Abstract Machine Models for Highly Parallel Computers

Abstract Machine Models for Highly Parallel Computers

EDITED BY

JOHN R. DAVY

*Lecturer in Parallel Processing, School of Computer Studies,
University of Leeds*

and

PETER M. DEW

*Professor of Computer Science, School of Computer Studies,
University of Leeds*

OXFORD   NEW YORK   TOKYO
OXFORD UNIVERSITY PRESS
1995

Oxford University Press, Walton Street, Oxford OX2 6DP
Oxford New York
Athens Auckland Bangkok Bombay
Calcutta Cape Town Dar es Salaam Delhi
Florence Hong Kong Istanbul Karachi
Kuala Lumpur Madras Madrid Melbourne
Mexico City Nairobi Paris Singapore
Taipei Tokyo Toronto
and associated companies in
Berlin Ibadan

Oxford is a trade mark of Oxford University Press

Published in the United States by
Oxford University Press Inc., New York

© The contributors, 1995

All rights reserved. No part of this publication may be
reproduced, stored in a retrieval system, or transmitted, in any
form or by any means, without the prior permission in writing of Oxford
University Press. Within the UK, exceptions are allowed in respect of any
fair dealing for the purpose of research or private study, or criticism or
review, as permitted under the Copyright, Designs and Patents Act, 1988, or
in the case of reprographic reproduction in accordance with the terms of
licences issued by the Copyright Licensing Agency. Enquiries concerning
reproduction outside those terms and in other countries should be sent to
the Rights Department, Oxford University Press, at the address above.

This book is sold subject to the condition that it shall not,
by way of trade or otherwise, be lent, re-sold, hired out, or otherwise
circulated without the publisher's prior consent in any form of binding
or cover other than that in which it is published and without a similar
condition including this condition being imposed
on the subsequent purchaser.

A catalogue record for this book is available from the British Library

Library of Congress Cataloging in Publication Data
(Data available)

ISBN 0 19 853796 4

Typeset by the authors and
editors using LaTeX

Printed in Great Britain by
Bookcraft (Bath) Ltd
Midsomer Norton, Avon

PREFACE

J. R. Davy and P. M. Dew

*Scalable Systems and Algorithms Group,
School of Computer Studies, University of Leeds,
Leeds LS2 9JT, W.Yorkshire, UK*

Introduction

Abstract models have played a profound though frequently unacknowledged role in the development of modern computing systems. They provide a precise definition of vital concepts, allow system complexity to be managed by providing appropriate views of the activity under consideration, enable reasoning about the correctness and quantitative performance of proposed problem solutions, and encourage communication through a common medium of expression.

Such models can be identified at various levels. Higher-level models generally provide a user-related approach to the description of problems and their possible solutions; lower-level models formalize the physical components which will be used to implement these solutions. In recent years there has been increasing recognition of the significance of bridging models between these levels; in particular the von Neumann model has played a vital role in bridging the gap between diverse high-level programming models and diverse machine architectures (Valiant 1990a). Some authors consider this unification to be a major enabler of the rapid development and application of sequential computing technology (McColl 1993, Valiant 1990a).

The world of parallel processing does not yet benefit from a similar, general-accepted unifying model. Rather there has been a proliferation of both higher-level (programming) and lower-level (machine) models. This diversity, contrasting markedly with the sequential world, has been a disincentive to the more rapid development and application of parallel processing technology, even though the potential performance benefits have long been understood.

Against this background, the vital role of abstract models has come into new focus. To date, models have tended towards undesirable extremes. On the one hand they may have desirable theoretical qualities but be unrealistic or very difficult to implement efficiently on current hardware. A classic example is the conceptually simple PRAM model (Fortune 1978), which enables elegant algorithm design and tractable performance analysis at a cost of ignoring many of the vital issues involved in hardware realization. The vigorous research efforts into parallel implementation of declarative languages provide further examples of elegant theoretical models whose practical implementation has not yet been demonstrated adequately for most purposes. At the other extreme, models may be efficiently implementable but undesirably complex or machine-dependent, restricting their long-term value. Even where the programming model is superficially similar (such as message passing, or a single-level shared memory), the underlying mechanisms for communication and synchronization may be substantially different. Commercially available parallel processing systems typically fall into this category.

The need for more appropriate models is therefore pressing. The feasibility of a single parallel equivalent to the von Neumann model may be a matter of debate; what is clear, however, is that the future development of parallel computing depends on the emergence of more usable models than have previously been available.

The first Workshop on Abstract Machine Models for Highly Parallel Computers was convened at the University of Leeds in March 1991 to address this vital area for progress. It was held under the auspices of the Parallel Processing Specialist Group (PPSG) of the British Computer Society. Academic and industrial delegates, from the UK, Europe, and the USA, came together to discuss wide-ranging issues at theoretical and practical levels. Though there was no overall consensus on the way forward, both the ongoing importance of the field and the desirability of a further workshop were acknowledged.

This second workshop, again under the auspices of the BCS PPSG, took place in April 1993 at Leeds University, with a similar international flavour. Eighteen plenary presentations, with both invited and submitted contributions, were interspersed with group activities exploring the current state of the art and potential future directions. Perhaps the main development since 1991 had been the increasing awareness of the claims made for Valiant's BSP (Bulk Synchronous Parallel) model (Valiant 1990a) as a potential unifying model for parallel computation, analogous to the von Neumann model and with the potential for efficient PRAM emulation (Valiant 1990b). Though these claims are far from universally accepted, there has been significant work in this area, as reflected in workshop presentations, and the vital underlying role of abstract models has again been emphasized.

This volume contains revisions of all the plenary papers from the workshop. They reflect the diversity of contributions, while highlighting major directions of current research activity.

The Papers

The absence of standard definitions and terminology is endemic to computing, and the field of abstract models is no exception. Two initial papers provide a foundation by addressing this issue, exploring the nature and role of abstract models. Heywood outlines three main levels of models, corresponding to languages, algorithms, and architectures, where the central level leads again to the concept of bridging model. Gurd and Snelling's paper proposes a comprehensive conceptual framework and corresponding vocabulary, covering the whole spectrum of computing activity from application description down to physical realization.

The next group of three papers covers the topical issues of the BSP model and PRAM emulation. McColl describes the philosophy and principles of BSP, highlighting its potential bridging role and consequent significance for the future development of parallel computing. Natvig discusses a language for synchronous MIMD programming on a PRAM, suggesting an implementation route through BSP. Nash et al. propose a modified PRAM model based on a BSP derivative; they emphasize the potential for realistic performance prediction at the algorithm design stage and demonstrate scalable simulated performance.

Parallel functional programming forms the subject-matter of the next group of three papers. The first, by Kumar et al., describes the CTDNET III graph reduction model, combining eager reduction with selective lazy features. It follows a 'traditional' method of automatically exploiting the implicit parallelism of functional programs. The performance problems associated with this approach have motivated some researchers to investigate alternative techniques for functional programming, using restricted algorithmic patterns with known efficient parallel implementations. The aim here is to obtain acceptable performance while preserving the theoretical benefits of functional languages. Both Rabhi's and Darlington and To's papers explore this 'skeleton' approach.

The skeleton philosophy amounts to using a limited set of computation and communication patterns, and is not restricted to the functional programming world. The next group of papers illustrates a variety of analogous methods in other contexts. Skillicorn uses the concepts of category theory to propose and illustrate an approach to building such skeletal models in a consistent and complete way. Shafarenko's paper discusses RETRAN, an array-based applicative language which allows general data parallel operations to be inferred and implemented in parallel. Sheffler introduces a model of parallel computing based on two fundamental architecture-independent operators for communication: **match** describes the communication pattern and **move** carries out the data movement.

A potential bridging model at a very different level from BSP is the subject of the next paper. The Open Software Foundation's ANDF (Architecture Neutral Distribution Format) is an open system intermediate compiler target. Lake's paper considers the implications of extending this format to parallel systems. Based on work carried out on TDF, the base technology selected for ANDF, it is a preliminary review of the area which takes into account some of the distinctive requirements of parallel computation.

Low-level standards for message-passing have been an important focus of the parallel processing community's efforts for some time. MacDonald's paper discusses many of the issues, illustrated by the CHIMP system. This then forms the basis for a further example of the skeleton approach through PUL, a modular suite of libraries supplying the program infrastructure for a variety of parallel programming patterns. Here, however, the principle is applied in an imperative context.

Simulation continues an important computational task. The paper by Muller *et al* defines four abstract models of simulation, and discusses their potential for parallelization. Links with other papers are established by the use of a functional language as a specification tool.

Performance modelling of parallel algorithms and systems is a key area of research, which is relatively under-represented in this workshop. Candlin and Phillips discuss an approach based on statistical modelling. A small number of program parameters affecting performance are identified; factorial experiments and analysis of variance enable their relative importance to be assessed, and conclusions for performance prediction and load balancing are drawn.

The next paper, by Barrett, illustrates the potential for abstract models to prove correctness of programming techniques. He draws a comparison between well-established compiler techniques for scheduling memory accesses on sequential computers and possible approaches for rescheduling communications on parallel computers. The notion of buffer tolerance is used to justify such rescheduling and leads to a semantics of communicating processes which is abstract with respect to synchronization.

Concurrent object-oriented programming brings new opportunities and challenges to parallel computing. The Warp protocol, introduced by Livesey and Allison, aims to provide a uniform approach to concurrency control in distributed object-oriented environments. The authors argue that it is also applicable to support other parallel abstract models with multiple concurrent object interactions.

The final paper, by Cornu and Vialle, describes the Cellular Abstract Machine (CAM), an abstract model dedicated to connectionist applications. It serves as a reminder of the extensive work in special-purpose parallel architectures, which may continue to play important niche roles outside the mainstream of general-purpose parallel computing.

Acknowledgements

We wish to thank the committee of the BCS Parallel Processing Specialist Group for their support throughout this venture. Several people helped with the organization of the workshop; special thanks go to Mr Charlie Brown for his handling of the mass of administration and to Mrs Maureen Hopkinson and Mrs Judith Thursby for secretarial support. The staff at Devonshire Hall, University of Leeds, led by Mrs Beverley Kenny, provided a welcoming environment which made the workshop all the more enjoyable.

We gratefully acknowledge the assistance of the Workshop Program Committee and others who gave their time to review all papers submitted to the workshop: Dr Z. Abdelouahab, Dr R. D. Boyle, Prof J. R. Gurd, Prof L. O. Hertzberger, Dr N. Jayaram, Prof C. Jesshope, Dr M. Kara, Dr T. Lake, Dr W. McColl, Dr J. M. Nash, Dr C. P. Wadsworth, Dr S. Winter.

Finally we wish to thank the staff of Oxford University Press for their help and encouragement.

References

S. Fortune and J. Wyllie. Parallelism in Random Access Machines. In Proceedings of the 10th Annual Symposium on Theory of Computing, pp 114-118, 1978.

W. F. McColl. General-purpose parallel computing. In Gibbons and Spirakis (eds), *Lectures on Parallel Computation*, volume 4 of *Cambridge International Series on Parallel Computation*, pp 337–391, 1993

L. G. Valiant. A bridging model for parallel computation. *Communications of the ACM*, **33(8)**, 103–111, 1990.

L. G. Valiant. General purpose parallel architectures. In *Handbook of Theoretical Computer Science*, volume A, pp 945-971, 1990.

CONTENTS

1. Models of Parallelism
   T. Heywood and C. Leopold — 1
2. A Terminology for (Parallel) SuperComputing
   J. R. Gurd and D. F. Snelling — 17
3. Bulk Synchronous Parallel Computing
   W. F. McColl — 41
4. General-Purpose Parallel Programming on the PRAM Model
   L. Natvig — 64
5. Parallel Algorithm Design on the WPRAM Model
   J. M. Nash, P. M. Dew, M. E. Dyer and J. R. Davy — 83
6. CTDNet III - An Eager Reduction Model with Laziness Features
   P. Kumar and J. P. Gupta S.C. Winter — 103
7. Exploiting Parallelism in Functional Languages: a 'Paradigm-Oriented' Approach
   F. A. Rabhi — 118
8. Building Parallel Applications Without Programming
   J. Darlington and H.W. To — 140
9. Categorical Data Types
   D.B. Skillicorn — 155
10. RETRAN: a Recurrent Paradigm for Data-Parallel Computing
    A.V. Shafarenko — 169
11. Writing Portable Parallel Programs with Match and Move
    T. J. Sheffler — 187
12. ANDF — Sequential to Parallel
    T. Lake — 209
13. A Framework for Portable Parallel Applications
    N. B. MacDonald — 225
14. Using a Functional Notation to Specify Abstract Simulation Models
    H.L. Muller P.H. Hartel and L.O. Hertzberger — 243
15. Statistical Modelling as a Tool for Studying the Performance of Parallel Systems
    R. Candlin and J. Phillips — 259

16. **Reschedulable communications**
    G. Barrett 281
17. **Some Practical Considerations for Object-Oriented Programming on Distributed Memory Parallel Computers**
    M. Livesey and C. Allison 295
18. **A Framework for Implementing Highly Parallel Applications on Distributed Memory Architectures**
    T. Cornu and S. Vialle 314

1
MODELS OF PARALLELISM

Todd Heywood

Department of Computer Science
University of Edinburgh
thh@dcs.ed.ac.uk

Claudia Leopold

LS Informatik I
Universität Würzburg
leopold@informatik.uni-wuerzburg.de

1.1 Introduction

Parallel computing systems can be considered to consist of languages, algorithms, architectures, and their interactions. The complexity of designing and analysing parallel systems requires that models be used at various levels of abstraction in order to provide a simplified view of a system. A *model* is simply an abstract view of a system, or more appropriately a part of a system, obtained by removing details in order to allow one to discover and work with the basic principles.

Recent years have brought a virtual explosion of proposed models. Their various purposes, their various levels of abstraction away from the hardware, and similar terminology can sometimes cause confusion. It seems necessary to clarify the roles of models in parallel computing.

In this paper we first outline a loose levelled framework, or hierarchy, of model types where each type corresponds to a component of a parallel system (as defined above: languages, algorithms, architectures). Interactions between components become mappings between model types. We then discuss a particular instance of this framework which we believe may result in scalable, cost-effective systems which balance ease of use and efficient performance. Lastly we consider where the important, and often ignored, issue of I/O is best represented in the framework.

It needs to be emphasized that we are not proposing any rigid taxonomy or classification, but just organizing key issues in parallel systems into a loose context. Boundaries between model types are often fuzzy.

1.2 Framework

The complexity of computing systems, partly due to the conceptual distance between the raw technology that one can build computers with and the problems that one wants to solve by using a computer, requires that the systems be designed and analysed via a division of labour. People working on different parts of a system will have different views of it, reflected by different *models*.

Low-level models are concrete about physical, technological and/or economic properties of the hardware components, but don't pay much attention to problem formulation techniques. High-level models are quite concrete about the ways people can best formulate and solve their problems, but don't (often) pay much attention to implementation details. In the middle levels, a good model will manage to balance attention to formulation and implementation.

There is a variety of frequently changing models at the high and low levels, as formulation and implementation details evolve in reaction to technology and problems, respectively. Due to their degree of abstraction from *both* implementation and formulation details, middle-level models may stay fixed for long periods of time. Further, their number should be small, or even one, so as to provide a basis for establishing a common environment necessary for the success of software and hardware industries. Valiant has called a model of this kind a 'bridging model' (Valiant 1990).

The von Neumann model of sequential computation is a bridging model. Though hardware and software industries have been in constant development, the changes have been subsumed by the degree of abstraction in the model for a long time. Lately though, the changes have become large enough that they cannot be subsumed:

- Technological advances have made massively parallel architectures possible (it does not seem possible to automatically parallelize sequential programs in an efficient manner).
- Problem sizes reached a threshold where I/O time must often be considered more important than computation time (hierarchical memories have hid this in the past, but it is becoming increasingly difficult), and there are important new interactive or real-time problems which require much I/O.

The ending of the central unifying standard resulted in a variety of diverging efforts to manage the new situation, which is much more complex due to the difficulty of managing explicit parallelism. Thus, recent years have brought a virtual explosion of proposed models of parallelism (less attention has been paid to I/O). Their differing purposes and levels of abstraction, along with coinciding terminology, sometimes seems to add to the confusion rather than providing a simplifying framework. It seems necessary to clarify the roles of models in parallel computation.

In this section, we outline a loose three-level hierarchical framework of model types. This is as opposed to the two level framework in sequential computing: the von Neumann architectural model (the RAM computational model is not really distinct but simply a formalization of it) and HOL programming models. The first two subsections discuss the model types that interact with the outside world: low-level architectural models and high-level programming models. The third subsection considers intermediate-level computational models, the key points being that they must be considered quite separate and distinct from the other two types, and that a model of this type should logically play the role of the bridging model.

Discussion of where I/O representation belongs in the framework is postponed to Section 1.4.

1.2.1 Architectural (technology) models

Architectural models are abstractions of the hardware and operating system that make up a machine. For example, an architectural model may describe the interconnection network and its purpose, e.g. to connect the processors to banks of shared memory (shared memory architecture) or to interconnect the processors with each other, where each has its own local memory (distributed memory architecture). Thus an architectural model describes how to perform communication, but not its implementation details. It also defines whether the computer is synchronous or asynchronous, SIMD or MIMD, and possibly other characteristics.

An architectural model may be even more abstract than this. For example, it may not specify the network, but only its performance characteristics (thus achieving some 'architecture/implementation-independence'). In this context, an architectural model may serve as a 'bridging model' overlying machines with a certain class of networks.

The advent of these higher-level architectural models (e.g. BSP and LogP; see below), in comparison to lower-level network architectural models, arose from the following situation. On the one hand, algorithms for network-level architectural models are designed to explicitly take advantage of locality because of the highly structured communication patterns imposed on designers by the network topology. However, this also imposes conceptual hardship by forcing them to conform to a rigid low-level structure. On the other hand there is the PRAM computational model, which imposes absolutely no structure on communication. Standard PRAM simulation techniques employing random hashing obliterate any natural locality that may be present in an algorithm. In other words, the PRAM can be seen as assuming a 'worst case' scenario for locality in algorithms.

The higher-level architectural models basically throw away the representation of locality (or proximity) in order to obtain simplicity of use via a global address space, and also some network-architecture independence (strictly speaking, they may allow the limited benefit of distinguishing between on-processor and off-processor communication, but this is a very restricted form of locality). These types of models may have different purposes though. Valiant's BSP model (Valiant 1990) is designed to specify a platform that can efficiently support a PRAM computational model, that computer engineers can aim for realizing in their machine designs. The LogP model (Culler 1993) is defined to correspond to common characteristics of current and near-term architectures, aiming to provide an architecture/implementation independent platform.

Some may see the BSP model, which embodies the philosophy of latency hiding for the purpose of efficient (work preserving) support of the PRAM as a computational model, as justification for ignoring locality. In our opinion, the cost-effective viability of this idea is questionable, especially with respect to scalability. The BSP requires very powerful, thus expensive, communication hardware in order to support the PRAM model and furthermore requires that the investment in communication hardware grow faster than that for computation as a machine scales up.

Another consideration has been raised recently by Bilardi and Preparata (Bilardi 1992). In this interesting paper, they argue that, ultimately, speed of light and physical device size limitations will make it effectively impossible to hide latency on a 'machine-wide' basis. Furthermore, speed of light considerations are already quite applicable to the wireability (thus scalability) of architectures, since 1 ns is already considered 'long' for today's technology and light travels just 30 cm in 1 ns.

It is our opinion that general locality (or neighbourhood locality, or proximity), represented by partitioning a model into sub-models, would combine well with the use of latency hiding (or slackness) to achieve efficient simulation of sub-models ('sub-PRAMs') on sub-networks of architectures. General locality would reduce the amount of slackness otherwise needed to achieve efficiency, resulting in *faster* efficient algorithms, due to the utilization of more processors in the architecture. It would allow the utilization of latency hiding on a limited basis such that a system could cost-effectively scale up (simply increase the number of 'sub-machines', reducing their sizes relative to total machine size, as the system scales up). The interplay and tradeoffs between general locality utilization and slackness, with respect to algorithm design and analysis, is the subject of current research. Note that, in effect, this philosophy involves employing a 'locality preserving BSP' architectural model to support a 'locality preserving PRAM' computational model. In Section 1.3 we outline an instance of the framework which could encompass this.

One can't disregard the representation of locality lightly. Locality is the prime resource that one can extract from problems to boost the performance of their solutions. At minimum, the *option* of locality needs to be provided.

LogP is a very recent proposed model which is somewhat similar in flavour to BSP. LogP is based on economic considerations which suggest that current and short-term architectures will most likely consist of a thousand or two off-the-shelf processing elements. Since LogP is designed for the short term, it does not appear to be a good choice for a bridging model. Long-term validity is too important, e.g. with respect to software investments, to be given up lightly.

We think the bridging model should be situated at the most steady point (over time) of the model hierarchy and believe it will be at a somewhat higher level. In other words, the bridging model should be a computational model, although a realistic one. The key principle separating architectural models from computational models is that the architectural model is defined in terms of technological, physical, and/or economical properties; the computational model is *designed* for a user's convenience, while respecting technological and physical practicalities. We return to this in more detail in Section 2.3.

Having said that, it also needs to be said that we do not intend to make the *hard* claim that the BSP and LogP are architectural models. The boundary between abstract architectural models and computational models is fuzzy and a matter of taste. Again, we are not proposing any rigid classification (it seems a waste of time to argue issues such as this).

More research into architectural models is required. First, it needs to be accepted that a computational model will overlay the architectural model and thus the architectural model will not be the 'user model'. The architectural model should be the theoretical equivalent of an experimental architecture CAD and simulation tool. By this we mean that the architectural model needs to be a flexible platform for studying the performance of mixes of man-made designs (algorithm structures as represented by an overlaying computational model) and architectural technology operating under physical laws and economic constraints. The tighter linking of architectural technology to physical limitations such as the speed of light and device size lower bounds is necessary so that the architectural model is not only relevant to current technology, but all possible future technology. The tighter linking to economic constraints should be self-evident.

Sub-topics of this topic are subjects of current work, but it appears that the following characteristics would be important for an architectural model (recall that since we are accepting that a computational model will be overlaying the architectural model, details which are not conducive to usability are permissible and necessary).

- Redefinition of 'unit time', normalized to the speed of light and distances in two- or three-dimensional space, not only between processors but within them.

- 'Upward' parameters to represent properties of algorithms at the computational model level, such as locality and data size (message size).
- 'Downward' parameters to abstractly represent physical properties of implementing technology, such as
  * processor size ('processor' = CPU, memory module, communication switches);
  * local bandwidth: wire size (length, width) and degree. Related to processor size via pin count limits;
  * global bandwidth: bisection width of network;
  * distances, which is an abstraction of hops in a network, or diameter of a network or sub-network;
  * various delay models (linear, logarithmic, constant) parametrized by sizes and distances.

The point is that, while a BSP or LogP augmented with general locality (or proximity) representation may prove to be adequate, there are other issues that also need to be studied in architectural models for supporting computational models before making that judgment.

An important application of a model along the above lines is to study contention, since this is the main limiting factor in supporting idealized computational models. We note that locality is known to significantly reduce contention in architectures, and thus bandwidth requirements (Agarwal 1991).

Finally, it seems essential that the architectural model represent the polynomial vicinity assumption implicit in Turing Machines: 'In n time units, a processor can access any one of a polynomial number of memory locations'.

1.2.2 Programming (semantical) models

Programming models are defined for the purpose of providing enough abstraction so that large programs can be organized, reasoned about, and verified correct (either through testing or formal proofs). Hence, their defining characteristic is simplicity of use, i.e. convenience in formulating solutions to problems and verifying that they are correct. Programming models are minimally defined in terms of the semantics of a particular implementation of a programming language.

The programming model is at the next higher level of abstraction over the computational model, which is discussed in the following subsection. One key difference between a programming model and a computational model is that a programming model describes memory in terms of data structures, while a computational model typically describes memory as a sequence of memory locations. Analogously, differences may exist in terms of control structures.

An ideal model would be a language-independent formal semantics, e.g.

a process calculus, in which case 'semantical model' would be a better term than 'programming model'. A semantics of control structures and data structures, for example, would have obvious benefits. The structures could potentially be mapped to lower-level structures in computational models, thus providing a link between the semantical and algorithmic worlds. Even more beneficial would be a semantics of 'programming paradigms' or 'common problem-solving techniques'. Section 1.3 discusses a possible path toward this.

There has been work on augmenting programming models with performance metrics. Even if they are well-defined and effective (a difficult task in a highly abstract model) with respect to some architectural model(s), this is not as simple and accurate as in a computational model where one just counts a few types of basic built-in operations, e.g. read, write, compute. Combining performance analysis with programming models seems a messy task with inherent tradeoffs between accuracy and architecture independence. Ideally, a programming model should stick to its role as defined above, and rely on mappings down to an underlying computational model which has the responsibility for performance analysis. This is not to say that a programming model should not have performance metrics attached, only that they should be defined in terms of an underlying computational model rather than architectural models.

1.2.3 *Computational (algorithmic) models*

Ideally, a computational model provides an abstract view of a class of architectures while accurately reflecting costs and resources of those architectures. It should be simple and general enough to make the design and analysis of algorithms, and the proving of lower bounds for problems, relatively easy. At the same time, complexity results of algorithms on the model should sufficiently predict their performance on architectures. In essence, this requires finding a balance between *simplicity* of use and *reflectivity* of realistic costs and resources.

Thus, computational models are at an intermediate level of abstraction, overlying architectural models and underlying programming models.

The report of the Purdue Workshop on Grand Challenges in Computer Architecture for the Support of High Performance Computing (Siegel 1992) gives a good description of what is needed in a computational model, and the necessary relationships to programming and architectural models.

The nature of a computational model is that it be *designed* for the user's (algorithm designer/analyser) *convenience*. Of course it needs to be realistic, but it cannot be too tied to specific architectural technologies; in other words, 'the [computational] model must have practical hardware realizations, but not dictate specifics of those realizations' (Siegel 1992). In particular, a computational model must not be made obsolete by advances in realizing technology. In comparison, the architectural model is defined primarily for analysing the performance effects of various combinations of technology with respect to physical and (sometimes) economical constraints.

For example, in our opinion the PRAM, H-PRAM (Heywood 1992a), LPRAM (Aggarwal 1990), YPRAM (de la Torre 1991), circuits, dags, and comparison networks are all computational models. They are all designed weighted mostly to the convenience and purposes of the user rather than enabling technology. Asynchronous PRAMs and the BPRAM (Aggarwal 1989) may be seen by some to be computational models, but we consider them architectural models since they are tailored to properties of architectural technology, and not to the user.

Since architectural models have elements of design in them and computational models have elements of reaction to the physical/technological world, the boundary between the two types is fuzzy and a matter of opinion. The key point here is simply that computational models are *primarily* prescriptive (designed for users) and architectural models are *primarily* reactive.

We believe that a computational model should play the role of the 'bridging model' (Valiant 1990), or 'idealized parallel computer model' (Siegel 1992), since the most steady point (least sensitive to changing technologies and problem formulation techniques) of the model hierarchy will be at a somewhat higher level than that of abstract architectural models such as BSP and LogP. Additionally, intuition suggests that the conceptual complexity of parallel computing over sequential computing means that programming models will have to be at higher levels of abstraction from the hardware. This would in turn mean that a bridging model for parallel computation would need to be at a higher level than sequential computing's bridging model.

It is difficult to find the right balance between the simplicity of use and reflectivity of costs and resources in underlying architectural models. It requires formulating compromises between contradictory requirements like asynchrony existent in architectures and synchrony desired by users, between the efficiency of explicit message passing and the simplicity of shared memory. In designing a computational model one has to make decisions on which properties of both technology and problem formulation/solution are so general and so decisive for the design of the whole system that they must be represented. We think such properties include the following:

- At least the option of using locality (in underlying architectural models, since this is the main resource from above that can be used to boost performance below).

- Synchronous communication (for the benefit of overlying programming models, giving determinate programs, i.e. simplicity). Note that this does not exclude asynchronous control.

- Modularity (of both software on the programming model and hardware under the architectural model).

The next section discusses an instance of the framework which is built around a computational model representing these properties.

In summary, we would denote those models being defined mainly in terms of technological or physical properties as architectural models. Programming models are defined mainly in terms of describing and reasoning about (large) problems and their solutions. Computational models are *designed* for the 'low-level user's' convenience, while respecting technological and physical practicalities, i.e. designed to balance simplicity of use and reflectivity of costs and resources in realistic machines. Further, as the intermediately abstract model in the hierarchical framework, the computational model logically should play the role of the bridging model.

1.3 An instance of the framework

In this section we suggest a possible instance of the hierarchical framework, i.e. a set of architectural, computational, and programming models that complement each other, in the context of cost-effectively scalable parallel computing.

Scalability minimally requires that systems be built around distributed memory machines, which may be constructed with either direct (point-to-point) or indirect (multistage) communication networks. Direct networks allow the exploitation of locality while indirect networks do not. Since locality is the main resource that can be extracted from applications to boost performance, it should not be thrown away lightly, particularly in a system employing direct networks. It seems paradoxical to build systems with direct networks which do not represent locality, at least optionally, in some way to a user. The framework instance given in this section is based on the use of direct network architectural models, and the increasingly abstract representation of general locality at the computational and programming model levels.

The Hierarchical PRAM (H-PRAM) (Heywood 1992a, Heywood 1992b, Heywood 1992c) is a model of computation which balances simplicity of use, and reflectivity of the costs and resources of architectural models employing direct networks (i.e. a 'good' H-PRAM algorithm will translate to a 'good' algorithm on a direct network model). It uses the PRAM as a submodel, allows the utilization of general locality, and enforces determinate (testable) computation in the face of asynchrony, thus providing a basis for bridging theory and practice of parallel computing. The H-PRAM satisfies almost all of the goals of the 'idealized parallel computer models' grand challenge of the Purdue Workshop on Grand Challenges in Computer Architecture for the Support of High Performance Computing (Siegel 1992). The remaining goal of cost-effective implementation of the model, although shown to be satisfied in principle by the H-PRAM, necessitates some further investigation since 'cost-effectiveness' is a subjective measure. Note though, that locality reduces bandwidth requirements in architectures, and thus their cost (Agarwal 1991).

The H-PRAM and its relationships to architectural models are covered in detail elsewhere (Heywood 1992a, Heywood 1992b) so this section mainly considers programming models within the framework instance. First, though, we emphasize that it would be beneficial to have a more abstract and complete architectural model to map the H-PRAM to, in order to investigate the details of cost-effective scalability and physical limitations in supporting a synchronous shared memory abstraction. This was discussed in Section 2.1.

Structurally, the H-PRAM is a PRAM which can partition itself (recursively) into sub-PRAMs, giving rise to a hierarchy of synchronous sub-PRAMs which operate asynchronously from each other. Communication is limited to be within sub-PRAMs, resulting in the representation of general degrees of locality.

The representation of general locality necessarily means taking responsibility for some memory management. The memory management paradigm of the H-PRAM lies between the extremes of totally automated (and 'non-local', i.e. the PRAM) and totally manual (architectures). One might see it as a compromise, where memory is seen as an array and there is responsibility for organizing it into groups by specifying permutations on it, but not for the details of implementing the organizing, i.e. for routing data to specific processors. Programming models overlaying the H-PRAM will have data structures giving an abstract view of the memory management, thus regaining the simplicity of the PRAM's automatic memory management given up to obtain reflectivity via general locality.

A hierarchy is formalized by a hierarchy relation. A straightforward *tree* hierarchy relation structures an H-PRAM computation in such a way that it can be represented by a graph. To see this, first note that the partition instruction acts like a kind of 'fork' operation, creating independent sub-PRAMs. Similarly, when sub-algorithms running on the sub-PRAMs terminate and the sub-PRAMs synchronize and combine it is as if a kind of implicit 'join' operation is being executed.

The forking and joining are structured by the hierarchy relation, giving rise to a high-level graphical representation of the general structure of an H-PRAM computation. The 's-graph' (structure graph) corresponding to the tree hierarchy relation is a series-parallel graph. This naturally raises the question of which other s-graphs would be useful for the high-level description of parallel programs, i.e. which other hierarchy relations would be useful. This in turn leads to consideration of combinations of different hierarchy relations in different portions of the same H-PRAM algorithm, i.e. embeddings of s-graphs in different larger s-graphs in the high-level representation.

The reasoning here is that there are classes of problems which submit to the same general solution technique (high-level control structure), and that large problems (structures) are normally expressed in terms of sub-problems (sub-structures). An ideal programming model would allow programmers to manipulate and reason about the high-level superstructures of H-PRAM programs, i.e. in terms of 'general algorithmic techniques'. For example, the tree hierarchy relation and series-parallel s-graph could be seen as embodying the general algorithmic technique of divide-and-conquer. A diamond dag s-graph would represent the dynamic programming paradigm. Specific algorithms would be obtained by inserting specific PRAM algorithms at the nodes of s-graphs. Note that in the s-graph descriptions of H-PRAM computations, nodes are seen as primitive operations and these primitive operations are actually (sub-) PRAM algorithms. In other words, it may be possible to have a programming model with a 'PRAM algorithm instruction set'. This is attractive since it shows a means of building on the body of work that has been done on PRAM computations.

Cole's Algorithmic Skeletons is a programming model which has the style of these ideas. Skeletons are templates which embody the high-level structures of computations that submit to efficient parallel implementation. In (Cole 1989), Cole provides divide-and-conquer, iterative combination, cluster, and task queue skeletons. A programmer working within this framework would look through various available skeletons and attempt to formulate a basic solution for the problem at hand in terms of one of the high-level structures. If a match is found then the programmer inserts code into the chosen skeleton to obtain a specific solution to the problem. Otherwise, a new skeleton (solution technique) may need to be developed. An interesting aspect to this philosophy, as Cole noted, is that as the body of work on these general algorithmic techniques grows, the lack of a suitable skeleton for a problem increasingly suggests that the problem may not submit to an efficient parallel implementation. It is also interesting to note that skeletons can be represented by higher-order functions, thus giving them a formalization. For more recent skeleton-based work, see (Darlington 1993, Rabhi 1993) and the references therein.

Other work that is similar in philosophy includes KIDS (Smith 1989) and P^3L (Pisa Parallel Programming Language) (Danelutto 1992). KIDS is a program derivation system that uses 'design tactics', which are high-level construction methods for designing 'algorithm theories', which represent the high-level structures common to a class of algorithms. P^3L provides high-level language constructors, or 'organizing forms', which embody common parallel algorithmic techniques.

We think something along these lines may eventually provide an ideal programming model overlying the H-PRAM. High-level s-graphs embodying general algorithmic techniques would correspond to hierarchy relations in the underlying model of computation. The abstraction in the programming model is much higher; reflectivity would be addressed in terms of s-graphs which are known to submit to efficient parallel solutions in the computational model, and the structures would be beneficial in the conceptual management of large programs. Clearly, this is a research topic in its own right.

In another paper elsewhere in this book (chapter 9), Skillicorn describes possible category theoretic methods of rigorously generating high level patterns (i.e. s-graphs), rather than relying on *ad hoc* means (Skillicorn 1993).

A programming model that deals with organized structures such as s-graphs should be conducive to obtaining formal proofs of correctness of programs, assuming that the sub-PRAM sub-algorithms comprising the nodes of the graphs have individually been proven correct (this is normal procedure in PRAM algorithm design). Note again that Algorithmic Skeletons are formalized by higher-order functions, and that algorithm theories in KIDS are formal representations of the structure of classes of algorithms.

This approach may address the 'problem' of Fortran being a commercial necessity in parallel computing. One way of reducing its influence could be to use data parallel Fortran in sub-PRAMs at the computational model level. The parallel Fortran would appear to the programming model analogously to how assembly language appears to sequential higher-order languages.

We have outlined a potential environment which attempts to maximize gaps between levels in the meta-model, without being so large as to lose reflectivity of the costs and resources in the next lower level. The key to retaining reflectivity through the levels of abstraction can be generally said to be the imposition of some structure on computations. However, the structure would become more abstract and easier to work with in successively higher-level models.

1.4 Where does I/O belong in the framework?

As discussed so far, all the model types in the framework focus on managing massive parallelism, without explicitly handling the second new reality of huge problem sizes. I/O will need to be dealt with somewhere in the framework, since parallel computers will need to deal with scientific problems too large to fit in primary memory, as well as with communications intensive applications such as HDTV, gigabyte networks, and visualization.

Abstracting from I/O in the computational and programming models may be justified as follows:

1. Virtual memory (which hides I/O) contributes significantly to simplicity.

2. Massive parallelism might make the distinction between processors and memory devices obsolete in the long run (compare the concept of spatial machines: 'Memory cells are essentially specifically programmed processors.' (Feldmann 1992)).

In any case, I/O will need to be addressed on the architectural level. We think an ideal way for this, at least in the higher-level models, would be to isolate the common characteristics of interprocessor communication in networks and data movement in memory hierarchies. These two tasks are closely related. In both cases, the programmer can boost performance via locality exploitation. Hence, a computational model representing communication cost and locality in an abstract way could model both interprocessor communication and I/O by the same means. This seems to be a way to achieve simplicity without giving up reflectivity.

A model aiming at bringing the two issues into a common framework is PMH (Alpern 1993). Other related models are P-HMM, P-BT (Vitter 1992b, Vitter 1992a), and P-UMH (Vitter 1993). These are basically distinguished from all the other models cited in this paper in that they are not based on the PRAM. Instead, they are based on models from the sequential domain which adapt the RAM to huge problem sizes, e.g. HMM (Aggarwal 1987a), BT (Aggarwal 1987b), and UMH (Alpern 1993). These models stress the blocking of data. We would classify them as abstract architectural models (necessitating an overlying computational model), as their explicit responsibility for data movement with the attached effects of asynchrony are not characteristics designed in for the user's convenience.

1.5 Conclusions

We have tried to place a general framework around current activity in modelling parallelism. It is not intended as a strict classification or taxonomy, but simply to try to provide some context to the modelling of parallel systems. Within this context, we have suggested a possible set of models that may lead to cost-effective, scalable parallel systems which balance ease of use and efficient performance.

The suggestion of the H-PRAM as the 'bridging computational model' of this framework instance needs to be be examined very carefully, since the establishment of a 'wrong' bridging model would be a costly mistake for the computer industry in the long run. Similarly, other proposed computational and abstract architectural models require very careful investigation. There is always some risk in predicting future developments. Nevertheless, the lack of a fixed bridging model obstructs commercial development. We think that the H-PRAM is a credible candidate.

We conclude by noting that parallel computing is just a specific instance of parallel activity. Ideally, the models of parallelism should apply to both the modelling of parallel systems and to modelling the world. Architectural models represent the materials we have to work with, and the physical laws governing them ('the world'). Computational (or algorithmic) models are abstractions for describing and designing the world, and analysing the performance of the designs. Programming (or semantical) models are frameworks for understanding and reasoning about the world.

REFERENCES

A. Agarwal, Limits on Interconnection Network Performance, *IEEE Transactions on Parallel and Distributed Systems*, 2(4), pp. 398–412, Oct. 1991.

A. Aggarwal, B. Alpern, A. K. Chandra, and M. Snir, A Model for Hierarchical Memory, *19th ACM Symposium on Theory of Computing*, pp. 305–314, 1987.

A. Aggarwal, A. K. Chandra, and M. Snir, Hierarchical Memory with Block Transfer, *IEEE Symposium on Foundations of Computer Science*, pp. 204-216, 1987

A. Aggarwal, A. K. Chandra, and M. Snir, On Communication Latency in PRAM Computations, *ACM Symposium on Parallel Algorithms and Architectures*, pp. 11-21, 1989

A. Aggarwal, A. K. Chandra, and M. Snir, Communication Complexity of PRAMs, *Theoretical Computer Science*, Vol. 71, pp. 3-28, 1990

B. Alpern, L. Carter, E. Feig, and T. Selker, The Uniform Memory Hierarchy Model of Computation, *Algorithmica* (in press)

M. Danelutto, R. DiMeglio, S. Orlando, S. Pelagatti, and M.Vanneschi, A Methodology for the Development and the Support of Massively Parallel Programs, *Future Generation Computing Systems*, 8(1-3), pp. 205–220, July 1992.

D. Bilardi and F.P. Preparata, Horizons of Parallel Computing, in *Future Tendencies in Computer Science, Control, and Applied Mathematics* (International Conference on the Occasion of the 25th Anniversary of INRIA), A. Bensoussan and J.P. Verjus (Eds.), pp. 155-174, Dec. 1992.

M.I. Cole *Algorithmic Skeletons: Structured Management of Parallel Computation*, Pitman & MIT Press, 1989.

D. Culler, R. Karp, D. Patterson, Abhijit Sahay, K.E. Schauser, E.Santos, R. Subramonian, and T. von Eiken, LogP: Towards a Realistic Model of Parallel Computation, *4th ACM SIGPLAN Symposium on Principles and Practice of Parallel Programming*, May 1993.

J. Darlington, A.J. Field, *et al*, Parallel Programming using Skeleton Functions, *Parallel Languages And Architectures, Europe:Parle '93*.

Y. Feldmann and E. Shapiro, Spatial Machines: A More Realistic Approach to Parallel Computation, *Commun. of the ACM*, Oct. 1992, pp. 61-73, 1992

T. Heywood & S. Ranka A Practical Hierarchical Model of Parallel Computation I: The Model, *Journal of Parallel and Distributed Com-*

puting, 16, pp. 212-232, 1992.

T. Heywood & S. Ranka A Practical Hierarchical Model of Parallel Computation II: Binary Tree and FFT Algorithms, *Journal of Parallel and Distributed Computing*, 16, pp. 233-249, 1992.

T. Heywood and S. Ranka, Sorting and List Ranking on the Hierarchical PRAM Model (Extended Abstract), *4th Symposium on the Frontiers of Massively Parallel Computation*, October 1992.

T. Heywood, K. Mehrotra, and S. Ranka, *Performance of Barnes-Hut N-body simulation in the abstract*, Technical Report CSR-27-93, Dept. of Computer Science, Univ. of Edinburgh.

F.A. Rabhi, Exploiting parallelism in functional languages: a paradigm-oriented approach, in *Abstract Machine Models for Highly Parallel Computers*, Oxford University Press, 1994.

H.J. Siegel, S. Abraham, *et al*, Report of the Purdue Workshop on Grand Challenges in Computer Architecture for the Support of High Performance Computing, *Journal of Parallel and Distributed Computing*, 16, pp. 199-211, 1992.

D.B. Skillicorn, Categorical data types, in *Abstract Machine Models for Highly Parallel Computers*, Oxford University Press, 1994.

D.R. Smith and M.R. Lowry, Algorithm theories and design tactics, *Mathematics of Program Construction*, Springer-Verlag Lecture Notes in Computer Science 375, 1989, pp. 379–398.

P. de la Torre and C. P. Kruskal, Towards a Single Model of Efficient Computation in Real Parallel Machines, *Parallel Architectures and Languages Europe (PARLE'91)*, Springer Verlag Lecture Notes in Computer Science 506, pp. 6-24, 1991

L.G. Valiant, A bridging model for parallel computation, *Commun. of the ACM*, Aug. 1990, pp. 103–111.

J.S. Vitter and M.H. Nodine, Large-Scale Sorting in Uniform Memory Hierarchies , *Journal of Parallel and Distributed Computing* 17, Jan. 1993, pp. 107–114.

J.S. Vitter and E.A.M. Shriver, *Algorithms for Parallel Memory I: Two-level Memories*, TR CS-92-04, Dept. of Computer Science, Brown Univ., Aug. 1992. (Also: Optimal Disk I/O with Parallel Block Transfer, *ACM Symp. on Theory of Computing*, 1990, pp. 159–169.)

J.S. Vitter and E.A.M. Shriver, *Algorithms for Parallel Memory II: Hierarchical Multilevel Memories*, TR CS-92-05, Dept. of Computer Science, Brown Univ., Aug. 1992.

2
A TERMINOLOGY FOR (PARALLEL) SUPERCOMPUTING

John R. Gurd and David F. Snelling

*Centre for Novel Computing
Department of Computer Science
University of Manchester
Oxford Road
Manchester M13 9PL, UK*

Abstract

We investigate the processes involved in 'mapping' large-scale computational problems onto high-performance (parallel) computing systems. Decisions in these processes are taken at points in a large and complex 'mapping space', and it is important to be able to describe 'where' each decision is made. The mapping space contains a wide variety of abstract machine models, but there is discernible structure which we attempt to describe. Appreciation of the nature of the mapping space may provide helpful insight when mapping specific problems onto specific real machines.

2.1 Introduction

This chapter addresses the general confusion surrounding high-performance (parallel) supercomputing by proposing a specific terminology for the subject and inviting reseachers to contribute to its refinement and eventual adoption as a standard. The reasons for attempting this are twofold:

- Firstly, there is a lack of *consistent* terminology in common use. Different words are used for the same concept, whilst the same word is used to refer to different concepts. This leads to confusion when communicating ideas, marketing products, etc.
- Secondly, there is a lack of *structure* in the study of the systems, techniques and tools of high-performance (parallel) computing. This has led to *ad hoc* (at best) and misleading (at worst) research. In either case, the potential benefits of research are not being realized as rapidly as they are needed.

Our presentation is intended to help alleviate these problems.

The paper starts, as near as possible, from first principles. It is split into three main sections, covering the following:

- general terms describing the processes involved in developing solutions to high-performance (parallel) computing applications;
- terms related to digital systems in general, and parallel computer systems in particular; and
- terms describing the quality and quantity of parallel activity in parallel systems.

In attempting to standardize vocabulary for an emerging technology, there is a danger that one will choose specific words in a way that upsets some part of the audience, thereby earning their displeasure or, worse, causing them to ignore the attempt to promote uniformity. It is our earnest hope that the reader will keep aloof from such parochial concerns, and judge this proposal solely on the basis of how well it describes and groups together *concepts* (i.e. judge it by the way it *defines* words, rather than the *choice* of words).

We have tried to maintain consistency, precision, and usability in our choice of terms, but we know that our concentration has wavered on occasion. We apologize in advance for any problems caused as a result of this.

2.2 General Terminology

2.2.1 *Principal Definitions*

This section introduces terms relevant to the development of solutions to high-performance computing applications. There is nothing specific to parallel supercomputing in these terms: they apply equally to traditional sequential computers, vector supercomputers, and so on.

2.2.1.1 *SuperComputer* A SuperComputer is any computer system that qualifies for the category of the *largest* (in terms of memory size) and *fastest* (in terms of processing rate and input/output bandwidth) available at any given time. This is obviously a 'moving target', since size and speed are continually increasing quantities*.

2.2.1.2 *Applications and Users* A problem whose computational, memory, or input/output requirements are such that a SuperComputer is *necessary* to solve it is termed a SuperComputer Application (Application, for short). A person who uses a SuperComputer to solve such a problem is termed a SuperComputer User (User, for short).

We intend no opprobrium to attach to these terms. It is clear that the reason for developing SuperComputers and SuperComputing techniques is primarily for the benefit of appliers of the technology and, through them, the world at large. We are aware that some 'Users' think it demeaning to be described as such, but we wish to make plain our view that Users are

---

*Similar definitions have been used by others, e.g. (Duff 1985).

the most important group in the wide range of people involved in Super-
Computing.

2.2.1.3 *SuperComputing* SuperComputing is the process of developing a
solution to a SuperComputer Application. In essence, this requires us to
move from a (possibly vague) 'application statement' to some form of 'corresponding' high-speed, digital activity. Applications are 'mapped' onto
SuperComputers using the techniques and tools of SuperComputing. Because of the large-scale nature of the Applications, and the many different
technical issues that are involved, these techniques are necessarily complex. Hence, the general process of 'mapping' involves many abstractions,
ranging from those concerned with 'capturing the Application' to those
concerned with the physical packaging of semiconductor devices.

2.2.1.4 *Levels of abstraction* In common with other forms of computational problem-solving, SuperComputing can be viewed as both working
from the Application towards the SuperComputer (i.e. 'top-down') and
vice versa (i.e. 'bottom-up'). In either approach, the development passes
through a number of stages, during which the major emphasis of the development changes (e.g. the first stage of a 'top-down' development is primarily concerned with describing the problem to be solved, whereas the final
stage is far more concerned with 'tuning' to achieve the highest possible
SuperComputer performance). Where we can identify a significant change
of emphasis, we define a Level of Abstraction.

The precise Levels of Abstraction involved may vary from Application
to Application. However, most SuperComputing developments currently
pass through five* major Levels of Abstraction, roughly as follows[†]:

- At the *Application Level*, a relatively simple *application model* is developed, describing the problem to be solved. This contains no information about the method(s) to be used for solution. Indeed, it may
often be a rather informal description of the problem (e.g. 'I want to
know what the weather will be like next Tuesday'; or 'I want to know
the fastest journey time between Manchester and Timbuktu on 27th
March 1993'). The major concern is to understand what the User
actually wants.

---

*There is an open question as to whether or not one should include the (re-)design of
SuperComputers themselves as part of the legitimate 'design space' (i.e. adding a sixth
Level of Abstraction at the 'lowest' end). Indeed, we might also consider adding other
Levels of Abstraction at the 'highest' end (see section 2.2.2.6).

[†]We first proposed these Levels of Abstraction in (Gurd *et al.* 1993). They follow
the general pattern of the phases of what has become known as the software development cycle, as described, for example, in (Peters 1988). There are interesting parallels
between this approach and the analysis presented in (Hoare 1987), which investigates
the mathematical nature and computational properties of different formal methods for
describing computational tasks.

- At the *Specification Level*, the application model is turned into a formal mathematical model, or *specification*, of the problem. The mathematical notation used will depend on the nature of the problem (but is not determined by it, since each problem can be described by more than one model). For example, the global weather system might be modelled by a set of partial differential equations, with appropriate boundary conditions (including definitions of the starting and finishing conditions); or the global travel problem might be modelled by the set of relations contained in the abstractly merged contents of *all* travel company databases.

  Initially, the specification is developed without paying heed to the kind of SuperComputer that might be used to compute a solution to the problem. The major concern is for *veracity* of the mathematical model.

- At the *Algorithm Level*, a systematic procedure, known as the algorithm, is developed, based on some *discrete data domain*. The algorithm is designed to solve the specified problem, either directly, or by converging towards a solution in a predictable fashion. The specified problem may be approximated by either or both of the particular choice of discrete data domain or the point at which a converging algorithm is terminated. For example, in the weather forecasting problem, the continuous variables in the differential equations may be approximated by a 'mesh' of suitably spaced, discrete 'grid-points' in space-time; or, in the travel problem, the overall situation might be approximated by a subset of the travel company databases (e.g. air-travel companies only, relying on the assumption that surface travel cannot compete for such a journey). Here again, the description is, at least initially, independent of the hardware that will perform the final computation.

  The algorithm is used to shift emphasis away from the abstract (and possibly continuous) domain associated with the specification towards the concrete (discrete) domain associated with computer storage. The algorithm also resolves any *implicit* parts of the specification, by defining appropriate *explicit* methods. The major concerns are with the trustworthiness (e.g. *accuracy* and *stability*) of any approximations made, and with the certainty of convergence.

- At the *Program Level*, the algorithm is expressed as a *program*, written in a (usually high-level) programming language. The data domains of the algorithm are mapped onto *data structures*, and the associated procedures onto appropriate *control structures**.

---

*We have used the language of 'imperative' programming to describe this mapping, but similar processes are necessary for the achievement of high-performance in 'declarative' programming. For example, state-of-the-art functional programming languages

At this Level of Abstraction, the emphasis moves away from the problem, towards the computer hardware: At present, even high-level programming languages are strongly based on specific (parallel) hardware vehicles. The major concerns are, firstly, *authenticity* of the program (i.e. certainty that it implements the algorithm) and, secondly, achievement of *acceptable performance*.

Of course, performance depends both on the algorithm and on the realization (see below), but the coupling between a program and its realization has traditionally been much stronger than that with the algorithm. Hence, it has become usual for poor realized performance to lead to adjustments to the program, rather than to modification of the algorithm. However, the use of parallel realizations is changing this situation, because the choice of algorithm is having a greater effect on performance (see, for example, (Dongarra *et al.* 1984)).

- At the *Realization Level*, the program is implemented on (parallel) computer hardware. The data structures of the program are mapped to specific *memory locations*, and the control structures become sequences of *instructions* for the processing element(s).

  For the vast majority of Users, the process of translating from program to realization is handled entirely by a *compiler*. However, highly parallel hardware has yet to be fully mastered by compilers, and the use of such parallelism has been made explicit at the Program Level, encouraging users to help the compiler do a better job. Once again, the major concerns are authenticity and performance.

It will be noted that many of the above Levels of Abstraction involve discrete models and systems. These are fundamental to SuperComputing (indeed, to any form of digital computing), and deserve additional attention. All such models and systems have the same basis, namely *state-transition*, and we take advantage of this in our later definitions (see Section 2.3).

2.2.1.5 *Systems* Each Level of Abstraction involves *nested* development of the detail of models or systems (which we shall henceforth call Systems). The following methodology is applied for this development:

1. The *outer* (or *enclosing*) Level of Abstraction specifies a *required behaviour* to be obtained from the System being developed.
2. Details of the System are then developed so as to conform to the required specification:
    - A System with a complex behaviour requirement can be constructed by *composing* together Systems with simpler behaviour ('bottom-up');

require that *annotations* be added to the functional 'code' to guide the compiler in the right direction for high-performance.

- Alternatively, a System with a complex behaviour requirement may be engineered by *decomposing* it into a system of Systems with simpler behaviour ('top-down').

3. The design process is recursive, in the sense that 'top-down' development of one of the 'Systems with simpler behaviour' might require further decomposition at the same Level of Abstraction.

In some respects, this echoes the 'top-down' and 'bottom-up' view of the Levels of Abstraction, introduced in Section 2.2.1.4, above. However, here we are working within a single Level of Abstraction, so our major concerns are constant. Instead, as we move up and down the 'levels' of the System design, we are intent on achieving *correctness* of the detailed design, preparatory to moving our attention to the 'next' Level of Abstraction.

2.2.1.6 *Levels of Design* Since the above definitions are recursive, then, for any complex System, there exists a Design Hierarchy (Hierarchy, for short), associated with the 'current' Level of Abstraction. At each distinct Level of Design in this Hierarchy, different behaviour is apparent.

When comparing Systems at adjacent Levels of Design, we refer to 'Systems with simpler behaviour' as Components. Thus, an (arbitrary) System is composed of multiple Components, which interact with one another, according to their specified behaviour, via a *communication medium*, developed as part of the System. We refer to the communication medium as Glue, holding the Components together. The (logical or physical) *structure* of the Components and Glue in a System constitutes its System Architecture (Architecture, for short).

The Components at any arbitrary Level of Design are viewed as Systems at the next 'lowest' Level of Design. The boundaries between Levels of Design are determined by the conventions of the particular Hierarchy associated with the 'current' Level of Abstraction (examples are given in Section 2.2.2).

Components at the 'bottom' of a Hierarchy (i.e. the Level of Design at which no further decomposition is *feasible* within that Level of Abstraction) are called Primitive Components (Primitives, for short). Components, elsewhere than the 'bottom' of the Hierarchy, whose implementation is already known (and for which, therefore, no further decomposition at that Level of Abstraction is *necessary*) are called Library Components. A 'top-down' designer can choose whether to decompose Library Components further in the 'current' Hierarchy, or to pick-up a ready-made implementation at the 'next lowest' Level of Abstraction.

2.2.2 *Examples*

The above definitions are overly general: To illustrate them further, we give the following examples, taken from the world of SuperComputing.

GENERAL TERMINOLOGY

2.2.2.1 *Packaging Hierarchy* This Hierarchy is concerned with the *physical* packaging of hardware Systems and Components at the Realization Level, usually in the kind of Hierarchy shown below:

Packaging Hierarchy	
'highest' level	multiple-site
:	site
:	computer room
:	cabinet
:	rack
:	printed circuit board
'lowest' level	integrated circuit

2.2.2.2 *Realization Hierarchy* Also at the Realization Level, this Hierarchy is concerned with the *logical* structuring of hardware Systems and Components. The most common basis for this logical structuring is communicating Finite State-Transition Automata (see section 2.3). A familiar Hierarchy for this is as follows:

Realization Hierarchy	
'highest' level	multi-cluster
:	multi-'processor' cluster
:	'processor' (see section 2.3)
:	functional unit
'lowest' level	gate / flip-flop

2.2.2.3 *Program Hierarchies* At the Program Level, different avenues have been opened up by previous experimentation. Presumably there is greater diversity here (compared to the Realization Level) because it is easier to change software paradigms than it is to develop novel hardware. To give an idea of the kinds of Hierarchy that may be encountered here, we give three fundamentally different examples:

Imperative Hierarchy	
'highest' level	multi-job
:	job
:	process / file
:	'code block' / structured variable
'lowest' level	statement / scalar variable

This is the traditional programming view of sequential computing, with which the reader is expected to be familiar.

Functional Hierarchy	
'highest' level	higher-order function
:	recursive function
:	non-recursive function
'lowest' level	primitive function

The hierarchical structure of functional programming is apparent in standard textbooks on the subject, e.g. (Reade 1989).

Object Hierarchy	
'highest' level	arbitrary object
:	library object
'lowest' level	primitive object

A full object hierarchy is available in most object-oriented systems: See, for example, the type system in Modula-3 (Nelson 1991).

2.2.2.4 *Algorithm Hierarchies* Once again, diversity is encountered at the Algorithm Level, mostly because different kinds of Application require different kinds of 'core' algorithm. Again, we give fundamentally different examples of associated Hierarchy:

Numerical linear algebra Hierarchy	
'highest' level	multi-System solver
:	linear System solver
:	matrix-matrix operation (BLAS 3)
:	matrix-vector operation (BLAS 2)
:	vector-vector operation (BLAS 1)
'lowest' level	arithmetic expression

For those who are not familiar with them, the Basic Linear Algebra Subprograms (BLAS) have been developed so as to provide Fortran programmers with a standard library of high-level routines for solving commonly occuring problems in linear algebra (Dongarra *et al.* 1988).

DataBase Hierarchy	
'highest' level	multi-database
:	database
:	relation
'lowest' level	datum

Here again, we expect the reader to be familiar with this area.

2.2.2.5 *Application Hierarchy* It is increasingly more difficult to give 'concrete' examples as the Level gets closer to the application. However, the same hierarchical view is usually taken at Application Level, where complex Applications are modelled as groups of simpler Application Components, in a Hierarchy similar to that shown below:

Application Hierarchy	
'highest' level	Application
'lowest' level	Application Component

2.2.2.6 *Organization Hierarchy* Abstracting still further, we could transcend the computer-oriented view of 'System', and move on to even higher level structures, such as a group of companies with an organization-wide Information Processing System, in the following type of 'organization-based' Hierarchy:

Organization hierarchy	
'highest' level	group of companies
:	company
:	division
:	group of people
'lowest' level	person

2.2.3 Encapsulation

2.2.3.1 Encapsulation Schema In the 'top-down' approach, the principal reason for structuring Systems in a Hierarchy is to *encapsulate* independent parts of the System design in such a way that later steps in their design do not interfere with one another. The properties of this structure can be succinctly described using the notion of an Encapsulation Schema, as follows:

- An Encapsulation Schema is a *directed acyclic graph* containing (up to) three kinds of *Node*, namely:
    * *Designer Nodes*, representing Systems that have been designed specifically by the User;
    * *Library Nodes*, representing Library Components;
    * *Primitive Nodes*, representing Primitive Components.
- Designer Nodes contain a description of the Architecture of the corresponding System/Component.
- Library Nodes and Primitives Nodes are the *leaf* Nodes of the Encapsulation Schema (i.e. they have no outgoing Arcs).
- Each *Arc* represents the conceptual link between the System represented by a Designer Node and one of its Components. Where appropriate, the Arcs can be annotated with parameters describing this link.

We use a 'modular' terminology for encapsulation: i.e. a Designer Node representing a Component is thought of as a *Module*, which is thus a unit of convenient encapsulation in a Hierarchy.

2.2.3.2 Physical versus Logical structuring Many of the above Hierarchies are 'soft', in the sense that the development of a System at one Level of Design can lead to at least one Component that needs to be developed further at the same, or even a higher, Level of Design. For an example of this, think of a Fortran *subroutine* which is itself expressed as a sequence of *calls* to other subroutines. We shall call these Soft Hierarchies.

On the other hand, there are also Hard Hierarchies, in which it is essential to move 'down' the Levels of Design at each stage of development. For an example of this, think of packaging, where a System such as a printed circuit board cannot be expressed other than in terms of Components that are physically smaller (e.g. integrated circuits).

In Soft Hierarchies, encapsulation into Modules is a matter of convenience for the designer. Hence, such Modules are usually co-incident with the Systems they implement, since there is no reason to break the modularity between design and implementation.

In Hard Hierarchies, encapsulation into Modules is related to external or physical constraints. Such Modules may bear little relationship to the Systems they implement, since they reflect constraints that are usually beyond the immediate concerns of the designer.

2.2.4 Overview

2.2.4.1 *Mapping Space* The preceding definitions delimit a SuperComputing Mapping Space in which solutions to Applications are developed by Users. Levels of Abstraction and Levels of Design form a nested Hierarchy defining the major concerns of the Mapping Space and the details of Systems, Components and Glue involved, respectively.

There is a natural ordering, from 'high' to 'low', of the Levels of Abstraction and Design. In 'top-down' development, design moves steadily 'down' from the 'higher' Levels of Abstraction, firstly towards 'lower' Design Levels in the same Hierarchy, then periodically shifting from one Level of Abstraction (and its set of major concerns) to another, 'lower' one. A shift from one Level of Abstraction to another occurs when we reach a leaf Node in the Encapsulation Schema (i.e. a Library Node or a Primitive Node). We use the generic term Level to refer to an arbitrary Level of Abstraction or of Design.

Hard Hierarchies, concerned with, for example, the physical packaging of hardware Systems and Components, are distinct from Soft Hierarchies, concerned with the logical relationships between software Systems and Components.

The totality of all possible Hierarchies constitutes the Mapping Space.

2.2.4.2 *Design Space* In any particular design, the range of Hierarchies that are considered will be limited to those with which the User is familiar. This group of Hierarchies forms the Design Space for that User. The Design Space is inevitably a subset of the Mapping Space, since the range of knowledge required to span *all* possible Hierarchies exceeds human capacity.

Where the Hierarchies of a Design Space are closely associated with one another, they form a specific SuperComputing Paradigm (e.g., the Hierarchies associated with SIMD array processing).

2.3 State-Transition Systems

This section re-focusses our discussion towards SuperComputers (as opposed to general 'Systems'). We first investigate abstracct models of discrete hardware Systems, since these are fundamental to all computing. We start with over-simplified models, and gradually work towards more realistic implementations. These models are exclusively based on the concept of State-Transition. We use a common characteristic to assess the 'computational power' of each model, as described in Section 2.3.1.1.

2.3.1 The Full Automaton

The fundamental model for digital computation is, by tradition, Finite State-Transition. A finite State is processed by an Automaton according to a well-defined State-Transition Cycle (Cycle, for short) in which the entire State is first *read*, in order to ascertain its status, and then *written*, according to that status (and, if interaction with the outside environment is required, according also to any external 'input'), in order to establish a new State. The user defines an Initial State, and then the Automaton is allowed to process the State, Cycle-by-Cycle, until it encounters a *halting condition*, at which point the Final State represents the outcome of the computation.

In this simplest version, all State is read (and later written) simultaneously. Hence, we call this arrangement the Read-All, Write-All Automaton (Full Automaton, for short), as illustrated in Fig. 2.1.

The behaviour of a Full Automaton is both sequential and parallel in nature. Sequentiality follows from the repetitive nature of its State-Transition Cycle. Parallelism arises from its ability to read and then write all parts of the State concurrently.

2.3.1.1 Processing-Capacity

The Processing-Capacity (in bits/Cycle) of any State-Transition System is the *maximum possible* size of its 'active' State*. The 'active' State is that part which is going to be overwritten at the end of the 'current' Cycle (see section 2.3.2.1. The Processing-Capacity is thus equal to the *maximum* number of bits of State that could ever be overwritten during a single State-Transition Cycle at the relevant Level.

This definition is analogous to the definition of Channel Capacity (Shannon and Weaver 1949), used in information theory to measure the maximum possible rate of data transfer across a noiseless channel. By focussing on bits *written* to State, it ignores the work done in *reading* data. However,

---

*In hardware, it is assumed that all State will be implemented by binary storage devices and that processing will be ultimately performed using binary digital logic. Hence, the basic unit of State is the binary digit, or *bit*. However, at higher levels of abstraction, it might be more convenient to use a coarser unit of State, such as the floating point operand.

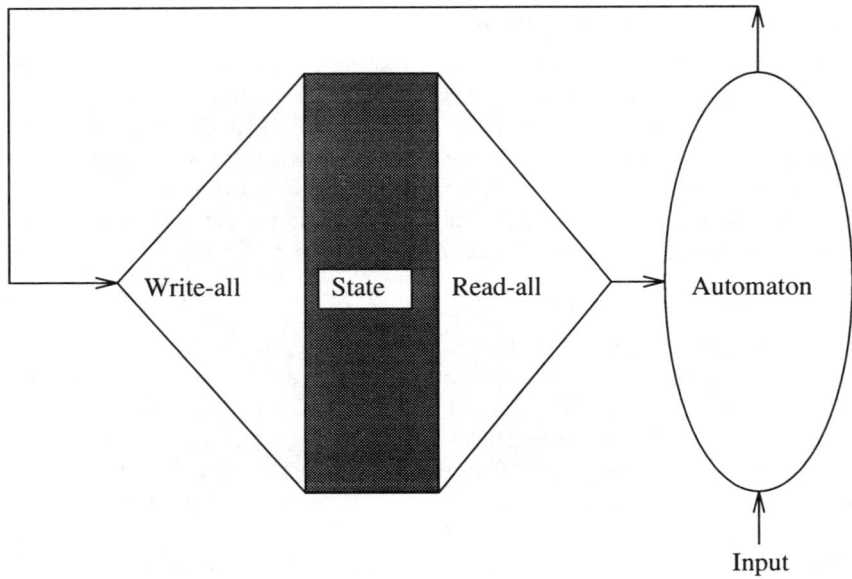

FIG. 2.1. Full Automaton

we believe this is justified by the fact that real progress can only be made in a computation by changing some part of State.

In practical terms, Processing-Capacity is a means of describing the *peak performance* of a State-Transition device, in many ways analogous to a hardware manufacturer's quoted peak MIPS or MFlop/s. A similar measure is described by (Hockney and Jesshope 1988).

The Processing-Capacity of a Full Automaton is equal to the *size* of its State, in bits (all State can be overwritten, every Cycle).

2.3.2 *Proto-Computers*

In practice, it is not feasible to construct large States that can be read and written in their entirety in a single Cycle. Hence, a modified version of the Full Automaton, namely the Read-Part, Write-Part Automaton, must be used to model a realistic computer. Because of its purpose, and also for the sake of brevity, we call a device of this nature a Proto-Computer.

2.3.2.1 *Uniform-State Proto-Computer* Compared with the Full Automaton, the amount of State that can be accessed at any one time is restricted in a Proto-Computer. Read and write accesses are usually restricted separately. Hence, in any one Cycle, the State is divided into the following three sections:

- The *Active State*: i.e. the part of the State that will be *overwritten* at the end of the current Cycle;

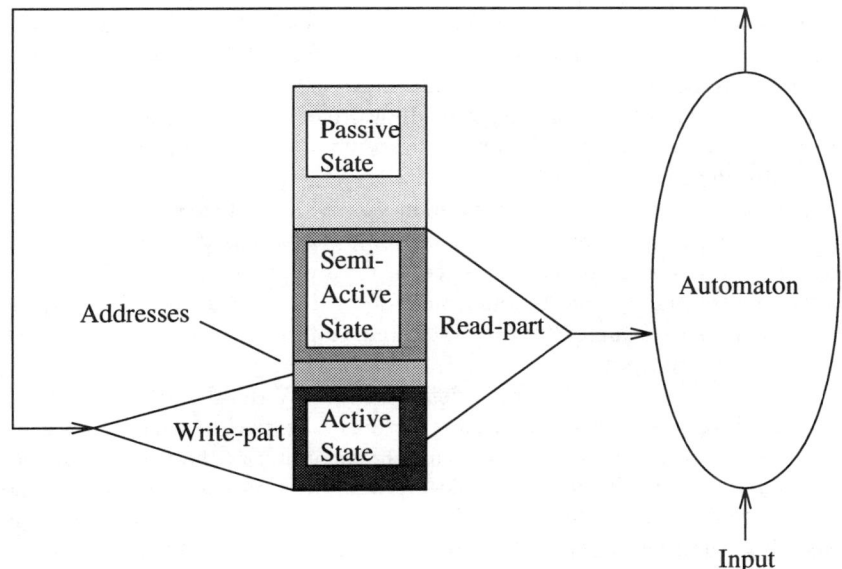

FIG. 2.2. Uniform-State Proto-Computer

- The *Semi-Active State*: i.e. any other part of the State (apart from the Active State) that is *read from* in order to start the current Cycle;
- The *Passive State*: i.e. the remaining part of the State that is neither written to nor read from during the current Cycle.

These different areas of the State are determined at the start of each Cycle, using a reserved part of the Active or Semi-Active State to indicate the Active State Addresses (Addresses, for short). Naturally, these Addresses can be overwritten, in full or in part, at the end of a Cycle, so that the Active and Semi-Active areas can move around the entire State over a period of time. By executing additional Cycles, and using additional State to hold the Addresses, a Proto-Computer can be programmed to emulate the behaviour of any Full Automaton.

In the Uniform-State Proto-Computer, Addresses are allowed to roam around the entire State without further constraint. This is akin to the classical 'von Neumann' computer arrangement, in which there is no distinction between 'code' and 'data'. Figure 2.2 illustrates this arrangement (omitting the details of how Addresses are formed).

The Processing-Capacity (in bits/Cycle) of a Uniform-State Proto-Computer is equal to the *maximum* size (in bits) of its Active State.

2.3.2.2 *Split-State Proto-Computer* In real computers, part of the State represents a *program* that is used to determine what actions are to be performed on the other parts of the State. Unconstrained ability to modify

30   SUPERCOMPUTING TERMINOLOGY

State, especially the ability to modify the 'code' of the program itself, leads to bad software engineering practices. Hence, it is usual now for real computers to separate 'code' from 'data'.

This practice is also prevalent in digital electronic engineering, where the 'code' for a specific function is 'hard-wired' in to a hardware State-Transition device.

In order to model this, it is convenient to divide the State into two parts, one fixed (the Fixed-State) and one variable (the Writeable-State). In the Fixed-State, a programmed sequence of data-dependent transitions that are to be executed by the Automaton is defined. The executed sequence is known as a *Code-Sequence**. The Writeable-State then records the progress (if any) in the computation.

Execution commences from a defined Initial Writeable-State, and the Automaton is then allowed to process the entire State (but without altering the Fixed-State), Cycle-by-Cycle, until it encounters a halting condition, at which point the Final Writeable-State represents the outcome of the computation. Each executed Code-Sequence is defined uniquely by the Fixed-State and the Initial Writeable-State, and is determinate†.

The concepts of Active State, Semi-Active State and Passive State extend naturally to form the following five categories of State during each Cycle:

- The *Semi-Active Fixed-State*: Instructions read at the start of (and probably being executed during) the current Cycle;
- The *Passive Fixed-State*: Instructions unused in the current Cycle;
- The *Active Writeable-State*: Data that will be overwritten at the end of the current Cycle;
- The *Semi-Active Writeable-State*: Data read at the start of the current Cycle;
- The *Passive Writeable-State*: Data unused in the current Cycle.

This arrangement is known as a Split-State Proto-Computer, and is illustrated in Fig. 2.3 (here again, the method of forming Addresses is omitted).

The Processing-Capacity of a Split-State Proto-Computer is equal to the maximum size of its Active Writeable-State.

2.3.2.3 *Comment*   It is important to remember that a Proto-Computer is *not* the same thing as a computer. Rather, it is an idealized (but relatively

---

*The Fixed-State defines a sequential 'program' of behaviour for the Proto-Computer. Depending on whether the associated Hierarchy is Hard or Soft, this program may be either *hard-wired* (e.g. microcode) or *programmable* (e.g. user-code).

†That is, the same initial conditions (of Fixed-State and Initial Writeable-State), and the same sequence of inputs, always cause exactly the same sequence of State-Transitions to be followed.

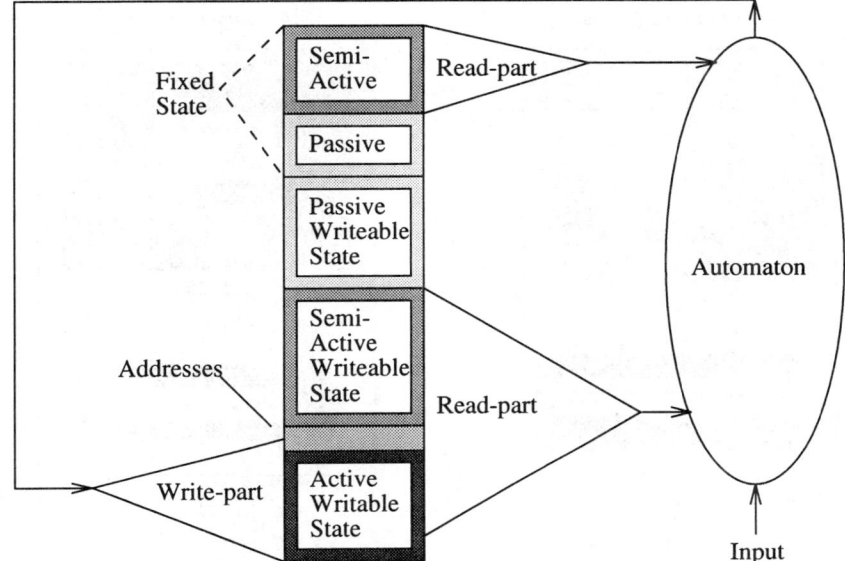

FIG. 2.3. Split-State Proto-Computer

practical) State-Transition device that is used to model discrete behaviour at many different Levels of Design, in both Hard and Soft Hierarchies.

Of course, the Proto-Computer model is often used to describe the instruction-level behaviour of a computer, and, for most modern computers, this behaviour can be modelled accurately by this means. However, it is only an approximate model of the hardware structure of a real computer.

The next two sections illustrate how a real computer can be implemented, in practice, using 'sub-systems' based on the State-Transition principles of the Proto-Computer.

2.3.3 *Processors*

A Processor is the simplest possible *implementation* of a Split-State Proto-Computer. It comprises a Sequence-Controller, which is responsible for executing coded instructions according to the defined State-Transition Cycle, plus a closely-coupled Memory, which is divided into a fixed part (the Fixed-Memory[†]) and a variable part (the Writeable-Memory).

Processors are often designed for a special purpose, and it is convenient to distinguish between[‡]:

[†]In this simple implementation, the Fixed-Memory corresponds to hard-wired behaviour, or microcode, rather than 'user-code'. The latter is 'stored' in the Writeable-Memory.

[‡]The first classification of this nature was the PMS notation, introduced by (Bell and Newell 1971).

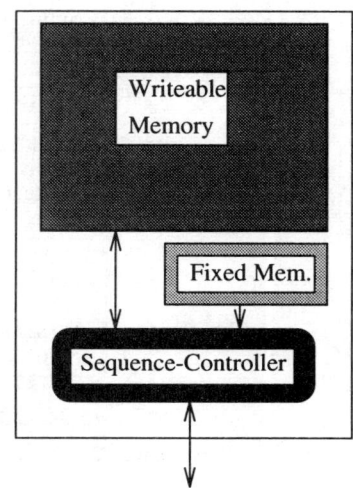

FIG. 2.4. Instruction-Processor and Memory-Processor structures

- *Instruction-Processors*, whose Fixed-Memory is used first to access parts of its (and other's) Writeable-Memory and to interpret and execute the resulting bits of data as a programmed Instruction;

- *Memory-Processors**, whose Fixed-Memory defines simple, pre-programmed responses to external requests (a typical Memory-Processor might respond to requests to *read* or *write* a location in its Writeable-Memory); and

- *Switch-Processors*, whose Fixed-Memory simply controls the routing of input data towards selected output data paths (possibly providing some form of 'buffering' for data *in transit*).

An Instruction-Processor and a Memory-Processor are illustrated in Fig. 2.4. It will be seen that the relative 'balance' between Fixed-Memory, Writeable-Memory, and Sequence-Controller is significantly different in these two devices.

The Processing-Capacity of a Processor is defined as the maximum amount of Writeable-Memory that can be overwritten in a single Cycle (i.e. the maximum size of the Active Writeable-State of the corresponding Split-State Proto-Computer).

---

*Often called *Memory Elements* or *Memories*.

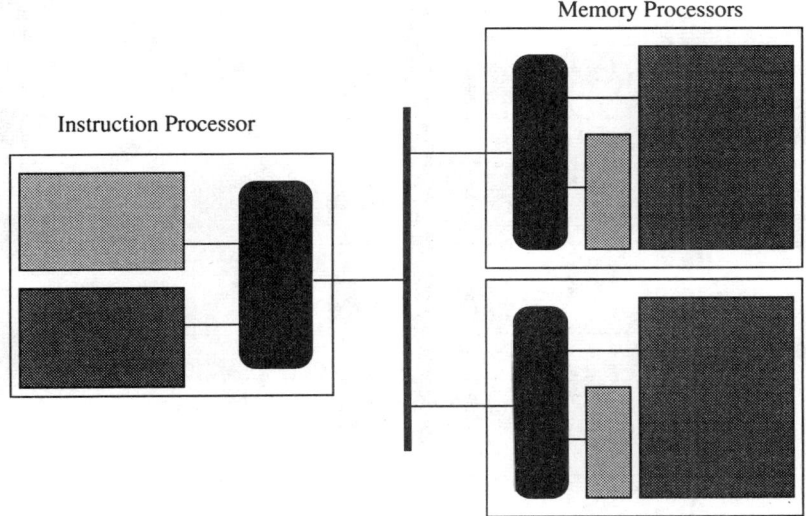

FIG. 2.5. Code-Sequential Computer implementation

2.3.4 *Computers*

A true Computer[†] is constructed using several Processors[‡]. It should be remembered that each Processor executes its hard-wired Code-Sequence sequentially, whilst accessing part of its Writeable-Memory, concurrently, during each Cycle. Each Processor can communicate with another Processor, and possibly change the content of the other Processor's Writeable-Memory, by sending a special form of *message* to the other Processor, either directly, or through a *network*.

The Processing-Capacity of *any* Computer is the sum of the Processing-Capacities of its Component Processors.

2.3.4.1 *Code-Sequential Computer*

The simplest form of Computer is the Code-Sequential Computer, which is a combination of one Instruction-Processor plus one-or-more Memory-Processors. This configuration is illustrated in Fig. 2.5.

Looked at as an entity, rather than as the sum of its Components, a Code-Sequential Computer exhibits the same kind of behaviour as a Proto-Computer. There is only one Instruction-Processor and, therefore, only one Code-Sequence visible at its highest-level. Hence, its overall behaviour is

---

[†] We use the generic term Computer to refer to any of the specific types of real Computer described in this section.

[‡] One is inevitably led to wonder why more than one Processor is *necessary*, but clearly it isn't. In fact, it is simply a consequence of the way that computer technology has *happened* to develop.

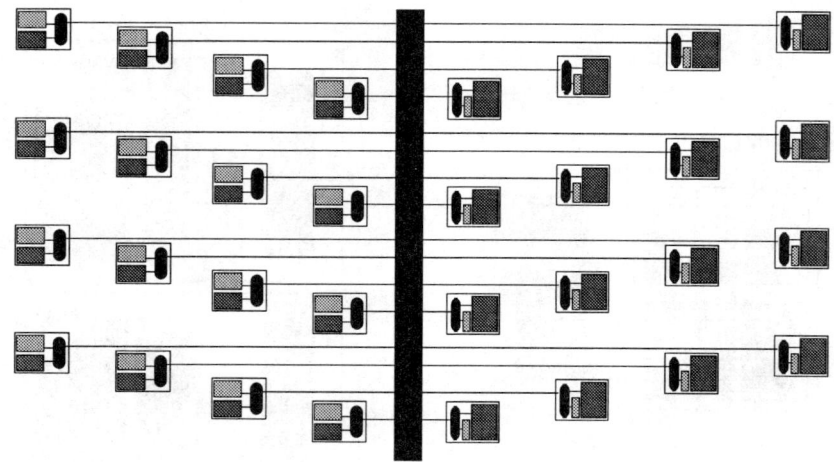

FIG. 2.6. Shared-Memory Code-Parallel Computer

both cyclical and determinate, just like the Split-State Proto-Computer[†]. The Memory-Processors act in a passive fashion, merely responding to the read and write requests generated by the one-and-only Instruction-Processor.

However, the location of its 'Fixed-State' and 'Writeable-State' is different, in that *both* parts are implemented in the Writeable-Memories of its constituent Memory-Processors. Equally, the notion of its Code-Sequence is different, since this is formed by the combined actions of the multiple Code-Sequences in the constituent Processors. Whilst the mapping between the two views is relatively straightforward, the reader needs to remember that this transformation has occurred. In the remainder of the paper we shall refer to the Fixed-State, Writeable-State and Code-Sequence of a Computer without further mention of these subtleties.

2.3.4.2 *Code-Parallel Computers* By adding more Instruction-Processors, it is feasible to construct a Computer in which multiple Code-Sequences are active. A device of this nature is called a Code-Parallel Computer. Unless its multiple Code-Sequence executions are strongly synchronized, it is possible to create non-determinate behaviour.

Examples of two kinds of Code-Parallel Computer are shown in Fig. 2.6 and Fig. 2.7. Both kinds comprise combinations of multiple Instruction-Processors and multiple Memory-Processors.

[†]We ignore the potentially non-deterministic effects of hardware interrupts.

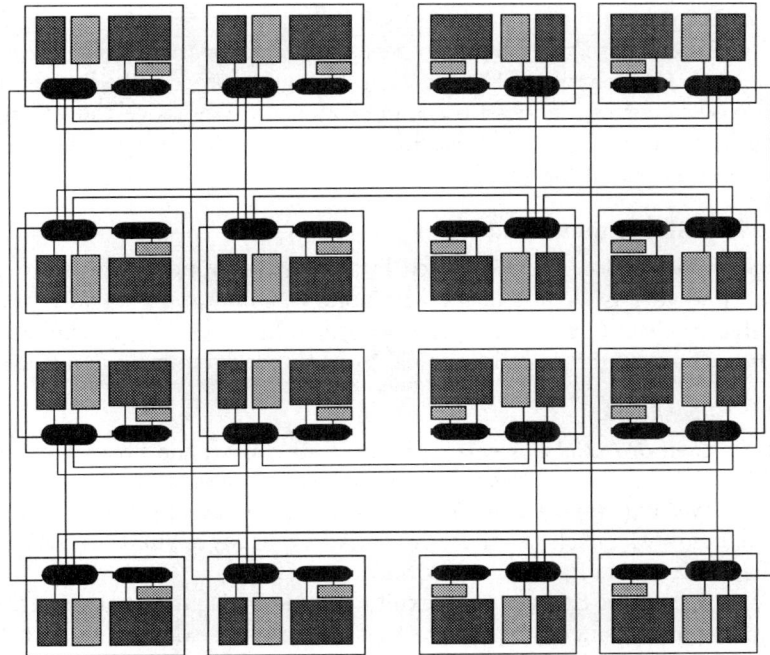

FIG. 2.7. Distributed-Memory Code-Parallel Computer

The configuration in Fig. 2.6 is known as a Shared-Memory, Code-Parallel Computer (Multi-Processor, for short)[††]. There is no particular relationship between the Instruction-Processors and the Memory-Processors of a Multi-Processor. Usually, the numbers of Instruction-Processors and Memory-Processors can be adjusted independently.

In contrast, each Instruction-Processor can be encapsulated with a Memory-Processor, as shown in the configuration of Fig. 2.7. Here, the numbers of each are equal. This configuration is known as a Distributed-Memory, Code-Parallel Computer (Multi-Computer, for short)[‡].

2.3.5 *Performance*

The Processing-Rate (in bits/Cycle) of a Computer is the *average* number of bits that are *actually* being written to its Writeable-Memory during a *typical* Cycle[§]. The Processing-Rate can also be measured in units other

---

[††] All Computers are, in a literal sense, Multi-Processors. The name has, however, become especially associated with Shared-Memory, Code-Parallel Computers.

[‡] The Multi-Computer gets its name because it is structured as a collection of multiple Code-Sequential Computers.

[§] Of course, the Processing-Rate can and should depend on the software being executed, according to the way we define 'typical' and 'average'.

than bits/Cycle[†].

The Processing-Rate effectively measures the *sustained performance* of a Computer for a particular Application. It is guaranteed to be less than, or equal to, the corresponding Processing-Capacity. We gauge the *efficiency* of the Computer for the Application by dividing the Processing-Rate by the Processing-Capacity.

2.4 Parallel SuperComputing

The previous section was motivated by the desire to model hardware Systems. However, as we noted, its terminology should apply equally well to Soft Hierarchies. This section proposes generalisations of the terminology, and how these might describe typical 'soft' SuperComputing activities.

2.4.1 *Classification*

2.4.1.1 *Code-Parallel Systems* As observed earlier, the State-Transition model can be used to describe Systems at many different Levels. A System at one Level is composed of Components that constitute the Systems at the next 'lower' Level. Activity at each Level may exhibit either Code-Sequential or Code-Parallel behaviour.

For example, a System that exhibits Code-Sequential behaviour, but is implemented using Components that collectively exhibit Code-Parallel behaviour is termed an Implicit Code-Parallel System. An example of this is the use of multiple function units in the implementation of the CDC 6600, which do not affect the appearance of the basic, Code-Sequential, 'von Neumann' instruction-cycle of this Code-Sequential Computer.

A System that exhibits Code-Parallel behaviour and is also implemented using Code-Parallel Components is termed an Explicit Code-Parallel System. An example of this is the implementation of Ada processes using a Multi-Processor. Here, both Levels exhibit Code-Parallel behaviour independently, but the required behaviour of the 'higher' Level System is forced onto the 'lower' Level System.

It is feasible to implement a System with Code-Parallel behaviour using Code-Sequential Components, by means of *time-sharing* the sequential state-transition implementation. We call this a Virtual Code-Parallel System. An example is the implementation of Ada processes using any Code-Sequential Computer.

Code-Sequential Systems implemented using Code-Sequential Components (e.g. Code Sequential Computers made from multiple Processors) do not have a special name.

2.4.1.2 *Data-Sharing* In a Code-Parallel Computer, the relevant parts of its Fixed-State must be availabie to each Code-Sequence so that each

---

[†]In particular, the floating-point Processing-Rate, in Flop/s or MFlop/s, is frequently quoted.

is capable of being executed in a physically disjoint Instruction-Processor, if so scheduled. Code-Sequences on disjoint Instruction-Processors *communicate*, so as to co-operate (or interfere) with one another, by means of *data-sharing* (i.e. by sharing access to all, or part of, the Writeable-State). Various mechanisms are used to allow Code-Sequences to synchronize their accesses to shared data and to constrain them so that they cannot easily run amok over the entire Writeable-State.

The highest degree of security is afforded by encapsulating each Code-Sequence in a protected environment, called a Process, and restricting communication at the program level so that it can only be achieved by *message-passing* between the Processes. In this scheme, the shared Writeable-State is distributed amongst the co-operating Processes, each part being directly accessible only to the 'owning' Process. Access to Writeable-State 'owned' by another Process is achieved via a run-time mechanism that implements a robust high-level communication protocol.

Compared with our models of real Computers (see Section 2.3.4), this represents another shift of Level, towards greater abstraction: Gradually, the hardware implementation is becoming less apparent. Although a Process behaves in the same fashion as a Code-Sequential Computer (and, hence, as a Proto-Computer), the relationship between its Fixed-State, Writeable-State and Code-Sequence and the Fixed-Memories and Writeable-Memories of its constituent Processors is highly complex, and perhaps even changeable *during* 'execution'.

2.4.1.3 *Data-Parallel Systems* Independently of Code-Parallel behaviour, it is usual (but not necessary) to process different parts of the Writeable-State concurrently. We term this Data-Parallel activity, and it is implemented in all modern Code-Parallel Computers.

The most common form of Data-Parallel activity is the parallel processing of the bits of a 'word' in a Code-Sequential Computer. The concept can be extended to allow concurrent processing of many words in parallel, as occurs in SIMD array processors, such as the AMT DAP*. Of course, in a practical SuperComputer System, there is a limit to the amount of Writeable-State that can be overwritten during a single State-Transition Cycle.

2.4.2 *Quantification of Parallelizm*

Perhaps more than anywhere else in parallel computing, confusion surrounds the terms used to describe the 'amount' of parallelism being considered in a parallel System. Qualitative terms, such as *low*, *medium* or *high*, have been used inconsistently in the literature to refer to many different degrees of parallelism. This section attempts to define such terms

---

*Distributed Array Processor (Reddaway 1973).

precisely.

It is important to be able to define these kinds of characteristic. Most parallelizing and vectorizing compilers incorporate some form of internal model of the target Architecture. With only occasional exceptions (Sarkar 1992), these models are rarely made explicit, and they usually include only the number of Processors and the times for various overheads.

2.4.2.1 *Measuring 'Parallel Size'* In order to measure the 'parallel size' of a System, we are essentially interested in the ratio between two quantities at different Levels of Design. The first quantity represents the absolute 'size' of the System, at the 'current' Level. The second quantity represents the absolute 'size' of its Sub-Systems (i.e. either its Components, at the next 'lowest' Level, or its sub-Components, at some other 'lower' Level). The ratio between these quantities is the amount of parallelism at the 'current' Level relative to the relevant 'lower' Level. We call this the System-Size, which must always be quoted relative to a named 'lower' Level.

One way of estimating the System-Size is simply to count the number of Sub-Systems at the relevant 'lower' Level. However, in doing this, there is an assumption that the Sub-Systems are all of equal absolute 'size'.

Another approach is to use the concept of Processing-Capacity, which we introduced in Section 2.3.1.1. Although Processing-Capacity is a measure of the *rate* of processing, it is clearly related to the physical size of the State*. Throughout Section 2.3, we gave specific examples of Processing-Capacity, demonstrating that it is relatively easy to measure in realistic hardware Systems. Computing the ratio of Processing-Capacities measured at different Levels should lead to a more precise assessment of System-Size.

Applied to Systems in Hard Hierarchies, the Processing-Capacity is an absolute property of the System Architecture and is not associated with any Application running on it. There is nothing to stop us using the same definition for Soft Hierarchies, but then its value will vary according to the Application, and it is not immediately obvious how it might be measured[†].

In summary:

- A Parallel System at a given Level is one whose Processing-Capacity exceeds the Processing-Capacity at some 'lower' Level.
- The System-Size (of the given Level *relative* to the 'lower' Level) is the ratio of the relevant Processing-Capacities.
- Arbitrarily, a Highly-Parallel System is defined as one in which the System-Size exceeds 100, and a Massively-Parallel System is one with

---

*Again, we believe the important part of State is that which gets written to, and we ignore the parts that get read from.

[†]The data-flow research community has developed a neat method for measuring the effective 'parallel width' of a data-flow program (Arvind *et al.* 1988, Gurd *et al.* 1987), which is an interesting example of what might be achievable in this area.

a System-Size greater than 10^4. We use the term Moderately-Parallel to refer to a System-Size of order 10.
- A Parallel Family is a set of related Parallel Systems of increasing System-Size.

2.4.2.2 *Scalability* An important property of Parallel Systems is the extent to which their System-Size can be increased:
- An *Extensible Family* is a Parallel Family with the property that the Processing-Capacity of its members increases as a strictly monotonic function* of the System-Size, up to some fixed limit known as the Maximum Family Size.
- A *Scalable Family* is an Extensible Family which also has the property that the Processing-Rate of the members of the Family increases as a strictly monotonic function of the System-Size†, up to the Maximum Family Size.

2.5 Conclusion

We have tried to introduce a consistent and comprehensive terminology for describing SuperComputing systems, techniques and tools. The immediate benefits of this should be:
- a clearer understanding of the Design Space for (parallel) SuperComputing;
- a common language in which to describe and discuss problems and solutions.

In addition, we can expect longer-term benefits as people realize that they are facing common problems, rather than continuing to work as if in isolation. In particular, we should anticipate a more systematic approach to future research and engineering practice.

2.6 Acknowledgements

We are endebted to the members of the Centre for Novel Computing and other colleagues at the University of Manchester for contributing to the discussions that led to this paper. In particular, we wish to record our thanks to Cliff Jones and Graham Riley.

---

*A strictly monotonic increase is one which continues for all possible values of the independent variable, in this case the System-Size: There is no asymptote.

†Of course, this is only possible for a program which exhibits a suitably large amount of exploitable parallelism, which we assume to be available. In practice, an asymptote will be imposed by any real program. The given definition is that of 'in-principle' scalability: Any Extensible Family for which some increase in the System-Size leads to a decrease in the Processing-Rate is not Scalable.

REFERENCES

Arvind, Culler, D.E., and Maa, G.K. (1988). *Assessing the Benefits of Fine-Grained Parallelism in Dataflow Programs*. Technical Report 279, Computational Structures Group, Laboratory for Computer Science, Massachussetts Institute for Technology, Cambridge, MA.

Bell, C.G., and Newell, A. (1971). *Computer Structures: Readings and Examples*. McGraw-Hill, New York.

Dongarra, J.J., et al. (1984). Implementing Linear Algebra Algorithms for Dense Matrices on a Vector Pipeline Machine. *SIAM Review*, **26**, 91–112.

Dongarra, J.J., et al. (1988). *A Set of Level 3 Basic Linear Algebra Subprograms*. Technical Report ANL-MCS-P1-0888, Argonne National Laboratory, Chicago.

Duff, I.S. (1985). The Use of Supercomputers in Europe. *Computer Physics Communications*, **37**, 15–25.

Gurd, J.R., et al. (1987). Performance Issues in Dataflow Machines. *Future Generation Computer Systems*, **3**(4), 285–297.

Gurd, J.R., et al. (1993). *A Framework for Experimental Analysis of Parallel Computing*. Technical Report UMCS-93-2-3, Department of Computer Science, University of Manchester.

Hoare, C.A.R. (1987). An Overview of some Formal Methods for Program Design. *IEEE Computer*, **20**(9), 85–91.

Hockney, R.W., and Jesshope, C.R. (1988). *Parallel Computers 2*. Adam Hilger, Bristol.

Nelson, G. (1991). *Systems Programming with Modula-3*. Prentice-Hall, Englewood Cliffs, NJ.

Peters, L. (1988). *Advanced Structured Analysis and Design*. Prentice-Hall, Englewood Cliffs, NJ.

Reade, C. (1989). *Elements of Functional Programming*. Addison-Wesley, Reading, MA.

Reddaway, S.C. (1973). DAP—A Distributed Array Processor. *Proceedings 1st ACM/IEEE International Symposium on Computer Architecture*.

Sarkar, V. (1992). Automatic Partitioning of a Program Dependence Graph into Parallel Tasks. *Proceedings 3rd Workshop on Compilers for Parallel Computers*, Technical Report ACPC/TR 92-8, Austrian Centre for Parallel Computation, Vienna, pp 70–95.

Shannon, C., and Weaver, W. (1949). *A Mathematical Theory of Communication*. University of Illinois Press, Champaign, IL.

3

BULK SYNCHRONOUS PARALLEL COMPUTING

W. F. McColl

Programming Research Group,
Oxford University,
11 Keble Road,
Oxford OX1 3QD, UK
mccoll@prg.oxford.ac.uk

Abstract

Bulk synchronous parallel architectures offer the prospect of achieving both scalable parallel performance and architecture independent parallel software. They provide a robust model on which to base the future development of general purpose parallel computing systems. In this paper we discuss some of the current issues involved in the development of parallel systems which support both fine-grain concurrency and global memory models. We describe the BSP and PRAM models, and demonstrate how they can be efficiently realized on distributed memory parallel architectures.

3.1 Introduction

For most of the 1980s, low-level hardware considerations have been the main driving force in parallel computing. Rapid progress in VLSI technology has permitted the development of a wide variety of distributed memory multicomputer architectures (Athas and Seitz 1988, INMOS Limited 1988, Seitz 1985, Seitz 1990, Whitby-Strevens 1985). These systems consist of a set of general-purpose microprocessors connected by a sparse network, e.g. array, butterfly, or hypercube. The relatively low speed and capacity of such networks forces the programmer to think in terms of a model in which one has multiple private address spaces connected in some complex way, e.g. in a hypercube structure, with explicit message passing by the programmer (Hoare 1985, Hoare 1991, Jones and Goldsmith 1988) for all non-local memory requests. The key to algorithmic efficiency in such systems is the careful exploitation of network locality. By minimizing the number of nodes through which a message has to travel one can substantially improve efficiency. Despite the programming difficulties inherent in this approach, a large amount of scientific and technical applications software has been developed for such systems. In positive terms, this work has demonstrated conclusively that for many important applications, scalable parallel performance can be achieved in massively parallel systems (Gustafson 1988a, Gustafson 1988b), despite the reservations expressed by Amdahl (Amdahl 1967). However, in this message-passing approach, most of the effort in software development tends to be devoted to the various low-level process mapping activities which need to be performed to achieve efficiency. Besides being extremely tedious in many cases, this usually produces software which cannot easily be adapted to another architecture. In a world of rapidly changing parallel architectures, this architecture-dependence has proved to be a major weakness, and it has inhibited the growth of the field beyond the area of scientific research.

An alternative approach, which has been extensively pursued by computer science researchers in the last decade, is to make software the driving force. A variety of approaches of this kind has been investigated. They differ in terms of the type of programming language considered, e.g. functional (Bird and Wadler 1988, Hudak 1989, Hudak et al. 1991) , single assignment, logic, mostly functional, and in the computational model which they adopt, e.g. graph reduction, rewriting, dataflow. However, they share a number of similarities, particularly in comparison to the framework proposed in this paper. One example of this approach is where one starts by noting that a high-level functional language (Bird and Wadler 1988, Hudak 1989, Hudak et al. 1991) (if properly used) can often expose a large amount of implicit parallelism in a computational problem. The decision to work with a functional language, for reasons of architecture-independence, naturally leads to a decision to adopt, say, graph reduction as the model of parallel computation. The technological (hardware) goal is then to develop a scalable massively parallel architecture for graph reduction (Peyton Jones 1989). This 'software first' approach has a great deal of merit given that hardware is changing rapidly and that the cost and time required to produce software makes architecture-independence in software a major goal. Unfortunately, however, the amount of progress which has been made on the development of efficient parallel architectures for graph reduction, dataflow or rewriting has not been particularly impressive so far, despite much effort. The experiences of the last decade suggest that, in the pursuit of efficiency, it is often necessary to compromise some of the elegance and simplicity of such approaches. To a large extent, this has already happened in dataflow implementations of functional languages, e.g. the implementation of Id (Nikhil 1991) on Monsoon (Papadopolous and Traub 1991). Modern dataflow architectures (Iannucci 1988, Iannucci 1990, Nikhil and Arvind 1989, Nikhil et al. 1991, Papadopolous and Traub 1991) are in many important respects quite close to those described in this paper. For example, they must achieve latency tolerance through multithreading since, in the dataflow model, memory accesses are *split transactions* (Nikhil et al. 1991). (In the dataflow model, a read may be requested before the value is computed.)

A third alternative is to have some model of parallel computation as the driving force. Around 1944, von Neumann produced a proposal (Burks *et al.* 1946, von Neumann 30 June 1945) for a general-purpose stored-program sequential computer which captured the fundamental principles of Turing's work (Turing 1936) in a practical design. The design, which has come to be known as the 'von Neumann computer', has served as the basic model for almost all sequential computers produced from the late 1940s to the present time. As noted in (Hennessy and Patterson 1990), 'The paper by Burks, Goldstine and von Neumann ((Burks *et al.* 1946)) was incredible for the period. Reading it today, one would never guess this landmark paper was written more than 40 years ago, as most of the architectural concepts seen in modern computers are described there.' For an account of the principles of modern general-purpose sequential (i.e. von Neumann) computer design, see (Hennessy and Patterson 1990). For sequential computation, the stability of the von Neumann model has permitted the development, over the last three decades, of a variety of high-level languages and compilers. These have, in turn, encouraged the development of a large and diverse software industry producing portable applications software for the wide range of von Neumann machines available, from personal computers to large mainframes. The stability of the underlying model has also allowed the development of a robust complexity theory for sequential computation, and a set of algorithm design and software development techniques of wide applicability. General-purpose sequential computing based on the von Neumann model has developed vigorously over the last four decades. The widespread adoption of the model has not proved to be a harmfully constraining influence, in fact, it has been quite the reverse. A variety of hardware approaches have flourished within the framework provided by the model. The stability it has provided has been invaluable for the development of the software industry.

No single model of parallel computation has yet come to dominate developments in parallel computing in the way the von Neumann model has dominated sequential computing (Gear 1991, Valiant 1990a, Valiant 1990b). Instead we have a variety of models such as VLSI systems, systolic arrays and distributed memory multicomputers in which, as we have noted, the careful exploitation of network locality is crucial for algorithmic efficiency. We will use the generic term 'special-purpose' to refer to this type of parallel computing. In (McColl 1993c) we describe a number of aspects of the work which has been done in recent years on the design, analysis, implementation, and verification of such special-purpose parallel computing systems. A major challenge for contemporary computer science is to determine the extent to which general-purpose parallel computing can be achieved. The goal is to deliver both scalable parallel performance and architecture independent parallel software. (Work on special purpose parallel computing (Mc-

Coll 1993c), and on architectures for declarative languages (Hudak 1989, Hudak et al. 1991, Nikhil 1991, Nikhil and Arvind 1989, Nikhil et al. 1991, Papadopolous and Traub 1991), having demonstrated that either of these alone can be achieved.)

Can we identify a robust model of parallel computation which offers the prospect of achieving the twin goals of general-purpose parallel computing – scalable parallel performance and architecture-independent parallel software ?

Success in this endeavour would permit the long overdue separation of software and hardware considerations in parallel computing. This separation would, in turn, encourage the growth of a large and diverse parallel software industry, and provide a focus for future hardware developments.

The achievement of these goals would have profound consequences for the future development of both the computing industry and the academic subject of computer science. Given this fact, one might suspect that this issue would be central to much of the current research in parallel computing. However, at present, relatively little work is being done with these goals directly in mind. Much of the practical work in massively parallel computing today is concerned with the development of scientific applications software, without particular regard for the development of a credible strategy which would permit portability of that software as new architectures appear.

The current situation in parallel computing is remarkably chaotic when compared with that of sequential computing. With no agreed model to provide a focus for technological innovation, parallel hardware suppliers continue to develop, and attempt to market, systems with widely differing characteristics. Those people with the unenviable task of choosing a parallel system for their organization are faced with the prospect of investing substantial resources in the purchase of such a machine, and in the development of software for it, only to find that the software quite quickly becomes obsolete. At the present time, the MFLOP performance of the processors used in parallel systems is increasing rapidly. Unfortunately, this is not being matched by corresponding increases in communications performance. This rising imbalance is likely to further increase the difficulty of achieving architecture-independence in software. An important general message of the results in this paper is that architecture-independence is more likely to be achieved in those parallel systems which invest more substantially in communications performance than in processor performance. It is striking that relatively few of the commercial parallel systems being produced today seem to reflect this basic idea.

The current chaos in parallel computing has led many to conclude that the answer to the above question, on the prospects for agreement on a model, is no. Advocates of 'heterogeneous parallel computing' (Kung 1991) take as their starting point the idea that no convergence on a model is likely to take place. They argue that a wide variety of designs which are to some extent 'special-purpose' will continue to be produced and marketed, and that the primary function of parallel computing should be to develop languages and communications networks for the coordination of these ensembles of devices. It is again striking that many in computing have already accepted the inevitability of this rather pessimistic scenario, especially as no serious theoretical impediments to the achievement of the goals of general-purpose parallel computing have yet been identified, despite much effort to find them. One can contrast this with the situation in complexity theory where the ideas of NP-completeness have demonstrated in a precise way that many desirable goals in terms of algorithmic performance, for problems in AI, scheduling, optimization, and so on, are unlikely to be achievable and that we must, in some way, limit our expectations. There is no compelling evidence that general-purpose parallel computing, as described above, cannot be achieved. We can be reasonably confident that, as future hardware developments alone fail to significantly increase the market for parallel systems, the manufacturers of those systems will see it as in their interests to seek convergence on a model, rather than to seek to avoid it. A major goal for computer science today is to develop the ideas and techniques which will provide the required solutions when that change in thinking comes about.

In this paper we will describe one possible way forward for parallel computing, based on the bulk synchronous parallel (BSP) model of computation (Valiant 1990a). While we believe that this approach has many strengths, we would not want to argue that it is the only viable approach. Two alternatives, which merit serious consideration, are the actor model (Agha 1990) and the dataflow model (Iannucci 1988, Iannucci 1990, Nikhil 1991, Nikhil and Arvind 1989, Papadopolous and Traub 1991). The most fundamental difference between these two approaches and the BSP model is that they both have at their core the idea of local (usually pairwise) synchronization events, whereas the BSP model, as well as various PRAM (Gibbons and Spirakis 1993, JáJá 1992, Valiant 1990b, Vishkin 1991b) and data parallel models (Blelloch 1990, Hillis and Steele Jr 1986), have the idea of global barrier synchronization as the basic mechanism. Another significant difference is that in the BSP, PRAM and data parallel approaches there is usually tight control of ordering and scheduling by the programmer. In contrast, a major attraction of the dataflow approach is that the programmer is freed from consideration of such issues. Although we have stressed the differences between these various approaches, there is reason to believe that, at the architectural level, the BSP, PRAM, actor, and dataflow models will require a number of similar mechanisms for efficient implementation, in particular, high-performance global communications, uniform memory access, and multithreading to hide network latencies. It is perhaps not unreasonable to summarize the current situation with respect to these various approaches as follows. Work on the actor and dataflow models is much more highly developed in the areas of programming languages and methodologies than it is in the area of algorithm design, analysis and complexity. In contrast, for the BSP and PRAM models we have a highly developed set of techniques for the design and analysis of algorithms, but we do not yet have an established framework for the programming of such systems.

3.2 Parallel Random Access Machine

Various idealized models of parallel computation have been used in the study of parallel algorithms and their complexity. Of these, the most widely studied has been the parallel random access machine. A *parallel random access machine (PRAM)* consists of a collection of processors which compute synchronously in parallel and which communicate with a common global random access memory. In one time step, each processor can do (any subset of) the following: read two values from the common memory, perform a simple two-argument operation, write a value back to the common memory. There is no explicit communication between processors. Processors can only communicate by writing to, and reading from, the common memory. The processors have no local memory other than a small fixed number of registers which they use to temporarily store the argument and result values. In a *Concurrent Read Concurrent Write (CRCW) PRAM*, any number of processors can read from, or write to, a given memory cell in a single time step. In a *Concurrent Read Exclusive Write (CREW) PRAM*, at most one processor can write to a given memory cell at any one time. In the most restricted model, the *Exclusive Read Exclusive Write (EREW) PRAM*, no concurrency is permitted either in reading or in writing. The CRCW PRAM model has a large number of variants which differ in the convention they adopt for the effect of concurrent writing. Three simple examples of such conventions are: two or more processors can write so long as they write the same value, one of the processors attempting to write will succeed but the choice of which one will succeed will be made nondeterministically, the lowest numbered processor will succeed (assuming some appropriate numbering.) In other CRCW models (Ranade 1989) one might have the possibility of concurrent writing in which the memory location is updated to the sum of the written values, or to the minimum of the written values. An important characteristic of the PRAM model is that it is a one-level memory (or shared memory) model, that is, all of the memory locations are uniformly far away from all of the processors, the processors have no local memory and there is no kind of memory hierarchy based on ideas of network locality. These simplifying properties of the PRAM model have made it extremely attractive as a robust model for the design, analysis, and comparison of algorithms, and we now have a large set of PRAM algorithms for important problems (Gibbons and Spirakis 1993, JáJá 1992, Kruskal *et al.* 1990, Vishkin 1991b).

In discussing the complexity of PRAM algorithms we normally refer only to the parallel time complexity and the number of processors required. However, the communication complexity of PRAM algorithms, in a simplified setting, has also been studied by a number of researchers: see, for example, (Jung et al. 1989, Papadimitriou and Ullman 1987, Papadimitriou and Yannakakis 1990). In their work, the computational problem to be solved is modeled as a dag, with nodes corresponding to the functions computed and arcs corresponding to functional dependencies. The task is to efficiently schedule the dag on a p processor parallel system which may have a large local memory at each processor, that is, to assign each node of the dag to one or more processors in the system which will compute that node. A schedule must satisfy the constraint that a node can only be computed at a given time step if its predecessors have been computed in previous time steps. Communication complexity is captured in an obvious way. If node v depends on node u, that is, there is an arc from u to v in the dag, and u, v are computed in distinct processors, then that arc is said to be a communication arc. The communication complexity c of a given schedule is simply the number of communication arcs in the dag. This measure captures an important practical cost in the implementation of parallel algorithms on a multiprocessor system, that is, the total message traffic generated. A number of important results have been obtained for this model, showing tradeoffs between the time required for a parallel computation and the total number of messages which must be sent. Algorithms for scheduling dags have also been developed. Such work may provide a theoretical basis for the future development of software tools which efficiently schedule shared memory parallel algorithms for implementation on distributed memory architectures.

Aggarwal, Chandra and Snir (Aggarwal et al. 1990) have also studied the communication complexity of PRAM algorithms. They consider the design of efficient algorithms for a model called the *local memory PRAM*, or *LPRAM*, which also captures both the communication and computation requirements of PRAM algorithms in a convenient way. An LPRAM is a CREW PRAM in which each processor is provided with an unlimited amount of local memory. Processors can simultaneously read from the same location in the global memory, but two or more are not allowed to simultaneously write into the same location. The input variables are initially available in the global memory, and the outputs must also be eventually there. The multiprocessor is a synchronous MIMD machine. In order to model the communication delay and computation time, it is convenient to restrict the machine such that, at every time step, the processors do one of the following:

- In one communication step, a processor can write, and then read a word from global memory.

- In a computation step, a processor can perform a simple operation on at most two values that are present in its local memory.

A computation is represented as a dag, and a schedule for a dag consists of a sequence of computation steps and communication steps. At a computation step each processor may evaluate a node of the dag; this evaluation can only take place at a processor when its local memory contains the values corresponding to all of the incoming arcs. After the computation step is completed the values for the outgoing arcs are held in the local memory. At a communication step, any processor may write into the global memory any value that is presently in its local memory, and then it may read into its local memory a value from the global memory. They analyse a number of important problems, in terms of the two LPRAM complexity measures, parallel time (number of computation steps) and communication delay (number of communication steps).

A major issue in theoretical computer science since the late 1970s has been to determine the extent to which the PRAM and related models can be efficiently implemented on physically realistic distributed memory architectures. A number of new routing and memory management techniques have been developed which show that efficient implementation is indeed possible in many cases (McColl 1993a, Valiant 1990a, Valiant 1990b). The efficient implementation of a single address space on a distributed memory architecture requires an efficient method for the distributed routing of read and write requests, and of the replies to read requests, through the network of processors. Consider the problem of packet routing on a p-processor network. Let an h-*relation* denote the routing problem where each processor has at most h packets to send to various points in the network, and where each processor is also due to receive at most h packets from other processors. We are interested in the development of distributed routing methods in which the routing decisions made at a node at some point in time are based only on information concerning the packets that have already passed through the node at that time. Using *two-phase randomized routing* (Valiant 1990b) we can show the following

Theorem 3.2.1 *With high probability, every* 1-*relation can be realized on a p processor cube-connected-cycles, butterfly, 2D array and hypercube in a number of steps proportional to the diameter of the network.*

Theorem 3.2.2 *With high probability, every* ($log\ p$)-*relation can be realized on a p processor hypercube in* $O(log\ p)$ *steps.*

Proofs of Theorems 3.2.1 and 3.2.2 can be found in (Valiant 1990b). (In addition to such theoretical results, randomized routing has also been shown, in many studies, to work extremely well in practice.) In order to show that we can efficiently simulate a shared address space on a distributed memory architecture we also need to show that we can deal with the problem of 'hot spots', that is, where a large number of processors simultaneously try to access the same memory module. One very effective method of uniformly distributing memory references, which has now been widely studied, is to hash the single address space. A detailed technical account of the role of hashing in achieving efficient general-purpose parallel computing on a distributed memory architecture can be found in (Valiant 1990b). We will only mention here the following two results which demonstrate that certain distributed memory architectures can efficiently simulate PRAMs. Let $EPRAM(p,t)$ $[CPRAM(p,t), HYPERCUBE(p,t), COMPLETE(p,t)]$ denote the class of problems which can be solved on a p processor EREW PRAM [CRCW PRAM, hypercube, completely connected network, respectively] in t time steps.

Theorem 3.2.3. ((Valiant 1990b))
With high probability, $EPRAM(p \ log \ p, t/log \ p) \subseteq HYPERCUBE(p,t)$.

Theorem 3.2.4. ((Karp et al. 1992))
With high probability,
$CPRAM(p \ log \ log \ p \ log^*p, t) \subseteq COMPLETE(p, t \ log \ log \ p \ log^*p)$.

Theorems 3.2.3 and 3.2.4 show that PRAM algorithms with a degree of parallel slackness can be implemented on distributed memory architectures in a way which is optimal in terms of the processor-time product.

Definition 3.2.5 *An m processor algorithm, when implemented on an n processor machine, where $n \leq m$, is said to have a* parallel slackness factor *of m/n for that machine.*

Parallel slackness is an idea of fundamental importance in the area of general-purpose parallel computing. If parallel algorithms and programs are designed so that they have more parallelism than is available in the machine, then the available parallel slackness can be effectively exploited to hide the kind of network latencies one finds in distributed memory architectures. The only requirement is that the processors provide efficient support for multithreading and fast context switching (Blumofe 1992, Nikhil et al. 1991, Papadopolous and Traub 1991). Latency tolerance via multithreading is likely to be more effective on large-scale general-purpose parallel computing systems than the use of complex caching schemes for latency reduction.

The idea of exploiting parallel slackness can even be carried over into the area of sequential computing. Much effort in recent years has been devoted to the development of complex heuristic techniques for the efficient prefetching of values from memory in sequential computations. A radical alternative to this approach is, instead, to design parallel algorithms for implementation on sequential machines. The parallel slackness of the algorithm can then be exploited to achieve efficient prefetching. For more on this topic, see (Vishkin 1991a, Vishkin 1992).

We have seen then that by achieving a degree of parallel slackness in program designs one can provide significant opportunities for the effective scheduling of those programs, by the programmer or by a compiler, to hide the various kinds of latencies which arise in both sequential and parallel computing. In recent years, parallel slackness, or overdecomposition, has come to be recognized as crucial, not only for efficient implementation of PRAM like models (Karp *et al.* 1992, McColl 1993a, Valiant 1990a, Valiant 1990b), but also for dataflow models (Nikhil *et al.* 1991). The prospects for 'autoparallelizing' sequential code, which may be regarded as the extreme opposite of this approach, appear very bleak indeed.

A general design principle which emerges from these studies is, therefore, that one should aim, at all times, to produce algorithms and programs which have more parallelism in them than is available in the machine. In the future we can expect to see the development of a variety of programming languages for general purpose parallel computing. A clear message from the above discussion is that such languages must permit, and indeed encourage, the development of programs which demonstrate a high degree of fine-grain concurrency. The GL programming language (McColl 1993b) is being developed with these ideas in mind.

3.3 Bulk Synchronous Parallel Computer

For a detailed account of the BSP model, and of the various routing and hashing results which can be obtained for it, the reader is referred to (Valiant 1990a, Valiant 1992a, Valiant 1992b). We concentrate here on presenting a view of (i) how a bulk-synchronous parallel architecture would be described, and (ii) how it would be used. A *bulk-synchronous parallel (BSP) computer* consists of the following:

- a set of processor-memory pairs
- a communications network that delivers messages in a point-to-point manner
- a mechanism for the efficient barrier synchronization of all, or a subset, of the processors

There are no specialized combining, replication or broadcasting facilities. If we define a time step to be the time required for a single local operation, that is, a basic operation on locally held data values, then the performance of any BSP computer can be characterised by the following four parameters:

- p = number of processors
- s = processor speed, that is, number of time steps per second
- l = synchronization periodicity, that is, minimal number of time steps between successive synchronization operations

- g = (total number of local operations performed by all processors in one second) / (total number of words delivered by the communications network in one second)

The parameter l is related to the network latency, that is, to the time required for a non-local memory access in a situation of continuous message traffic. The parameter g corresponds to the frequency with which non-local memory accesses can be made; in a machine with a higher value of g one must make non-local memory accesses less frequently. More formally, g is related to the time required to realize h-relations in a situation of continuous message traffic; g is the value such that an h-relation can be performed in gh steps.

A BSP computer operates in the following way. A computation consists of a sequence of parallel *supersteps*, where each superstep is a sequence of steps, followed by a barrier synchronization at which point any memory accesses take effect. During a superstep, each processor has a set of programs or threads which it has to carry out, and it can do the following:

- perform a number of computation steps, from its set of threads, on values held locally at the start of the superstep
- send and receive a number of messages corresponding to non-local read and write requests

The BSP computer is a two-level memory model (McColl 1993a), that is, each processor has its own physically local memory module; all other memory is non-local, and is accessible in a uniformly efficient way. By uniformly efficient, we mean that the time taken for a processor to read from, or write to, a non-local memory element in another processor-memory pair should be independent of which physical memory module the value is held in. The algorithm designer/programmer should not be aware of any hierarchical memory organization based on network locality in the particular physical interconnect structure currently used in the communications network. Instead, performance of the communications network should be described only in terms of its global properties, for example, the maximum time required to perform a non-local memory operation, and the maximum number of such operations which can simultaneously be in the network at any time. The complexity of a superstep S in a BSP algorithm is determined as follows. Let L be the maximum number of local computation steps executed by any processor during S, h_1 be the maximum number of messages sent by any processor during S, and h_2 be the maximum number of messages received by any processor during S. The cost of S is then $max\{l, L, gh_1, gh_2\}$ time steps. (An alternative is to charge $max\{l, L+gh_1, L+gh_2\}$ time steps for superstep S. The difference between these two costs will not, in general, be significant.) When g is small, for example, $g = 1$, the BSP computer corresponds closely to a PRAM, with l determining the degree of parallel slackness required to achieve optimal efficiency. For a BSP computer of this kind, that is, with a low g value, we can use hashing to achieve efficient memory management (Valiant 1990a). The case $l = g = 1$ corresponds to the idealized PRAM, where no parallel slackness is required. In designing algorithms for a BSP computer with a high g value, we need to achieve a measure of *communication slackness* by exploiting thread locality in the two-level memory, that is, we must ensure that for every non-local memory access we request, we are able to perform approximately g operations on local data. To achieve architecture independence in the BSP model, it is therefore appropriate to design parallel algorithms which are parametrized not only by n, the size of the problem, and p, the number of processors, but also by l and g. The following example of such an algorithm appears in (Valiant 1990a). The problem is the multiplication of two $n \times n$ matrices A, B on $p \leq n^2$ processors. The standard $O(n^3)$ sequential algorithm is adapted to run on p processors as follows. Each processor computes an $(n/p^{1/2}) \times (n/p^{1/2})$ submatrix of $C = A.B$. To do so it will require $n^2/p^{1/2}$ elements from A and the same number from B. For each processor we thus have a computation requirement of $O(n^3/p)$ operations, since each inner product requires $O(n)$ operations, and a communications requirement of $O(n^3/p)$ for the number of non-local reads, since $p \leq n^2$. If we assume that both A and B are distributed uniformly amongst the p processors, with each processor receiving $O(n^2/p)$ of the elements from each matrix, then the processors can simply replicate and send the appropriate elements from A and B to the $2p^{1/2}$ processors requiring them. Therefore, we also have a communications requirement of approximately $n^2/p^{1/2} = O(n^3/p)$ for messages sent. We thus have a total parallel time complexity of $O(n^3/p)$, provided $l = O(n^3/p)$ and $g = O(n/p^{1/2})$. An alternative algorithm, given in (Aggarwal *et al.* 1990), that requires fewer messages altogether, can be implemented to give the same optimal runtime, with g as large as $O(n/p^{1/3})$ but with l slightly smaller at $O(n^3/p \log n)$.

The BSP model can be regarded as a generalisation of the PRAM model which permits the frequency of barrier synchronization to be controlled. By capturing the network performance of a BSP computer in global terms using the values l and g, the model enables us to design algorithms and programs which are parametrized by those values, and which can therefore be efficiently implemented on a range of BSP architectures with widely differing l and g values. It therefore provides a solution to the problem posed at the start of the paper. We have a simple and robust model which permits both scalable parallel performance and a high degree of architecture independence in software. Its simplicity also offers the prospect of our being able to develop a coherent framework for the design and analysis of parallel algorithms.

The use of the parameters l and g to characterize the communications performance of a BSP computer contrasts sharply with the way in which communications performance is described for most distributed memory architectures on the market today. We are normally told many details about local network properties, for example, the number of communications channels per node, the speed of those channels, the graph structure of the network, and so on. The way in which such descriptions emphasise local properties of the network, rather than its global properties, reflects the fact that most of those machines are designed to be used in a way where network locality is to be exploited. Those customers who have highly irregular problems, for which such exploitation is much more difficult, are often much less impressed by such machines when they are told about the global performance of the network in situations where network locality is not exploited. A major feature of the BSP model is that it lifts considerations of network performance from the local level to the global level. We are thus no longer particularly interested in whether the network is a 2D array, a butterfly, or a hypercube, or whether it is implemented in VLSI or in some optical technology. Our interest is in global parameters of the network, such as l and g, which describe its ability to support non-local memory accesses in a uniformly efficient manner. As an aside, we note that it might be an interesting and instructive exercise to benchmark the various parallel architectures available today, in terms of such global parameters.

In the design and implementation of a BSP computer, the values of l and g which can be achieved will depend on (i) the capabilities of the available technology, and (ii) the amount of money that one is willing to spend on the communications network. As the computational performance of machines, that is, the performance captured by p and s, continues to grow, we will find that to keep l and g low it will be necessary to continually increase our investment in the communications hardware as a percentage of the total cost of the machine. A central thesis of the BSP and PRAM approaches to general-purpose parallel computing is that if these costs are paid, then parallel machines of a new level of efficiency, flexibility, and programmability can be obtained. On the basis of Theorems 3.2.1 and 3.2.2 we might expect to be able to achieve the following values of l and g for a p processor BSP computer, by using the network shown.

Network	l	g
2D Array	$O(p^{1/2})$	$O(p^{1/2})$
Butterfly	$O(log\ p)$	$O(log\ p)$
Hypercube	$O(log\ p)$	$O(1)$

These estimates are based entirely on the asymptotic degree and diameter properties of the graph. In a practical setting, the use of techniques such as wormhole routing (Leighton 1992, Seitz 1990), rather than store and forward routing, would also have a significant impact on the values of l and g which could be achieved.

In the BSP model described above, communication is point-to-point. There are no specialized combining, replication, or broadcasting facilities.

In contrast to this, it has often (Abolhassan *et al.* 1991, Ranade 1989) been proposed that one should use more complex (and costly) routing networks, containing combining hardware, in order to efficiently support a much broader class of communications (and, in some cases, that one should even support simple forms of computation, for example, the computation of prefix sums, within the 'router'). In a recent paper, Valiant (Valiant 1992a) describes a mechanism for recirculating messages in a simple point-to-point routing network so that the added functionality of a network with combining hardware, for arbitrary communication patterns, can be efficiently achieved by the more basic device. There would not appear, therefore, to be a very strong case for modifying the architectural requirements of the BSP model in this direction.

Although we have described the BSP computer as an architectural model, one can also view bulk synchrony as a programming model or, indeed, as a kind of programming methodology. The essence of the BSP approach is the notion of the superstep and the idea that the input/output associated with a superstep (or reading/writing, depending on how one views it) is performed as a global operation, involving a whole set of individual sends and receives. Viewed in this way, a 'BSP program' is simply one which proceeds in phases, with the necessary global communications taking place between the phases. The BSP approach can, therefore, be regarded as a programming methodology which is applicable to all kinds of parallel architecture, for example, shared memory multiprocessors, distributed memory architectures, or networks of workstations. (The values of g in such different forms of architecture would, of course, vary enormously.) It would appear then, that the BSP approach provides a consistent, and very general, framework within which to develop portable parallel software for the wide range of parallel architectures which are likely to emerge in the future.

3.4 Challenges

In the previous sections we have seen that there are a variety of theoretically and practically efficient solutions to the problem of supporting a single address space on a distributed memory architecture. In this section we briefly describe some of the main issues which need to be addressed in the future in order to continue the development of this framework for general-purpose parallel computing based on fine-grain concurrency in a shared address space.

3.4.1 *Architecture*

Most distributed memory architectures are based on conventional microprocessors (Hennessy and Patterson 1990). We need alternative processor designs which can support a very large number of lightweight threads simultaneously, and can provide fast context switching, message handling, address translation, hashing, and so on. (Boothe and Ranade 1992, INMOS Limited 1988, Whitby-Strevens 1985). If such designs are not produced then we may find that the processors, and not the communications network, will be the bottleneck in the system.

We need to continue to develop improved networks for communication (Dally 1990, Leighton and Maggs 1989) and synchronization (Birk *et al.* 1989, Kruskal *et al.* 1988). There is currently great emphasis in parallel computing on various 'Grand Challenge' applications in science and engineering. While not doubting the importance of these applications, we would suggest that perhaps the most important challenge for parallel architectures at the present time is to develop systems for which global 'inefficiency parameters', such as l and g in the BSP model, are as low as possible. The use of optical technologies may prove to be extremely important in this respect (McColl 1993a). In focusing our attention on the reduction of global parameters such as l and g, we should note that it may not necessarily be cost-effective to try to obtain the extreme case of the PRAM, where l and g are both 1. At any given point in time, the capabilities and economics of the technologies available will determine the most cost-effective values of such parameters. An important advantage of the BSP model (Valiant 1990a) over the PRAM (Abolhassan *et al.* 1991, Ranade 1989) is that it provides an architecture-independent framework which allows us to take full advantage of whichever values of l and g are the most cost-effective at a given point in time.

Large general-purpose parallel computer systems will inevitably suffer hardware faults of various kinds during their operation. We need to develop efficient techniques which can provide a degree of fault tolerance for processors, memories, and communications links. An interesting approach to this problem is to use the idea of information dispersal (Lyuu 1992, Rabin 1989), where a space-efficient redundant encoding of data is used to provide secure and reliable storage of information, and efficient fault tolerant routing of messages. Other approaches to the problems of fault tolerance are described in (Kedem *et al.* 1991, Shvartsman 1991).

The communications architecture of the Inmos T9000 transputer has been designed to support the kind of high performance global communications required to efficiently implement BSP computer systems.

3.4.2 *Algorithms*

Although the potential for automating memory management via hashing is a major advantage of the BSP model, the BSP algorithm designer may wish to retain control of memory management in the two-level memory to achieve higher efficiency, for example, on a BSP computer with a high value of g. A systematic study of direct bulk-synchronous algorithms remains to be done. Some first steps in this direction are described in (Bisseling and McColl 1993, Gerbessiotis and Valiant 1992, Valiant 1990a).

3.4.3 *Languages and Software*

The PRAM model was developed to facilitate the study of parallel algorithms and their complexity. In that context it has proved to be extremely useful. However, as we have pursued the design and implementation of parallel architectures based on the PRAM model, it has become clear that we have no well-developed framework for the programming of such architectures. This can be contrasted with other approaches to general-purpose parallel computing such as the actor and dataflow models, where there has been an intensive effort to develop a programming framework, although rather less on the investigation of parallel algorithms and their complexity. It is vital for the success of the approach described in this paper that we develop programming languages and methodologies for the kinds of parallel architectures proposed. Of the various challenges mentioned, this is perhaps the most important, and in many respects the most difficult one. The apparent unwillingness of many programmers of parallel machines to use anything other than minor variants of the sequential languages FORTRAN and C is widely perceived to be a major impediment to the continuing development of parallel computing. Another impediment is, of course, the 'dusty decks' of old FORTRAN codes which many organizations are unwilling, or unable, to abandon. Many new parallel programming languages have been proposed and rejected over the last decade or so. Nevertheless, we must continue to seek a programming model which will provide a means of achieving the architecture-independence sought, while permitting scalable parallel performance on the kinds of architectures described. Some preliminary work in this direction can be found in (McColl 1993b). It is to be hoped that as such a programming framework is developed we will also be able to provide a strategy for the migration of the dusty decks to the new architectures.

3.5 Other Approaches

A large number of approaches are currently being proposed as the basis of a framework for general-purpose parallel computing. In this paper I have presented the case for the BSP/PRAM approach. In this section I will briefly mention, and comment on, some of these other approaches. Perhaps the most conservative of the alternatives is SIMD or data parallelism. Although a number of interesting algorithms have been developed for such architectures (Blelloch 1990, Hillis and Steele Jr 1986, Steele Jr and Hillis 1986), the model does not appear to be sufficiently general, even when extended to its SPMD form. Another conservative approach is simply to continue with architectures based on message passing across a fixed set of channels (Hoare 1985, Hoare 1991, Jones and Goldsmith 1988). Although such a model is adequate for the development of many special-purpose parallel systems, and for low-level systems programming, it does not appear to offer enough in terms of architecture-independence. An approach related to message passing which appears to be more attractive is the actor model (Agha 1986), which we might think of as message passing using names rather than a fixed set of channels. The names are first class objects and can be passed in messages. The graph of possible interactions between actors can thus change dynamically. The actor model provides a convenient framework for concurrent object-oriented programming (Agha 1990). Dally has developed an interesting parallel architecture, called the J Machine (Dally et al. 1989, Dally and Wills 1989), which supports the actor model.

The dataflow model has evolved considerably over the last decade. Modern designs for dataflow architectures (Iannucci 1988, Iannucci 1990, Nikhil and Arvind 1989, Nikhil et al. 1991, Papadopolous and Traub 1991) emphasize the importance of ideas such as efficient multithreading and the exploitation of parallel slackness, in the same way as the PRAM architectures do. There are, of course, major differences between the two approaches in terms of synchronization control, scheduling control, and so on. It is not yet clear whether the freedom which the dataflow model offers the programmer has a cost to be paid in terms of scalable parallel performance.

Other approaches to general-purpose parallel computing which have been suggested in recent years include asynchronous PRAMs (Cole and Zajicek 1989, Gibbons 1989), block PRAMs (Aggarwal et al. 1989), hierarchical PRAMs (Heywood 1991), the LogP model (Culler et al. 1993), tuple space (Carriero and Gelernter 1989), graph reduction (Peyton Jones 1989), rewriting, and shared virtual memory.

3.6 Conclusion

The goals of general-purpose parallel computing are to achieve both scalable parallel performance and architecture-independent parallel software. Despite much effort to find them, no serious theoretical impediments to the achievement of these goals have yet been found. We have argued that the BSP computer is a robust model of parallel computation which offers the prospect of achieving both requirements. The main challenge at the present time is to develop an appropriate programming framework for the BSP model.

Two other models which appear to offer the required architecture-independence are the actor and dataflow models. The most fundamental difference between these two approaches to parallel computing and the BSP, PRAM, and data parallel models, is that they both have at their core the idea of local (usually pairwise) synchronization events, whereas the BSP, PRAM and data parallel models have the idea of global barrier synchronization as the basic mechanism. Another significant difference is that in the BSP, PRAM, and data parallel approaches there is usually tight control of ordering and scheduling by the programmer. In contrast, a major attraction of, for example, the dataflow approach, is that the programmer is freed from consideration of such issues. It is not yet clear whether the actor and dataflow models can offer the same scalability in parallel performance as we have demonstrated can be obtained for the BSP model. It is also unclear at present whether they can offer a convenient framework for the investigation of parallel algorithms and their complexity. Nevertheless, by virtue of their attractiveness in programming terms, they merit serious consideration.

Although we have stressed the differences between these various approaches, there is reason to believe that, at the architectural level, the BSP, PRAM, actor, and dataflow models will require a number of similar mechanisms for efficient implementation; in particular, high-performance global communications, uniform memory access, and multithreading to hide network latencies.

REFERENCES

F Abolhassan, J Keller, and W J Paul. On the cost-effectiveness of PRAMs. In *Proc. 3rd IEEE Symposium on Parallel and Distributed Processing*, pages 2–9, 1991.

A Aggarwal, A K Chandra, and M Snir. On communication latency in PRAM computations. In *Proc. 1st Annual ACM Symposium on Parallel Algorithms and Architectures*, pages 11–21, 1989.

A Aggarwal, A K Chandra, and M Snir. Communication complexity of PRAMs. *Theoretical Computer Science*, 71,3–28, 1990.

G Agha. *Actors : A Model of Concurrent Computation in Distributed Systems*. MIT Press, Cambridge, MA, 1986.

G Agha. Concurrent object-oriented programming. *Communications of the ACM*, 33(9),125–141, September 1990.

G M Amdahl. Validity of the single processor approach to achieving large scale computing capabilities. In *Proc. AFIPS Spring Joint Computer Conference 30*, pp 483–485, 1967.

W C Athas and C L Seitz. Multicomputers : Message-passing concurrent computers. *IEEE Computer*, 12(8),9–24, August 1988.

R S Bird and P Wadler. *Introduction to Functional Programming*. Prentice Hall, 1988.

Y Birk, P B Gibbons, J L C Sanz, and D Soroker. *A simple mechanism for efficient barrier synchronization in MIMD machines*. Research Report RJ 7078, IBM Research, October 1989. Also appears in *Proc. 1990 IEEE International Conference on Parallel Processing*, Volume II Software, pages 195-198.

R H Bisseling and W F McColl. Scientific computing on bulk synchronous parallel architectures. Technical report, Shell Research (KSLA, Amsterdam) and Programming Research Group, Oxford University, 1993. (In preparation).

G E Blelloch. *Vector Models for Data-Parallel Computing*. MIT Press, Cambridge, MA, 1990.

R D Blumofe. Managing storage for multithreaded computations. Technical Report (M.Sc. Thesis) MIT/LCS/TR-552, Laboratory for Computer Science, Massachusetts Institute of Technology, September 1992.

B Boothe and A Ranade. Improved multithreading techniques for hiding communication latency in multiprocessors. In *Proc. 19th Annual International Symposium on Computer Architecture*, pages 214–223, 1992.

A W Burks, H H Goldstine, and J von Neumann. *Preliminary discussion of the logical design of an electronic computing instrument. Part 1,*

Volume 1. The Institute of Advanced Study, Princeton, 1946. Report to the U.S. Army Ordnance Department. First edition, 28 June 1946. Second edition, 2 September 1947. Also appears in *Papers of John von Neumann on Computing and Computer Theory*, W Aspray and A Burks, editors. Volume 12 in the Charles Babbage Institute Reprint Series for the History of Computing, MIT Press, 1987, pp 97-142.

N Carriero and D Gelernter. How to write parallel programs: A guide to the perplexed. *ACM Computing Surveys*, 21(3),323–358, September 1989.

R Cole and O Zajicek. The APRAM : Incorporating asynchrony into the PRAM model. In *Proc. 1st Annual ACM Symposium on Parallel Algorithms and Architectures*, pp 169–178, 1989.

D Culler, R M Karp, D A Patterson, A Sahay, K E Schauser, E Santos, R Subramonian, and T von Eicken. LogP: Towards a realiztic model of parallel computation. In *Proc. 4th ACM SIGPLAN Symposium on Principles and Practice of Parallel Programming*, pp 1–12, May 1993.

W J Dally and D S Wills. Universal mechanisms for concurrency. In E Odijk, M Rem, and J-C Syre, editors, *Proc. PARLE 89 : Parallel Architectures and Languages Europe. LNCS Vol.365*, pp 19–33. Springer-Verlag, 1989.

W J Dally, A Chien, S Fiske, W Horwat, J Keen, M Larivee, R Lethin, P Nuth, and S Wills. The J-Machine : A fine-grain concurrent computer. In G X Ritter, editor, *Proc. Information Processing 89*, pp 1147–1153. Elsevier Science Publishers, B. V., 1989.

W J Dally. Network and processor architecture for message-driven computers. In R Suaya and G Birtwistle, editors, *VLSI and Parallel Computaion*, pp 140–222. Morgan Kaufmann, San Mateo, CA, 1990.

C W Gear, editor. *Computation and Cognition. Proceedings of the First NEC Research Symposium*. SIAM Press, 1991. Panel Session - The Future of Parallelism, pp 153-168.

A V Gerbessiotis and L G Valiant. Direct bulk-synchronous parallel algorithms. Technical Report TR-10-92 (Extended version), Aiken Computation Laboratory, Harvard University, 1992. Shorter version appears in Proc. 3rd Scandinavian Workshop on Algorithm Theory, July 8-10, 1992. LNCS Vol. 621, pp 1-18, Springer-Verlag.

A M Gibbons and P Spirakis, editors. *Lectures on Parallel Computation*, volume 4 of *Cambridge International Series on Parallel Computation*. Cambridge University Press, Cambridge, UK, 1993.

P B Gibbons. A more practical PRAM model. In *Proc. 1st Annual ACM Symposium on Parallel Algorithms and Architectures*, pp 158–168, 1989.

J L Gustafson. Development of parallel methods for a 1024-processor hypercube. *SIAM Journal on Scientific and Statistical Computing*, 9(4),609–638, July 1988.

J L Gustafson. Reevaluating Amdahl's Law. *Communications of the*

ACM, 31(5),532–533, May 1988.

J L Hennessy and D A Patterson. *Computer Architecture : A Quantitative Approach*. Morgan Kaufmann, San Mateo, CA, 1990.

T H Heywood. A practical hierarchical model of parallel computation. Technical Report SU-CIS-91-39, School of Computer and Information Science, Syracuse University, November 1991.

W D Hillis and G L Steele Jr. Data parallel algorithms. *Communications of the ACM*, 29(12),1170–1183, December 1986.

C A R Hoare. *Communicating Sequential Processes*. Prentice Hall, 1985.

C A R Hoare. The transputer and occam : A personal story. *Concurrency : Practice and Experience*, 3(4):249–264, August 1991.

P Hudak, S Peyton Jones, and P Wadler, editors. Report on the Programming Language Haskell - A Non-Strict, Purely Functional Language. Version 1.1 , 1991.

P Hudak. Concept, evolution, and application of functional programming languages. *ACM Computing Surveys*, 21(3),359–411, 1989.

R A Iannucci. Toward a dataflow/von Neumann hybrid architecture. In *Proc. 15th Annual International Symposium on Computer Architecture*, pp 131–140, 1988.

R A Iannucci. *Parallel Machines : Parallel Machine Languages*. Kluwer Academic Publishers, Dordrecht, 1990.

INMOS Limited. *Transputer Reference Manual*. Prentice Hall, 1988.

J JáJá. *An Introduction to Parallel Algorithms*. Addison-Wesley, 1992.

G Jones and M Goldsmith. *Programming in occam 2*. Prentice Hall, 1988.

H Jung, L Kirousis, and P Spirakis. Lower bounds and efficient algorithms for multiprocessor scheduling of dags with communication delays. In *Proc. 1st Annual ACM Symposium on Parallel Algorithms and Architectures*, pp 254–264, 1989.

R M Karp, M Luby, and F Meyer auf der Heide. Efficient PRAM simulation on a distributed memory machine. In *Proc. 24th Annual ACM Symposium on Theory of Computing*, pp 318–326, 1992.

Z M Kedem, K V Palem, A Raghunathan, and P G Spirakis. Combining tentative and definite executions for very fast dependable parallel computing. In *Proc. 23rd Annual ACM Symposium on Theory of Computing*, pp 381–390, 1991.

C P Kruskal, L Rudolph, and M Snir. Efficient synchronization on multiprocessors with shared memory. *ACM Transactions on Programming Languages and Systems*, 10(4),579–601, 1988.

C P Kruskal, L Rudolph, and M Snir. A complexity theory of efficient parallel algorithms. *Theoretical Computer Science*, 71,95–132, 1990.

H T Kung. New opportunities in multicomputers. In C W Gear, editor, *Computation and Cognition. Proceedings of the First NEC Research Symposium*, pages 1–21. SIAM Press, 1991.

F T Leighton and B M Maggs. Expanders might be practical: Fast algorithms for routing around faults on multibutterflies. In *Proc. 30th Annual IEEE Symposium on Foundations of Computer Science*, pp 384–389, 1989.

F T Leighton. *Introduction to Parallel Algorithms and Architectures : Arrays, Trees, Hypercubes*. Morgan Kaufmann, San Mateo, CA, 1992.

Y-D Lyuu. *Information Dispersal and Parallel Computation*, volume 3 of *Cambridge International Series on Parallel Computation*. Cambridge University Press, Cambridge, UK, 1992.

W F McColl. General-purpose parallel computing. In Gibbons and Spirakis (1993), pp 337–391.

W F McColl. GL : An architecture independent programming language for scalable parallel computing. Technical Report 93-072-3-9025-1, NEC Research Institute, Princeton, June 1993.

W F McColl. Special-purpose parallel computing. In Gibbons and Spirakis (1993), pp 261–336.

R S Nikhil and Arvind. Can dataflow subsume von Neumann computing? In *Proc. 16th Annual International Symposium on Computer Architecture*, pp 262–272, 1989.

R S Nikhil, G M Papadopolous, and Arvind. *t : A killer micro for a brave new world. Computation Structures Group Memo 325, Laboratory for Computer Science, Massachusetts Institute of Technology, July 1991.

R S Nikhil. Id - Language Reference Manual. Version 90.1. Computation Structures Group Memo 284-2, Laboratory for Computer Science, Massachusetts Institute of Technology, July 1991.

C H Papadimitriou and J D Ullman. A communication-time tradeoff. *SIAM Journal on Computing*, 16(4),639–646, August 1987.

C H Papadimitriou and M Yannakakis. Towards an architecture-independent analysis of parallel algorithms. *SIAM Journal on Computing*, 19(2),322–328, 1990.

G M Papadopolous and K R Traub. Multithreading : A revisionist view of dataflow architectures. In *Proc. 18th Annual International Symposium on Computer Architecture*, pp 342–351, 1991.

S L Peyton Jones. Parallel implementation of functional programming langauges. *The Computer Journal*, 32(2),175–186, 1989.

M O Rabin. Efficient dispersal of information for security, load balancing, and fault tolerance. *Journal of the ACM*, 36(2),335–348, April 1989.

A G Ranade. Fluent parallel computation. Ph.D. Thesis, Department of Computer Science, Yale University, May 1989.

C L Seitz. The Cosmic Cube. *Communications of the ACM*, 28(1),22–33, January 1985.

C L Seitz. Concurrent architectures. In R Suaya and G Birtwistle, editors, *VLSI and Parallel Computaion*, pages 1–84. Morgan Kaufmann, San Mateo, CA, 1990.

REFERENCES

A A Shvartsman. Achieving optimal CRCW PRAM fault-tolerance. *Information Processing Letters*, 39(2),59–66, 1991.

G L Steele Jr and W D Hillis. Connection Machine Lisp : Fine-grained parallel symbolic processing. In *Proc. ACM Conference on Lisp and Functional Programming*, pp 279–297, 1986.

A M Turing. On computable numbers, with an application to the Entscheidungsproblem. *Proceedings of the London Mathematical Society. Series 2*, 42:230–265, 1936. Corrections, *ibid.*, 43 (1937), pp 544-546.

L G Valiant. A bridging model for parallel computation. *Communications of the ACM*, 33(8),103–111, 1990.

L G Valiant. General-purpose parallel architectures. In J van Leeuwen, editor, *Handbook of Theoretical Computer Science : Volume A, Algorithms and Complexity*, pp 943–971. North Holland, 1990.

L G Valiant. A combining mechanism for parallel computers. Technical Report TR-24-92, Aiken Computation Laboratory, Harvard University, November 1992.

L G Valiant. Why BSP computers? Technical Report TR-26-92, Aiken Computation Laboratory, Harvard University, November 1992. To appear in Proc. 7th International Parallel Processing Symposium, April 1993.

U Vishkin. Can parallel algorithms enhance serial implementation? Technical Report UMIACS-TR-91-145, Institute for Advanced Computer Studies, University of Maryland, 1991.

U Vishkin. Structural parallel algorithmics. In J Leach Albert, B Monien, and M Rodriguez Artalejo, editors, *Proc. 18th International Colloquium on Automata, Languages and Programming, LNCS Vol.510*, pp 363–380. Springer-Verlag, 1991.

U Vishkin. Methods in parallel algorithmics and who may need to know them? In T Ibaraki, Y Inagaki, K Iwama, T Nishizeki, and M Yamashita, editors, *Algorithms and Computation, Third International Symposium, ISSAC 92, LNCS Vol. 650*, pp 1–4. Springer-Verlag, December 1992.

J von Neumann. *First draft of a report on the EDVAC*. Moore School of Electrical Engineering, University of Pennsylvania, 30 June 1945. Contract No. W-670-ORD-4926 between the United States Army Ordnance Department and the University of Pennsylvania. Reprinted in *Papers of John von Neumann on Computing and Computer Theory*, W Aspray and A Burks, editors. Volume 12 in the Charles Babbage Institute Reprint Series for the History of Computing, MIT Press, 1987, pp 17-82.

C Whitby-Strevens. The transputer. In *Proc. 12th Annual International Symposium on Computer Architecture*, pp 292–300, 1985.

4

GENERAL-PURPOSE PARALLEL PROGRAMMING ON THE PRAM MODEL

Lasse Natvig

Department of Computer Systems and Telematics
Norwegian Institute of Technology
University of Trondheim
N-7034 Trondheim, Norway
lasse@idt.unit.no

Abstract

Wyllie (Wyllie 1979) outlined a high-level notation called parallel pidgin algol for expressing algorithms on the PRAM model. Inspired by this, we have developed a high-level parallel programming language providing synchronous MIMD programming. The language makes it possible to implement parallel algorithms directly from the descriptions found in theoretical computer science literature. Implemented algorithms are executed on a CREW PRAM simulator that offer extensive facilities for measuring and gathering of statistics. The simulator has been used to evaluate the practical value of some of the many famous parallel sorting algorithms developed within the theory community (Natvig 1990a, Natvig 1990b, Hagaseth 1991).

Our experience with synchronous MIMD programming is promising. It provides a combination of the easier programming and debugging implied by synchronous operation and the flexibility and expressiveness of MIMD programming. Practitioners have generally regarded the PRAM model as purely theoretical, since it is very difficult to implement with existing technology. However, Valiant in his work on the bulk synchronous parallel (BSP) model (Valiant 1990), suggests that a 'PRAM language would be ideal' for programming on this model. The PRAM language would be transformed through an extensive compilation and runtime mapping process—to bulks of instructions for executing on a 'BSP-machine'. Valiant's work, together with the rapid progress in both hardware and compiler technology, makes us believe that synchronous MIMD programming is a new and promising paradigm for general-purpose parallel programming that should be further explored.

The existing version of the synchronous MIMD language has been used in connection with teaching of parallel algorithms, and we are currently working on a new Pascal-based version building on the gained experience.

4.1 Introduction, background, and motivation

4.1.1 *The gap between theory and practice*

There are at least two main directions of research in parallel algorithms. In theoretical computer science parallel algorithms are typically described in a high-level mathematical notation for abstract machines, and their performance is typically analysed by assuming infinitely large problems. On the other side we have more practically oriented research, where implemented algorithms are tested and measured on existing computers and finite problems.

The gap between theory and practice is evident, and it seems natural to ask the following questions: Are the theoretically fast parallel algorithms simply not known by the practitioners, is the case that they are not understood, or are they in general judged as without practical value?

Most parallel algorithms developed within theoretical computer science are described for the PRAM model (Fortune 1978, Wyllie 1979). One way to learn more about the possible practical value of such algorithms is to implement them and compare their performance with well known algorithms solving the same problem. This approach, that is, *implementing* algorithms executed on an *abstract machine* for solving *finite* problems, is possibly a new approach. Experience about the practical value of two famous parallel sorting algorithms have been gathered and reported using this approach (Natvig 1990b, Natvig 1990a, Hagaseth 1991).

4.1.2 *Evaluation of the practical value of theoretical parallel algorithms implies synchronous MIMD PRAM programming*

This paper concentrates on *synchronous MIMD programming*, a paradigm for parallel programming that was implied by the wish to implement PRAM algorithms. This style of programming was outlined by Wyllie as early as in the late 1970s, but has been given surprisingly little attention. The reason *may* be that the topic does not directly fit into either the theoretically or the practically oriented research on parallel algorithms.

Algorithms described for the PRAM model are most easily implemented on a PRAM machine. Using more realistic machine models will introduce a larger conceptual gap between the description and the implementation, implying more difficult programming and less portable results. Worse though, implementing an algorithm described at a high-level for the PRAM model on a more realistic machine model will necessarily introduce a larger degree of machine adaptation and programmer-dependent design choices. This makes it much more difficult to obtain a fair comparison of algorithms.

Practical algorithms developed for more realistic machine models are easily adapted to the PRAM model with little modifications, since these models in general are contained as a subset of the PRAM model. The practical algorithms will typically exploit a smaller part of the computational power* inherent in the PRAM model than the theoretical algorithms. Implementing both algorithms on the PRAM model will then favour the theoretical algorithm—strengthening the conclusion when a theoretical algorithm is found inferior to a practical algorithm, but not vice versa (Natvig 1990b).

After only a few months of programming in this new paradigm, using only simple prototype programming tools, we found that it provides remarkably easy parallel programming! We do now believe that this may be a viable way for providing easy and efficient high-level general-purpose parallel programming. Only the future can prove this to be correct or not, but there is no doubt that further research is needed.

4.1.3 About the paper

The paper starts by giving a brief description of the PRAM model and how its specification has been extended to a (CREW) PRAM simulator well suited for developing and measuring PRAM programs. We then advocate the view that the PRAM model implies a *synchronous MIMD* programming style, and this paradigm is then explained using examples on the pseudocode level.

Our experience so far with synchronous MIMD programming on the PRAM model is very promising, and we describe some of its main benefits—easy programming and debugging, and portability. An evident drawback of using the PRAM model as a programming model is that it is difficult to realize with existing technology in a way that provides efficient execution. However, Valiant's proposal of the BSP (Bulk Synchronous Parallel) model for parallel computing (Valiant 1990) is encouraging in this context. Valiant argues that a bridging model between hardware and software designers is needed for general-purpose parallel processing to become widespread. He mentions a PRAM language as an ideal high-level language for programming in the context of the BSP model, and suggest how the BSP model may be realized with existing technology. At the end of the paper we sketch some ongoing projects that are inspired by Valiant's work.

4.2 CREW PRAM programming

4.2.1 The CREW PRAM Model—Summary and Extensions

The *P-RAM (parallel random access machine)* was first presented by Fortune and Wyllie (Fortune 1978). It was further elaborated in Wyllie's Ph.D. thesis (Wyllie 1979). The P-RAM is based on random access machines (RAMs) operating in parallel and sharing a common memory. Thus

---

*Such as the ability for many processors to communicate simultaneously through the shared memory.

it is in a sense the model in the world of parallel computations that corresponds to the RAM model, which certainly is the prevailing model for sequential computations.

Today, the mostly used name on the original P-RAM model is probably CREW PRAM. CREW is an abbreviation for the very central concurrent read exclusive write property; several processors may at the same time step read the same variable (location) in global memory, but they may not write to the same global variable simultaneously. (When we use the unqualified term PRAM, we mean the original P-RAM described by Fortune and Wyllie.)

A CREW PRAM is a very simple and general model of a parallel computer. It has an unbounded number of equal processors. Each has an unbounded local memory, an accumulator, program counter, and a flag indicating whether it is running or not. A CREW PRAM has an unbounded global memory shared by all processors. An unbounded number of processors may write into global memory as long as they write to different locations. An unbounded number of processors may read any location at any time. All processors execute the same program. At each *global* time step in the computation, each running processor executes the instruction given by its own *local* program counter in one unit of time.

The CREW PRAM simulator extends the original PRAM specification by a few necessary details. It assumes a simple instruction set for each processor, where the time consumption of each instruction type is defined as a parameter. Further, each processor knows which processor set (see below) it is member of, and its logical number in this set. The unbounded quantities in the PRAM specification are represented by constants defined in the simulator (memory), or by dynamic allocation (processors). Thus, the memory size of the computer executing the simulator implies an upper bound on these quantities. This has not been a severe limitation till now. Further details are given in (Natvig 1990b).

4.2.2 *Why Synchronous MIMD Programming?*

Several recent textbooks (Akl 1989, Quinn 1987, Gibbons 1988) describe the PRAM model as a SIMD model according to Flynn's taxonomy, and this seems to be a *de facto* understanding of the PRAM model. Our view, that it is a MIMD model, is motivated by a wish to follow the original description of the PRAM model, and an understanding which seem to be more widespread in the theory community.

It may be that a large part of the algorithms presented for the (CREW) PRAM model is SIMD style, and this may be explained by the fact that all available synchronous computers have been SIMD. However, using the PRAM as a SIMD model is a choice made by the programmers, not a limitation enforced by the model. The CREW PRAM model implies a *single program*, which does *not* necessarily imply a single instruction *stream*.

The crucial point here is the local program counters as described in (Wyllie 1979) page 17:

'... each processor of a P-RAM has its own program counter and may thus execute a portion of the program different from that being executed by other processors. In the notation of (Flynn 1966), the P-RAM is therefore a 'MIMD' ... '

Cook (Cook 1981) expresses the same view:

'The P-RAM of Fortune and Wyllie ... different parallel processors can be executing different parts of their program at once, so it is 'multiple instruction stream'.

It should also be explicitly stated that the PRAM model is synchronous. In (Sabot 1988) the author compares his own model for parallel programming with the PRAM model. He starts by stating that the processors in the PRAM model operates asynchronously—which certainly is *not* true for the original PRAM model. Further, he writes (about the PRAM model) that it '... is confusing and hard to program.'. This view may also be explained by the history of old computers—all MIMD computers build so far have been asynchronous. Sabot states that the (asynchronous) MIMD power is overwhelming, and that PRAM programmers deal with this power by writing SIMD style programs for it; he claims that '*all* parallel algorithms in the proceedings of the ACM Symposia on Theory of Computing (STOC) 1984, 85 and 86—are of data parallel form' (i.e. SIMD). This is *not* true. Algorithms in the proceedings of the STOC (and other similar) Symposia are typically described in a *high-level mathematical notation—where the choice between SIMD or MIMD programming style to a large extent is left open to the programmer*. Parts of an algorithm are most naturally coded in SIMD style, while other parts most naturally are expressed in MIMD style.

4.2.3 Parallel Pseudo Pascal—PPP

In his thesis (Wyllie 1979), Wyllie outlined a high-level pseudo-code notation, called *parallel pidgin algol*, for expressing algorithms on a CREW PRAM. The notation introduced here, called Parallel Pseudo Pascal (PPP), is intended to be a modernization of Wyllie's notation. PPP is an attempt to combine the pseudo-language notation (called 'super pascal') used for sequential algorithms by Aho, Hopcroft and Ullmann in (Aho 1982), with the ability to express parallelism and processor allocation as it is done in parallel pidgin algol. Since the notation is *pseudo* code it should only be used as a *framework* for expressing algorithms. Only a very brief description is given here.

A statement may be any of the alternatives shown in Fig. 4.1, or a list of statements separated by semicolons and enclosed within **begin** ... **end**. Variables will normally not be declared, their definition will be implicit from the context or explicitly described. It is crucial to separate *processor-*

1. (a) *Processor-localVariableName* := Expression
   (b) SharedVariableName := Expression
2. **if** Condition **then** Statement1 { **else** Statement2 }
3. **while** Condition **do** Statement
4. **for** Variable := InitialValue **to** FinalValue **do**
   Statement
5. **procedure** *ProcedureName* (FormalParameterList) Statement
6. *ProcedureName* (ActualParameterList)
7. **assign** ProcessorSpecification {, **to** ProcSetName }
8. **for each processor in** ProcSetName { **where** Condition } **do**
   Statement
9. any other well defined statement

FIG. 4.1. Informal outline of the Parallel Pseudo Pascal (PPP) notation.

local variables, which are placed in the local memory of a processor, from shared variables. *Shared variables* are placed in the common global memory of the PRAM, and are therefore accessible by all the processors. Shared variables are emphasized in PPP by underlining. Processor-local variables and procedure names are written in *italics*.

Statement 1 to 6 should be self-explaining, and statement 9 should be familiar to all who have experienced pseudo-code. Here, only statement 7 and 8 need some further explanation.

4.2.4 *Processor sets, processor allocation, and activation*

The **assign** statement is used to allocate a set of processors as a named entity, a *processor set*. The 'ProcessorSpecification' is a description or expression that specifies the number of processors that should be in the set. It may contain variables to provide dynamic allocation at runtime. (The **to** ProcSetName clause may be omitted in simple examples when only one set of processors is used.)

The **assign** statement is realized by an $O(\log n)$ time algorithm using the standard PRAM FORK instruction (Fortune 1978) in a binary tree-structured 'chain reaction'. n is the number of processors in the set. The **assign** statement does also assign local processor numbers to the processors. A *processor set* is thus a named collection of processors numbered $1, 2 \ldots n$. For each processor, this number is stored in the processor-local variable p. This has shown to be a very nice feature when expressing parallel algorithms, because it in general provides a natural way for a processor to select its own part of the problem to be solved.

Processor activation (statement type 8) is the way to specify that code should be executed in parallel by a set of processors. The optional **where** clause is used to specify a condition that must be true for a processor in the processor set for the statement to be executed by that processor.

(1) **for each processor in** Aprocs **where** p is odd **and** $p < n$ **do begin**
(2) $temp := \underline{A}[p+1]$;
(3)   **if** $temp < \underline{A}[p]$ **then begin**
(4) $\underline{A}[p+1] := \underline{A}[p]$; $\underline{A}[p] := temp$;
      end;
end;

FIG. 4.2. Example of PPP notation for a part of an odd-even transposition sort algorithm

Processors in the set evaluating this condition to false will not be activated by the statement. A (rather low-level) PPP example demonstrating the **where** clause is shown in Fig . 4.2. The code stub may be a central part of an odd-even transposition sort implementation (See for instance (Akl 1989) or (Quinn 1987).) $\underline{A}$ is an array of n elements stored in the PRAM shared memory. The elements are numbered $1, 2, \ldots n$, and Aprocs is a set of n processors. The processor activation in the figure says that the odd numbered processors up to, but not including, processor n should change its 'own element'* with its 'right neighbour', if and only if these two elements are out of order.

Note that a crucial property of the high-level PRAM programming outlined by Wyllie is that statements such as the **if** statement in this example (3) are executed synchronously by the processors. This implies that the processors that evaluate the condition in statement (3) to false should not end the **if** statement before the processors that evaluate the same condition to true. The result is some wasted idle-time, but also the *important advantage that all the processors are synchronous when they start at the next statement.* This is further elaborated in the next section.

4.2.5 *The importance of synchronous operation*

Parallel programs are in general difficult to understand. However, requiring that the processors operate synchronously may often make it much easier to understand parallel algorithms. Consider the simple program in Fig . 4.3. This example is taken from (Wyllie 1979).

The processor activation statement (line (1)) describes that the **if** statement shall be executed in parallel by all processors in the set called ProcSet. Assume that ProcSet contains two processors. Further assume that one processor evaluates the condition 'cond(p)' to **true** while the other evaluates it to **false**, so that the **then** and the **else** clause are executed concurrently

---

*We may think that processor i is assigned to element no. i in the array, even though all elements are accessible by all processors due to the PRAM definition.

(1) **for each** processor **in** ProcSet **do**
(2)   **if** cond(p) **then**
(3) $\underline{x} := f(\underline{y})$;
    else
(4) $\underline{y} := g(\underline{x})$;

FIG. 4.3. Example program with unpredictable behaviour.

(1) **for each** processor **in** ProcSet **do begin**
(2)   **if** cond(p) **then**
(3)   tx $:= f(\underline{y})$;
    else
(4)   ty $:= g(\underline{x})$;
(5)   wait for both processors to reach here;
(6)   **if** cond(p) **then**
(7) $\underline{x} :=$ tx;
    else
(8) $\underline{y} :=$ ty;
  end;

FIG. 4.4. Example program transformed to predictable behaviour by using explicit resynchronization (line (5)). (Line (5) may be omitted from a PRAM program—see the text.)

by two processors. (This may happen because the condition may refer to variables local in each processor, or to the processor number p). $\underline{x}$ and $\underline{y}$ are global variables. Depending on the relative times required to compute the functions f and g, either the old or new value of $\underline{x}$ will be used in the **else** clause. It is therefore difficult to argue about the precise behaviour of this program.

As Wyllie points out (Wyllie 1979), the code generated from the program in Fig. 4.3 is a legitimate program for the P-RAM model—but this kind of programming is not recommended.

A program with more self-evident behaviour can easily be obtained from Fig. 4.3, as shown in Fig. 4.4. (It is here assumed that 'cond(p)' is unaffected by lines (3) and (4).) Line (5) describes an explicit *synchronization* that is necessary to guarantee that the old values of $\underline{x}$ and $\underline{y}$ are used in the computation of f and g. Such synchronizations are in general a valuable tool to restrict the space of possible behaviour and make it easier to understand parallel programs. This observation probably led Wyllie to define

what is the most important property of (high-level) programming of the
P-RAM;
Definition: *Synchronous statements*
*If each processor in a set P begins to execute a statement S at the same
time, all processors in P will complete their executions of S simultaneously.*

This property is crucial in making PRAM programs easy to understand
and easy to analyse, but may at first sight seem to put a lot of burdens on
the programmer. However, a proper compiler for a PRAM language with
this property may to a large extent produce code that automatically provides *synchronous statements*. More difficult cases may be handled by the
compiler in co-operation with the programmer or solely by the programmer.
From now on, we will assume the existence of a compiler that, whenever
possible, automatically ensures that the statements in our PRAM language
are executed as *synchronous statements*. Therefore we may omit statement
(5) in Fig. 4.4 since the compiler on reading the **if** statement in line (2–4)
will produce code so that all processors in **ProcSet** simultaneously will start
executing the **if** statement in line (6–8). This is easily done by so called
compile time padding if the compiler may calculate the time consumption
of f and g.

4.2.6 Compile-time padding (CTP)

Assuring the *synchronous statements* property may in most cases be done
by so called *compile-time padding*. When the compiler knows the execution
time of the **then** and **else** clauses of an **if** statement—inserting an appropriate amount of waiting ('NOP's') in the shortest (w.r.t. execution time)
of these will make the whole **if** statement to a synchronous statement.

Automatic compile-time padding is a necessity for making it possible
to express algorithms in a synchronous *high-level* language. The exact
execution times of the various constructs should be hidden in a high-level
language. It is not desirable, nor should it be possible, that the property
of 'synchronous statements' is entirely preserved by the programmer.

Compile-time padding should be done automatically wherever it is possible by the compiler. Relatively simple computations at compile time will
reveal where NOP's should be inserted in the generated code to produce
a synchronous statement. This will in most cases produce more efficient
code than may be achieved by the simpler 'brute force' method, which
is to let the compiler insert general synchronization code at the end of
non-synchronous statements. General synchronization is relatively time-consuming, so a little increase in compile-time spent on padding will significantly reduce the run-time of the program. Note that if synchronous
statements are required, the compile-time padding will not produce any
overhead at run-time. The reason is that it is only the uncritical execution
paths which are padded until they have the same length as the critical
(longest) execution path. The compile-time padding may be done at least

three levels of sophistication:

1. The simplest cases are those that resemble the example above with an **if** statement where the **else** or the **then** clause may be expanded with NOP's to ensure that all processors finish the statement simultaneously. Sequential code is most easily handled. Loops are easily handled if the number of iterations is known at compile-time. In cases where this is not possible the possibility of computing the number of iterations at run-time should be investigated by the compiler. If this is possible, the compiler may generate code that at run-time calculates and produces the appropriate amount of waiting.

   A proper language will of course contain nested statements and procedure calls—thus the analysis must be hierarchical. An **if** statement may therefore be difficult to make synchronous if one of its clauses contain program fragments that are difficult in this context.

2. In cases where the compiler is unable to ensure synchronous statements with compile-time padding (CTP), the programmer should be informed by a warning explaining the problem. In many cases the programmer will be able to modify the program in a way that makes CTP possible. (See Natvig 1990, page 74). This process should be integrated in the program development cycle in the PRAM programming environment.

3. There will of course be some cases where the compiler and the programmer are unable to ensure synchronous operation. The most typical case will be when the program represents or requests an inherently asynchronous operation, such as an I/O operation. The only remedy in these cases is to insert general code for resynchronization. If the compiler knows the actual number of processors at compile-time it may choose a synchronization algorithm that will be efficient for the actual case*. When the number of processors is unknown at compile-time, a general synchronization algorithm should be used.

   In the earliest phases of program development the programmer will normally concentrate on program structure, readability and correctness. The insertion of explicit synchronization should be done silently by the compiler. Later, the programmer may be focusing on performance, and switching on a compiler option should force the compiler to inform the programmer about explicit synchronization.

The compile-time padding techniques discussed above should be relatively straightforward to integrate with other tasks performed by a compiler. The hierarchical analysis can be done in a syntax tree that should

---

*An $O(\log n)$ synchronization algorithm will be fast for medium and large number of processors (n), but a simpler linear synchronization algorithm may be faster for a small number of processors.

```
            ...
            { Each processor holds one object }
(1)         if type of object is Rectangle then
                Compute area of Rectangle object
            else if type of object is Triangle then
                Compute area of Triangle object
            else
                Compute area of Trapezium object;
(2)         Compute total sum; { $O(\log n)$ standard parallel prefix* }
            ...
```

FIG. 4.5. Part of synchronous MIMD program for computing the total area of a compound surface

be kept 'alive' till the end of the code generation phase. Program-libraries should store information about the expected time consumption of each routine to integrate CTP with separate compilation of program modules. Note also that space consuming long sequences of NOP instructions may be avoided by using an appropriate number of iterations in a NOP-loop.

The following example is included to illustrate various aspects of synchronous MIMD programming. Assume that we are given the task of making a synchronous MIMD program that computes the total area of a complex surface consisting of various objects.

Example: case-1
Assume that the surface consists of only three kinds of objects, Rectangle, Triangle or Trapezium. Fig. 4.5 shows a part of the program executed by a processor set containing one processor for each object. Statement (1) demonstrates a multiway branch statement that utilizes the multiple instruction stream property to give faster execution and better processor utilization than on a SIMD machine. If the time used to compute the area of each kind of object is roughly equal, MIMD execution will be roughly three times faster than SIMD execution for this case with three different objects. Since statement (1) is synchronous, the processors should leave the statement simultaneously. This is easily achieved by compile-time padding since the time used to compute the area of each kind of object may be determined by the compiler. If we assume that the most lengthy operation is to calculate the area of Triangle the effect of compile-time padding can be illustrated as in Fig. 4.6.

Example: case-2
Now, assume that each object is a polygon with from 3 to 10 edges. As

---
*See for instance (Hillis and Steele 1986) or (Akl 1989).

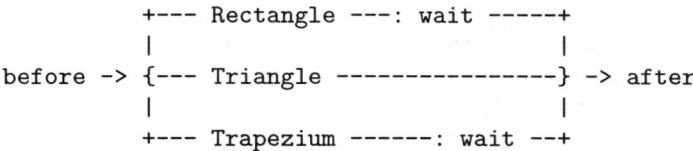

FIG. 4.6. Compile-time padding, i.e. inserting waiting in uncritical paths

```
         ...
         { Each processor holds one polygon }
(1)      Area := ComputeArea(this polygon);
         { All processors should be here at the same time unit }
(2)      Compute total sum; { O(log n) standard parallel prefix }
         ...
```

FIG. 4.7. Synchronous MIMD program, case-2.

shown in Fig. 4.7 the area of each object (polygon) is computed by a function called *ComputeArea*. Knowing the maximum number of edges in a polygon, the compiler may calculate the maximum time used to execute the procedure, and produce code that will make all calls to the procedure use this maximum time. Note that this is a way of increasing the amount of determinism in the program execution—directly providing advantages such as easy debugging and analysis. However, the solution may at first sight seem inefficient because of the wasted waiting time. It *may* be inefficient for cases with a relatively small number of processors where all the polygons have significantly less than 10 edges, but not in general. The alternative, which is to let each processor compute the area for its 'owned' object as fast as possible, and then let all the processors in the set do a general synchronization, will in general be more time-consuming for a large number of processors. The reason is that the time needed for general synchronization grows with the number of processors,[†] while the synchronization provided by the CTP has a constant time consumption.

Example: case-3
We now consider the case that each object is an arbitrary complex polygon (that is, with an unknown number of edges), and that the size of the most complex polygon is not known by the processors. (Knowing the size of

---

[†]The CREW PRAM simulator (Natvig 1990b) offers a general synchronization algorithm which use $O(\log n)$ time for synchronizing n processors. A similar $O(\log n)$ time algorithm is the *synchronization barrier* reported in (Lubachevsky 1989).

the most complex polygon would make it possible to use the techniques described for case-2.) In this situation it is impossible for the compiler to know the maximum execution time of *ComputeArea*. The best it can do is to insert an explicit synchronization just after the return from the procedure. In this case the use of general synchronization does not influence the time complexity of the computation, since the succeeding statement (2) in Fig. 4.7 also requires $O(\log n)$ time.

4.2.7 *Parallelism should make programming easier!*

There are at least two main reasons that the use of parallelism in future computers may make programming easier. The main motivation for using parallel processing is to provide more computing power to the user. This power may in general make it possible for the programmer to use simpler and 'dumber' algorithms—to some extent. It is well-known that coding to achieve maximum efficiency is difficult and time-consuming. Consequently, *parallelism may increase the number of cases where we can afford to use simple 'brute force' techniques.* Also, increased computing power (or a large number of processors) may make it more common to use 'redundancy' in problem-solving strategies—which again may imply conceptually simpler solutions.*

Many of our programs try to reflect various parts of, or processes in, the real world. There is no doubt that the real world is highly parallel. Therefore, *the possibility of using parallelism will in general make it easier to reflect real-world phenomena and natural solution strategies.* One example is given in (Natvig 1990b).

For traditional, sequential programming, there is far from any consensus about which of the various programming paradigms being most successful in providing easy programming. We must therefore expect similar discussions for many years about the corresponding parallel programming paradigms.

4.2.8 *Synchronous MIMD programming is easy*

Though we have experienced synchronous MIMD programming only through simple prototype tools, we are convinced that this programming style has several very nice features:

Easy programming
Synchronous statements imply that the benefits of synchronous operation are obtained at the 'source code level', and not only at the instruction level. It gives the advantages of SIMD programming, that is, programs that are easier to reason about (Steele 1990). However, we have kept the

---

*An example is that a parallel processor with a sufficiently large number of processors may allow the application of odd-even transposition sort which is faster, *and much simpler and more predictable* than for instance quicksort on an uniprocessor.

flexibility of MIMD programming. This was illustrated by the example above where Fig. 4.5 shows a MIMD program that may compute the area of three different kinds of objects simultaneously. Such natural and efficient handling of conditionals is not possible within the SIMD paradigm.

There certainly are cases where asynchronous parallelism is most desirable. Many real-world phenomena are most naturally represented as co-operating asynchronous processes. Such situations may easily be handled if the used language has some sort of compiler directive for switching on/off the 'synchronous statements property' and the associated compile-time padding. The only restriction will be that the time-difference between any two events will be 'discretized' to a whole number of CREW PRAM time units. This should not be a problem in practice, since this time unit will be small.

Easy analysis

Synchronous programs exhibit in general a much more deterministic behaviour than asynchronous programs. This implies easier analysis. The fact that the high-level statements by default are synchronous also simplifies analysis. This is shown by *case-1* of the preceding example where compile-time padding was used to assure that a computational task would use an equal amount of time for three different cases. Here, redundant operations (that is, wait-operations) simplify analysis and programming.

The requirement of synchronous high-level statements may be a good reason for coding a procedure such that its time consumption is made *independent* of its input parameters. A natural approach is then to code the procedure for solving the most general case, and to use this code on all cases. This removes the possibility of saving some processor time units on simple cases—but more important, it simplifies the programming of the procedure. Also, a fixed known time consumption may eliminate the need for general synchronization, and therefore possibly give a faster overall solution. A consequence is that the time consumption of an algorithm will typically be less dependent on the actual nature of the problem instance, such as presortedness (Manilla 1985) for sorting. This makes it easier to use the algorithm as a substep of a larger synchronous program.

As indicated above during the discussion of compile-time padding, a proper compiler for a PRAM language will produce information about the time consumption of various parts of the code. This should be available through a programming environment for helping the programmer to analyse the performance of the algorithms. In many cases, an intelligent programming environment may produce a performance model that in turn will make it easier for the programmer to develop a large software system with predictable performance.

Easy debugging

'*The main problems associated with debugging concurrent programs are increased complexity, the 'probe effect', nonrepeatability, and the lack of a synchronized clock.*' (McDowell 1989).
The debugging of the synchronous MIMD programs developed on the CREW PRAM simulator described in (Natvig 1990b) has been quite similar to debugging of uniprocessor programs. The synchronous operation and the execution of the programs on a simulator have eliminated most of the problems with concurrent debugging. The CREW PRAM model implicitly provides a synchronized clock known by all the processors. Synchronous operation will in general reduce the amount of nondeterminism, and therefore imply an increased degree of repeatability. Full repeatability is possible on the simulator, and it also monitors a program without any disturbance of the program being evaluated — that is, no 'probe effect'.

Not all these benefits would be possible to obtain on a 'CREW PRAM machine'. However, a proper programming environment for such a machine should contain a simulator for easy debugging in 'early stages' of the software testing.

4.3 Synchronous MIMD Programs Executed Efficiently on a BSP-Machine

Our current research has a very ambitious goal; to provide easy and efficient, high-level general-purpose parallel programming! A new language, called PRAM pascal (see (Dale 1992)), is being developed using the experience from PIL and the CREW PRAM simulator (documented in (Natvig 1990b)). The most ambitious part of this goal is the wish to *efficiently* execute PRAM pascal. A large number of processors operating synchronously towards a global shared memory are very difficult to realize with existing technology in an efficient way. However, the BSP model (Valiant 1990) may give a solution to this problem. Further, we believe that development of large and reliable parallel processing applications with good performance requires programs that are easy to develop and understand.

4.3.1 *Efficient execution on a BSP-machine*

'*What is the right division of labour between programmer, compiler and on-line resource manager? What is the right interface between the system components? One should not assume that the division that proved useful on multiprogrammed uniprocessors is the right one for parallel processing.*' (Snir 1989).

Valiant has recently described (Valiant 1990) the need for a bridging model for parallel computation, and proposes the *bulk-synchronous parallel* (BSP) model as a candidate for this role. He attributes the success of the von Neumann model of sequential computation to the fact that it has been a bridge between software and hardware. On the one hand the software designers have been producing a diverse world of increasingly useful and resource demanding software assuming this model. On the other hand the hardware designers have been able to exploit new technology in realizing more and more powerful computers providing this model. Valiant claims that a similar standardized *bridging model* for parallel computation is required before general-purpose parallel computation can succeed. It must imply convenient programming so that the software people can accept the model over a long time. Simultaneously, it must be sufficiently powerful for the hardware people to continuously provide better implementations of the model. See Valiant's paper (Valiant 1990) for further details.

A computer based on the BSP model may be programmed in many styles, but Valiant claims that a PRAM language would be ideal. The programs should have so-called *parallel slackness*, which means that they are written for v virtual processors run on p physical processors, $v > p$. This is necessary for the compiler to be able to 'massage' and assemble *bulks* of instructions executed as so-called supersteps giving an overall efficient execution. Using the PRAM model as programming model will typically result in programs with a large degree of parallel slackness. Further, a compiler with compile-time padding will produce code with associated information about the time consumption of its various parts. This is exactly the information that is needed to assemble the above-mentioned bulks of instructions. The increased degree of determinism found in synchronous programs will also make the 'bulkifying' easier.

4.3.2 *Ongoing research and future work*

Much research remains until it will be possible to offer efficient execution of PRAM pascal. This section explains why it may be worthwhile to strive for this ambitious goal even if we do not succeed.

We are currently working on PRAM pascal and a new PRAM simulator building on the experience gained through the prototype tools (Natvig 1990b), see Fig. 4.8. Execution of PRAM pascal on a simulator will clearly be an useful tool for research and teaching in the field of parallel algorithms.

In the context of project and diploma work in computer architecture various projects will be started to learn more about how a PRAM program may be executed on the BSP-model, and how the BSP-model may be implemented with existing and foreseeable technology. Valiant's BSP model is a totally new computing concept requiring drastically new designs of compilers and runtime systems. The use of the model raises many questions. How should PRAM programs be mapped to the BSP model? How much of the mapping should be done at compile-time, and what should be left to runtime? How is the BSP model best implemented? A BSP simulator may be the most appropriate investment in the start to learn more about the BSP model. Later on we hope to execute PRAM pascal transformed to BSP code on real hardware.

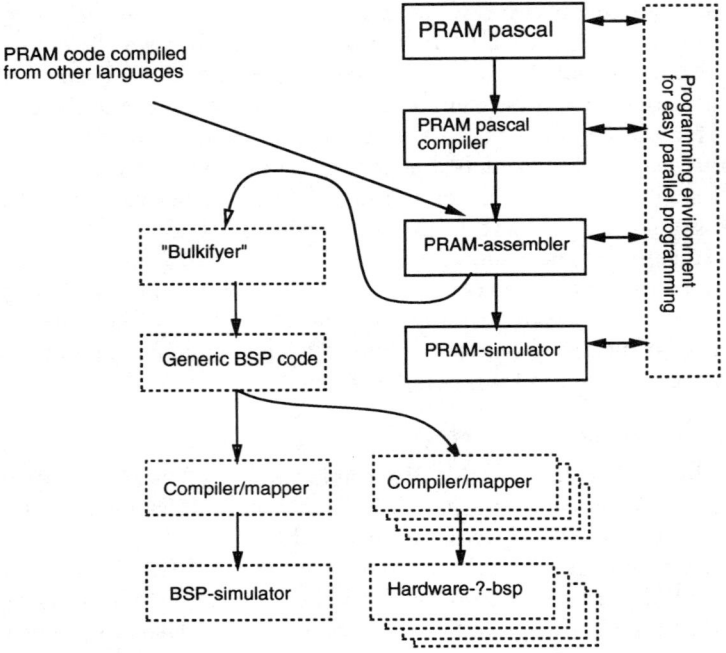

FIG. 4.8. Overview of various projects centred around the PRAM and the BSP model. The main goal is to provide easy parallel programming. Solid lines indicate ongoing projects, while dashed lines represent planned projects.

REFERENCES

Aho, A V, Hopcroft, J E and Ullman, J D (1982). *Data Structures and Algorithms*, Addison-Wesley Publishing Company, Reading, Massachusetts.

Akl, S G (1989). *The Design and Analysis of Parallel Algorithms*, Prentice Hall International, Inc., Englewood Cliffs, New Jersey.

Cook, S A (1981). Towards a complexity theory of synchronous parallel computation. *L'Enseignement Mathematique*, **XXVII**, 99–124.

Dale, B T (1992). *PRAM Pascal Prototype Compiler*, Master Thesis, Department of Computer Systems and Telematics, Norwegian Institute of Technology, University of Trondheim, Norway.

Flynn, M J (1966). Very High-Speed Computing Systems. *Proceedings of the IEEE*, **54**(12), 1901–9.

Fortune, S and Wyllie, J (1978). Parallelism in Random Access Machines. *Proceedings of the 10'th ACM Symposium on Theory of Computing (STOC)*, pp 114–8.

Gibbons, A and Rytter, W (1988). *Efficient Parallel Algorithms*, Cambridge University Press, Cambridge.

Hagaseth, M (1991). *Polylogarithmic Sorting on the CREW PRAM Model*. Master thesis, Department of Computer Systems and Telematics, Norwegian Institute of Technology, University of Trondheim, Norway",

Hillis, W D and Steele, G L Jr. (1986). Data parallel algorithms. *Communications of the ACM*, **29**(12), 1170–83.

Lubachevsky, B D (1989). Synchronization Barrier and Tools for Shared Memory Parallel Programming. *Proceedings of the 1989 International Conference on Parallel Processing, vol. II*, pp 175–9.

Manilla, H (1985). Measures of Presortedness and Optimal Sorting Algorithms. *IEEE Transactions on Computers*, **C-34**(4), 318–325.

McDowell, C E and Helmbold, D (1989). Debugging Concurrent Programs. *ACM Computing Surveys*, **21**(4), 593–622.

Natvig, L (1990). Logarithmic Time Cost Optimal Parallel Sorting is Not Yet Fast in Practice!. *Proceedings of SUPERCOMPUTING'90*, pp 486–94.

Natvig, L (1990b). *Evaluating Parallel Algorithms: Theoretical and Practical Aspects*, PhD thesis, Department of Computer Systems and Telematics, The Norwegian Institute of Technology, The University of Trondheim, Norway.

Quinn, M J (1987). *Designing Efficient Algorithms for Parallel Computers*. McGraw-Hill Book Company, New York.

Quinn, M J (1989). Analysis and Benchmarking of Two Parallel Sorting Algorithms: Hyperquicksort and Quickmerge. *BIT* **29**, 239–250.

Sabot, G W (1988). *The Paralation Model Architecture-Independent Parallel Programming*. The MIT Press, Cambridge, Massachusetts.

Snir, M (1989). Parallel Computation Models—Some Useful Questions. *Proceedings of the NSF - ARC Workshop on Opportunities and Constraints of Parallel Computing*. pp 139–45.

Steele, G L Jr. (1990). Making Asynchronous Parallelism Safe for the World. *Proceedings of POPL-90*, pp 218–27.

Valiant, L G (1990). A Bridging Model for Parallel Computation, CACM **33**(8), 103–111.

Wyllie, J C (1979). *The Complexity of Parallel Computations*. PhD thesis, Dept. of Computer Science, Cornell University.

5
PARALLEL ALGORITHM DESIGN ON THE WPRAM MODEL

J. M. Nash, P. M. Dew, M. E. Dyer and J. R. Davy

Systems Architecture Group,
School of Computer Studies, University of Leeds,
Leeds, LS2 9JT, UK
nash@scs.leeds.as.uk

Abstract

This paper describes the WPRAM programming model, supported on a derivative of the Bulk Synchronous Parallel (BSP) architecture. The WPRAM provides a weakly coherent shared address space with asynchronous concurrent requests for shared variables. All operations have an associated time cost, and are scalable with respect to increasing machine size. This provides a basis for designing scalable parallel algorithms whose time complexity accounts for all overheads. A simulator for the WPRAM enables absolute performance to be estimated, using machine parameters from Inmos T9000 and C104 components.

We outline a programming methodology in which an algorithm is first expressed in terms of the more abstract Strong PRAM model. This algorithm is then transformed onto the WPRAM model by considering practical issues of data locality, process synchronization and granularity. The scalability characteristics of the algorithm can then be analyzed; a new approach to assessing scalability is proposed. Experimental results from the simulator may be used to validate this analysis.

These ideas are demonstrated using the Simplex method for linear programming as a case study.

5.1 Introduction

To be able to make intelligent design decisions when writing parallel algorithms requires the existence of a well-defined computational model with predictable performance characteristics. Performance should be modelled in terms of the complexity of each operation, including overheads such as process management, data access, and process synchronization. The model should enable the algorithm designer to find out how the performance of an algorithm changes with varying data locality and granularity characteristics.

The PRAM model (Fortune 1978) represents a very abstract view of a parallel machine in which processors operate synchronously (at no cost) and have access to a global memory at unit cost. It has provided a simple and elegant framework for the design and analysis of parallel algorithms, but its idealization has limited its value for implementing programs on real parallel systems. Hence a number of more practical models have been suggested. For instance, the Block PRAM (Aggarwal 1989) includes a 2-level memory hierarchy accounting for network latency in global memory requests. Gibbons (Gibbons 1989) discusses a family of asynchronous PRAMs with costed barrier synchronization operations to maintain memory coherency. One member of this family, the Phase LPRAM, also costs latency for global memory accesses and enables such requests to be pipelined. Valiant defines an XPRAM model (Valiant 1990a) which barrier synchronizes all processors at fixed intervals, and uses it as an intermediate model in simulating standard PRAM models on distributed memory architectures. However these augmented PRAM models are still effectively restricted to executing SPMD algorithms.

This paper defines the WPRAM model, which permits a greater degree of asynchrony by allowing multiple process groups to synchronize independently and by permitting pairwise synchronization of arbitrary processes. It provides a weakly coherent CREW address space whose coherency is maintained by the various synchronization operations. This memory model is augmented by a set of asynchronous, combinable fetch-and-op operations (Gottlieb 1985).

All WPRAM operations are scalable, with a specified time complexity. Scalability is guaranteed because the WPRAM is supported on a BSP-like machine model (Valiant 1990), which essentially consists of a distributed memory MIMD machine using point-to-point message delivery. A simulator for the WPRAM has been written and can be used to predict absolute performance, using machine parameters derived from Inmos T9000 and C104 components.

A program development methodology is described which begins by designing an algorithm for the more abstract Strong PRAM model (Kruskal 1990). This algorithm is then transformed for the WPRAM by accounting for practical issues of data locality, process synchronization, and granularity. Performance can then be assessed both analytically and using the simulator, and the algorithm can be modified if appropriate. A new practical measure for scalability is proposed for this purpose.

The structure of the paper is as follows. We first describe the original BSP model and outline the modifications introduced here. The WPRAM model is then defined and an overview of the simulator is given. After a brief description of the Strong PRAM model, a case study is presented to illustrate the programming methodology. This is based on the Simplex method for linear programming. The scalability of the resulting program is analyzed using results from the simulator.

5.2 The architectural model

5.2.1 *The BSP model*

Valiant put forward the Bulk Synchronous Parallel (BSP) model (Valiant 1990) as a candidate for a *bridging model* which aims to provide some stability between the rapidly changing software and hardware areas of parallel computing. The model essentially consists of:

- a set of processor and memory pairs, or *nodes*;
- point-to-point message delivery between nodes;
- the ability to barrier synchronize all or a subset of the nodes.

Following Heywood's definitions elsewhere in this volume (see chapter 1) we classify this as an *architectural model*, since it is implementation-independent.

A program written directly for such a model consists of asynchronous *supersteps* separated by barrier operations. Each superstep involves the processors carrying out some local computation and sending or receiving some messages. All message delivery needs to be completed before the next superstep can begin. In particular, a processor should not attempt to read data which another processor is updating, unless the two operations are separated by a barrier.

A predictable latency between the sending and receipt of a message across the network can be achieved through the use of *randomized routing* (Valiant 1983). This removes any hot-spots which could potentially arise in the network by effectively decoupling the data access patterns of the processors from the resulting paths taken by the messages through the network.

However, if many processors wish to simultaneously access data that is on the same memory module then performance would degrade due to the serialization of the requests. The solution is to place each datum on a randomly chosen memory module using a randomized hash function (Mehlhorn 1984), so that each node (probabilistically) receives an even number of requests. Although theoretically the hash function needs to be a polynomial of degree $log_2(P)$ for P nodes, practical results (Jones 1992) show that this can be reduced to a low degree, giving an approximately constant calculation time. It has been shown that the use of such hash functions results in only a small constant loss of efficiency compared to the manual mapping of data to memory modules (Valiant 1990, Valiant 1992a).

The use of *parallel slackness* in an algorithm (creating more processes than there are processors) can be used to effectively hide remote access latencies and provide an efficient implementation of the user's algorithm (Valiant 1990).

5.2.2 *A modified architectural model*

The architectural model used in this paper differs from the BSP model in that it does not directly support a barrier operation. Instead, a barrier (or any other synchronization mechanism) can be implemented at a higher level (in our case in the WPRAM model) by using message passing. This has the advantage of making the model more general than BSP. The barrier mechanism was not included as a fundamental part of the machine model because it was felt that this would limit the functionality of the model (for instance, Valiant only allows a barrier operation across all processors). Also most parallel machines do not directly support this function. Similar reasoning is used in the LogP model (Culler 1993).

A second difference is that the architectural model used here assumes the presence of an asynchronous combining mechanism. This supports the scalable concurrent access of data by combining groups of similar requests before they reach their destination. This may be simply to allow data to be concurrently read by many processors, or to allow atomic updates such as integer addition (see next section). Asynchrony here implies that processors do not need to synchronize (using a barrier operation) when combining is required.

Combining can be accomplished either in hardware by augmenting the network or by randomized software algorithms. Hardware combining (Gottlieb 1985) requires that a switching element in the network can recognize combinable requests for the same destination when they arrive at approximately the same time. As well as recognizing that the requests can combine, the switches must also support the combining operation itself (integer addition for example). This necessarily means that the variety of such operations will be limited. The switch will need to keep a note that a combining operation has occurred (together with any relevant intermediate data) so that the result can be uncombined as it passes back through the switch. Asynchronous combining could also be implemented in software using a variant of the synchronous combining algorithm proposed by Valiant (Valiant 1992).

The last departure from the BSP model is that our model provides a more detailed set of cost parameters. A summary of these costs is shown in Table 5.1 below, where P is the number of processors.

The original BSP model aims to provide enough information to allow a user to write scalable and portable algorithms. The model provides costs for completing a barrier operation (L) and for the bandwidth of the machine, defined as the ratio of the total computational throughput of the processors to the total throughput of the network (g). The architectural model used here also allows the user to predict the time complexity of an

Table 5.1 Architectural model costs

Operation	Cost
network latency	$D = O(log(P))$
network bandwidth	$O(P.log(P))$
machine granularity	$g = O(1)$
access of X remote words	$O(D + X)$
local operations	$O(1)$

algorithm by specifying costs for all primitive operations. Constant costs can be assumed for such operations as context switching, arithmetic calculations, message handling, and local process management. In addition the model includes a constant cost for the frequency with which a processor can access the network (the g parameter in the BSP model, here called the *machine granularity*). This is derived from the fact that the bandwidth of the network must increase as $O(P.log(P))$ (Valiant 1990).

It is expected that this architectural model will be implemented using modern components such as the T9000 and C104 from Inmos Ltd. The hidden constants in the performance costs can then be derived.

5.3 The WPRAM model

We now define the WPRAM computational model. This is motivated by Valiant's XPRAM but is modified to support MIMD programming by permitting much greater asynchrony. In particular, barrier operations are under software control rather than at fixed intervals. The model provides a set of scalable operations, each with an associated cost derived from the underlying architectural model on which it is implemented.

5.3.1 *Process management*

The WPRAM model is based on a set of P processors which operate asynchronously. Each processor supports multiple asynchronous processes, allowing sharing of local data and the exploitation of parallel slackness. Parallelism is based on the creation of groups of identical processes which may be placed on the same processor or arbitrarily across the machine. In addition, the WPRAM allows the creation of the same process on each processor, guaranteeing parallel rather than merely concurrent execution (and also allowing the programmer to explicitly control load balancing for example). The complexity cost of creating a group of n processes across the machine is given as $O(D.log(min(n, P))+n/P)$. The logarithmic term represents the depth of a balanced binary tree, rooted at the creating process and spanning the machine, used to distribute the processes. Each branch of this tree connects two processors in the machine and so has a cost of

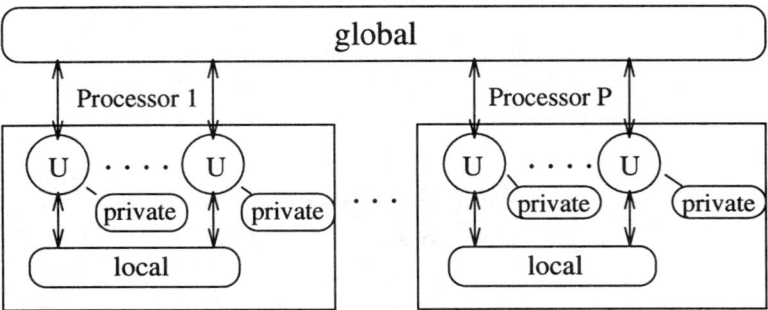

U - User process

FIG. 5.1. WPRAM Shared address space model

D, the diameter of the network, which is $log(P)$. Finally, the term n/P represents the serial component of creating the processes at each processor.

5.3.2 Data access

The WPRAM model supports a single flat shared address space with three data storage types (Fig. 5.1):

- *Private* data is placed on the processor of the allocating process. The data can only be accessed by this process and reflects the data locality in an algorithm on a single process basis. The cost of accessing x words is $O(x)$ since there is no use of the network.
- *Local* data is again placed on the allocator's processor but can be accessed by any process in the machine. This can be used to take advantage of data locality in an algorithm when used with local process groups (see above). Remote access of x words from local memory costs $O(D + x)$ due to the network delay.
- *Global* data is again accessible to any process but is not guaranteed to be placed on the same processor as the allocator. This is used when data locality is not an important issue and can provide shared data access with predictable performance without the need for the user to think about memory mapping issues. This has the same cost as the access of local data remotely.

Private storage is implemented by placing the data in the process's private address space or in the local cache of the processor. Implementation of the global storage type uses randomized hashing as described earlier. Local storage can be supported using a hashing function which always hashes an address to the local memory module (Jones 1992). The memory model of the WPRAM is termed *weakly coherent* since updates to local and global data are not guaranteed to be visible to other processes until they have synchronized. Synchronization may take the form of either creating process group or an explicit synchronization operation as given below.

Concurrent access of shared data is provided by a limited set of scalable atomic combining operations shown in Table 5.2. These are supported by the combining mechanisms of the underlying architectural model. As an example, *read&add (v,1,t)* enables each process $p_1 \ldots p_n$ to add an integer constant (one in this case) to a shared variable v (the result before the update being returned in a private variable t). Assuming v was initially zero, each process will receive a unique result in the range *0...n-1* into its private variable t, according to some arbitrary serial ordering of the requests, and v will hold the value n. A similar set of primitives called *void&op()* (where *op* represents *add*, *and*, or *or*) perform the same functions but do not return a result. Concurrent access takes place asynchronously and has the same performance as exclusive data access.

Table 5.2 Combining operations

Operation	Meaning
read&add (v,i,t)	add integer i to v
read&swap_int (v,i,t)	place integer i into v
read&swap_addr (v,a,t)	place address a into v
read&and (v,i,t)	bitwise AND integer i with v
read&or (v,i,t)	bitwise OR integer i with v

Asynchronous combining was chosen as opposed to synchronous combining because it allows the processes to operate asynchronously, preserving the maximum amount of parallelism in an algorithm. Combining is probabilistic using this method, but the probability of it happening increases when the network is more heavily loaded (Wilson 1992). Synchronous combining shares the general feature that more efficient use is made of the hardware resources (since combining is guaranteed to occur), but this requires a much tighter coupling between the processes, which may limit the parallelism available.

5.3.3 *Process synchronization*

Process synchronization in the WPRAM takes a number of forms. One is the creation of a process group, guaranteeing the coherency of data produced by the creating process to the processes within that group. Another is the blocking of the creating process until all of the processes in a specified group have terminated, guaranteeing the coherency of data produced by the group to the creating process. This can be generalized to the first process group from a set of groups (derived from the *alternation* facility, available in Occam2 (Pountain 1987) for example). We also permit a process group to *barrier synchronize*, as in the BSP architecture, guaranteeing data visibility within the group. This has the same complexity cost as the creation of a process group since the operation consists of first sending messages to the root of the associated spanning tree and then returning messages to the leaves.

Finally, the WPRAM includes *tag* variables which can synchronize one process with an event generated by another process. This provides a simple mechanism for the rescheduling of a blocked process and can similarly be used to block a process on one of a number of possible events. Tags provide more flexibility than barriers and cost only the same as exclusive data access. However, synchronization by tags guarantees only the coherency of data between the participating processes. A straightforward implementation of tags uses exclusive data access and combining operations.

5.3.4 *The Scalability of the WPRAM model*

Costs of the main WPRAM operation are shown in Table 5.3. Their scalability relates to the fact that they contain no cost components which grow linearly with the number of processors. For example, the creation of a process group is logarithmic in both the number of processes being created and the number of processors (ignoring the serial process creation component which might dominate when a large process group is created on a small number of processors). The same is true for the barrier operation. Most other operations in the WPRAM can be implemented using point-to-point message passing and take advantage of the combining operations to atomically update data.

5.3.5 *The WPRAM simulator*

A simulation of the WPRAM model which includes the performance model has been written so that the parallel execution time of an algorithm can be found. The simulator is based on the interaction of processes which represent the nodes of the target machine and the user processes in the WPRAM model. Practical costs for the operations in the machine model have used measured performance figures for the T9000 transputer and simulations of the C104 packet router.

Each user process has an associated independent clock which tracks its parallel execution time. A process also has a *slackness* variable which keeps a count of the total amount of processing carried out on the processor since that process was last context switched (i.e. the amount of parallel slackness). A process will only be context switched if it requires the result of a remote access or if it blocks when synchronizing with one or more other processes.

A remote operation is costed using four components:
- $local_1$: the preprocessing carried out before the remote access;

THE STRONG PRAM MODEL

Table 5.3 Costs of WPRAM operations

Operation	Cost
Process Management	
Creation of N local processes	$O(N)$
Creation of N global processes	$O(D.log(min(N,P))) + N/P)$
Termination of N local processes	$O(N)$
Termination of N global processes	$O(D + N/P)$
Data Access	
Access of X private words	$O(X)$
Access of X local words locally	$O(X)$
Access of X local words remotely	$O(D + X)$
Access of X global words	$O(D + X)$
Process Synchronization	
Barrier between N local processes	$O(N)$
Barrier between N global processes	$O(D.log(min(N,P))) + N/P)$
Local tag operation locally	$O(1)$
Local tag operation remotely	$O(D)$
Global tag operation	$O(D)$

- *remote*: the incurred delay of the remote access due to the network and the protocol on the destination processor;
- $local_2$: the further local computation which can be executed which the remote access is in progress;
- $local_3$: the local work carried out when the remote access returns.

The values $local_1$ and $local_2$ can be added to the *slackness* variables of each local process ($local_3$ being added when the operation completes). The clock of the process will be updated by the maximum of the incurred delay and the amount of parallel slackness generated in the meantime.

Further implementation details can be found in the first author's PhD thesis (Nash 1993).

5.4 The Strong PRAM model

The Strong PRAM (Kruskal 1990) is a more abstract model of parallel computation with a potentially infinite number of sequential processors operating synchronously. All processors share a CRCW (concurrent-read, concurrent-write) address space accessible at unit cost, enabling an arbitrary number of processors to read or write to a variable concurrently. There is also a set of concurrent atomic combining operations.

```
double A[n][d], c[n], b[d], column[d+1];
int pivot, index;
loop
    pivot takes the value of the index of the most negative c item;
    if pivot is defined
    {
        index takes the value of ratio (A[pivot]);
        if index is not defined
            exit;                                   /* solution unbounded */
        else
        {
            column[] takes the value of A[pivot] : c[pivot];
            simplex (b,column,index);
            carry out simplex (A[j]:c[j],column,index)
                            for each column j of A;
        }
    }
    else
        exit;                                       /* solution found */
end loop;

int function ratio (double vector[])
    return index i with smallest +ve ratio b[i]/vector[i];

void function simplex (double vector[], column[], int index)
    vector is updated using column[] and index;
```

FIG. 5.2. Sequential algorithm for simplex method

The Strong PRAM is an attractive model for a first stage of WPRAM algorithm design since it abstracts from many of the detailed concerns of the WPRAM while maintaining a similar memory model. In order to reflect the WPRAM model more accurately we have not used the full power of the Strong PRAM, limiting ourselves to the more restrictive CREW memory model. In addition, we have assumed a set of combining operations for the Strong PRAM which match those of the WPRAM.

An algorithm written for the Strong PRAM can be mapped to the WPRAM by considering issues of data locality, synchronization, and granularity. The following case study gives an example of this approach.

5.5 Case study : the Simplex method

We now show how the Strong PRAM model may be used to derive an initial parallel algorithm for the Simplex method, which is then mapped to the WPRAM model by taking into account various practical considerations.

CASE STUDY : THE SIMPLEX METHOD

5.5.1 Outline of the problem studied

The Simplex method can be used to derive the values of a set of n variables $x_1..x_n$ which minimize some function z of these variables subject to a set of d constraints. This can be summarized by the following formulae :

Minimize $z = \sum_j (c_j.x_j) \quad 1 \leq j \leq n$
Subject to
$$\sum_j (a_{ij}.x_j) \leq b_i \quad 1 \leq i \leq d$$
$$x_j \geq 0 \quad 1 \leq j \leq n$$
$$d \ll n$$

where $A = (a_{ij})$, b_i, and c_j represent the given constants.

The solution presented here assumes that the vector b contains only positive values. This removes the need for the use of *artificial variables*. The constraints are also represented by inequalities, removing the need for *surplus variables*. We also assume that the number of constraints is much smaller than the number of variables. Such problems occur very frequently; for example the problem of finding all of the vertices of a polyhedron (Chvatel 1983). This assumption allows us to exploit the parallelism within the algorithm by spreading the columns of the matrix across the processors.

5.5.2 The sequential algorithm

Fig. 5.2 outlines the sequential solution to the Simplex method. The algorithm consists of three main phases. The first is the selection of the most negative element of the c vector. If none exists then the algorithm terminates since a solution has been reached, otherwise the index of this value is held in the variable *pivot*. The corresponding column of the A matrix, $A[pivot][]$, is then searched for the smallest positive ratio of $b[i]/A[pivot][i]$. The row number holding this value in A is then assigned to the variable *index*. If all ratios are negative then the algorithm terminates since the solution is unbounded. This column of A (concatenated with the corresponding value of c) is then copied to the variable *column[]*. The rest of the algorithm then updates b and the columns of A (concatenated with the c vector), setting the chosen column $A[pivot][]$ to the unit vector. The complexity of the algorithm is $O(ndr)$ where r represents the number of iterations.

5.5.3 The Strong PRAM algorithm

The Strong PRAM algorithm for the Simplex method (Fig. 5.3) assigns a column of A and element of c to each processor. The chosen column for this iteration is determined by each processor incrementing the integer variable *race* if its c element is negative. So rather than the *most* negative c element being chosen, an element which is simply negative is used. The combinable operation read&add() accomplishes this, with the value of *race* before the update occurs being returned. This operation happens concurrently, with

```
int race, tmp;

for all j in 1..P do parallel
    loop
        race takes the value 0;
        if c[j] is negative
        {
            read&add (race, 1, &tmp);
            if tmp equals zero
            {
                index takes the value of ratio (A[j]);
                if index is defined
                {
                    column[] takes the value of A[j] : c[j];
                    simplex (b,column,index);
                }
            }
        }
        if index is not defined or race is 0
            exit;      /* solution unbounded or solution found */
        else
            simplex (A[j]:c[j],column,index);
    end loop;
end;
```

FIG. 5.3. Strong PRAM algorithm for Simplex method

one processor successfully retrieving the value zero. This processor can then update the *index* variable as before and set the *column[]* variable and update the *b* vector. It should be noted that there is an implicit suspension of the other processors while the successful processor completes this stage. The other processors then update their own columns if the final result has not yet been found (if *race* is not zero), or the result is not unbounded (if *index* is defined). The complexity of the algorithm is simply $O(dr)$ since each column is given to a unique processor.

5.5.4 *The WPRAM algorithm*

An outline of the solution of the Simplex Method for the WPRAM is shown in Fig. 5.4. This takes takes into account the important practical factors of locality, granularity, and synchronization.

The first factor, locality, is easily accounted for. Since the access of common data incurs a latency, each processor of the WPRAM stores a number of matrix columns locally, in the private variables *local_A* and *local_c*, using static data partitioning.

The second factor, granularity, is potentially more complex. The number of columns per processor is given by the *cols_per_proc* constant in the

CASE STUDY : THE SIMPLEX METHOD

```
#define cols_per_proc ? ( set to n/P )

global double A[n][d], c[n], b[d], column[3][d+1];
global int pivot, index[3], race[3];

for all j in 1..P do parallel

  private int tmp, double local_A[][], local_c[];

  copy 'cols_per_proc' columns of A[][] and elements of c[]
      into local_A[][] and local_c[];
  loop

    if any elements of local_c[] are negative
    {
      read&add (race[iter], 1, &tmp);
      if tmp equals 0
      {
        pivot takes the value of a negative element;
        index[iter] takes the value of ratio (local_A[pivot]);
        if index[iter] is not defined
            exit;                         /* solution unbounded */
        else
        {
          column[iter][] becomes local_A[pivot] : c[pivot];
          simplex (b,column[iter],index[iter]);
        }
      }
    }
    barrier_sync();                                            (i)

    if index[iter] is not defined or race[iter] is 0
        exit;       /* solution unbounded or solution found */
    else
        race[(iter+2) mod 3] takes the value of 0;             (ii)
                                 /* executed by processor 1 */
        simplex (local_A[]:local_c[],column[iter],index[iter]);
    iter takes the value of (iter+1) mod 3;

  end loop;
  copy local_A[][] and local_c[] back to A[][] and c[];
end;
```

FIG. 5.4. WPRAM solution of Simplex method

algorithm, and is simply calculated as n/P. Allocating a processor multiple columns allows it to execute other work so as to hide the latency of a remote data fetch. An example of this can be seen at the resetting of *race* at point (ii) (by processor 1) with the updating of the local matrix columns. This is related to the parallel slack requirement since it recognizes the advantage of overlapping communication with computation. Strictly, parallel slackness as specified by Valiant (Valiant 1990) requires that the algorithm create $O(P.log(P))$ processes in total. However, the efficiency benefits of this in the case of the Simplex algorithm are not obvious since the only remote accesses taking place are to distribute the pivot column and pivot row, most of which is carried out by a single process. Moreover, the amount of computation carried out to compute an updated column is quite small. Hence there is no advantage in allocating multiple processes to each local set, so the simple solution adopted in Fig. 5.4 seems adequate.

The final factor is the number and type of synchronization points in the algorithm, since the WPRAM operates asynchronously. In this example, the only synchronization operation required is the barrier synchronization of the processors at certain points in each iteration to guarantee the values of certain common variables. If there were only one copy of the *race* variable then there would need to be three synchronization points in the algorithm. The first is actually shown in the algorithm at point (i) and guarantees the coherency of the variables *race*, *index*, and *column*. Another two synchronization points would also have to be placed either side of the re-initialization of the variable *race*. The algorithm presented here removes these two extra synchronizations by providing three copies of *race*, indexed by the private variable *iter*. While the copy indexed by *iter* is being used by the algorithm, the previous copy of *(iter+2) mod 3* is being re-initialized and the next copy of *(iter+1) mod 3* can be used immediately in the next iteration. The same applies for the variables *column[]* and *index[]*. In fact there only need be two copies of these two variables since they are simply overwritten in the next iteration, but for coding simplicity three have been used.

Two important changes would be required if the WPRAM combining operations were synchronous. The first change requires a barrier synchronization after the *read&add()* operations so that the processors may be returned their results. One would be required at the access of *race* in order to choose the pivot column. A second barrier would need to be inserted at the concurrent reading of the variables *race[iter]*, *index[iter]* and *column[iter][]* after the original barrier in the WPRAM algorithm. So there now exist three barriers in the new algorithm using software combining, as opposed to the original one. This highlights the problem of the increased coupling of processes, imposing extra synchronization requirements on the algorithm. It also removes the advantage of executing combinable requests as early in the algorithm as possible, since the requests need to wait for the next synchronization point before they can be executed. Thus it provides further justification for the choice of asynchronous combining.

5.6 Performance analysis

5.6.1 *Method of analysis*

Part of the method for the design of an WPRAM algorithm is to determine how *successful* we have been in parallelizing the problem by deriving a scalability measure for the algorithm. The term *scalable* is commonly used for parallel algorithms to denote an algorithm which is able to make effective use of larger numbers of processors, given some increase in the problem size.

Suppose $t(N,P)$ is the parallel execution time of an algorithm (worst-case or amortized) solving a problem of size N on a P processor machine. This can be used to derive the measures for speedup

$$s(N,P) = t(N,1) / t(N,P)$$

and efficiency

$$e(N,P) = t(N,1) / P.t(N,P) = s(N,P) / P.$$

Speedup compares the advantage of a parallel execution of an algorithm as opposed to a sequential one, whereas efficiency measures the proportion of the time that processors do useful work. Efficiency will always decrease with P (for a fixed value of N) due to the increased communication overheads. The disadvantage of both measures is that they are functions of two variables. We would like a scalability measure that is either a simple number or a function of a single variable, so that it may be easily understood.

The widely used *isoefficiency* measure (Kumar 1990) measures the scalability of an algorithm by the rate of increase of problem size (in terms of the number of processors) required to maintain a constant efficiency. This measure has the advantage of describing the scalability of an algorithm as a function of a single variable, the number of processors in the machine. However, although the measure provides useful information on the *worst-case* increase in problem size required (since maintaining constant efficiency is very hard to accomplish for most algorithms), in our view the measure does not reflect what most algorithm designers require. This requirement is that the algorithm make good use of the machine resources as the number of processors increase, but allowing for the fact that larger numbers of processors will probably not be able to use the machine quite as efficiently as smaller numbers. This can be reflected by allowing the efficiency to decline slowly as the number of processors increases.

The measure in this paper uses a *scaling function*, $S(P)$, to describe at what rate the problem size is increased with P. Scaling functions will have the general form $S(P) = c.P^j . \log^k P$ for constants c, j, and k. An algorithm may be considered to have good scalability if a linear $((S(P) = c.P)$ or quasi-linear $(S(P) = c.P.\log^k P)$ scaling function leads to constant or

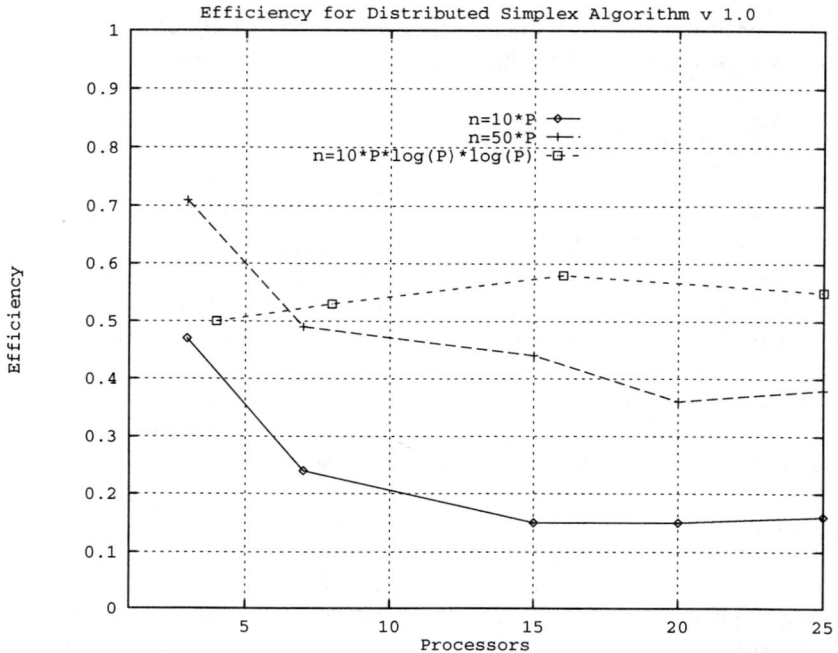

FIG. 5.5. Simplex algorithm efficiency results

slowly declining efficiency. 'Slowly' in this context is somewhat subjective and we have not currently quantified it more precisely. The time complexity $t(N, P)$, derived from the WPRAM cost model, can be used to determine the efficiency $e(S(P),P)$ of the algorithm, and hence investigate the effects of different scaling functions. Experimental results from the simulator can be used to test the validity of the scalability predictions.

5.6.2 Results

A theoretical analysis of the cost of the WPRAM algorithm given the complexity results described earlier gives the result $t(n, P) = O(r(nd/P + log^2(P)))$, since the columns, n, of the matrix are spread evenly across the processors and a barrier operation is carried out within each iteration. The variable r is the number of iterations of the algorithm and d the number of rows of the matrix. This results in the efficiency function $e(n, P) = O(n/(n + P.log^2(P)))$, assuming that the number of rows of the matrix, d are fixed. Chosing the scaling function to be $n = S(P) = c.P$ (that is, the number of columns of the matrix increases linearly), for some constant c, gives the result of $e(c.P, P) = O(c/(c + log^2(P)))$. How well this works in practice can then be determined by altering the value of c experimentally (given below). It should also be noted that chosing the scaling function to be $c.P.log^2(P)$ results in $e(c.P.log(P), P) = O(c)$. This quasi-linear scaling function is the constant efficiency result which would be produced by the isoefficiency method and is described as *near optimal* by (Rao 1989).

The simulator was used to assess how well the Simplex algorithm scaled up to 25 processors, for a linear scaling function with the constant $c = 10$ and $c = 50$. Results were also obtained for the quasi-linear scaling function with $c = 10$. The figures given are for the average of multiple simulated runs carried out for a number of randomly generated matrices.

Figure 5.5 shows the results obtained. Predictably, increasing the constant in the linear scaling function produces a noticeable increase in the efficiency over the range of processors. We would expect the efficiency to continue to decline slowly for much larger numbers of processors. However, limited computing resources and time constraints have not yet allowed us to explore this possibility more fully. The quasi-linear scaling function produces an approximately constant efficiency as expected.

5.7 Conclusions and future work

The major contributions of the WPRAM are:

- it provides a parallel programming model in which all practical overheads can be costed, allowing realistic complexity and scalability analysis;
- it provides scalable performance, with underpinning theoretical results showing that this can realistically be delivered.

A priority of our future work is therefore to build an implementation of the WPRAM model, using hardware components such as the Inmos T9000 and C104.

A scalable architecture is of little benefit if it is not exploited by scalable algorithms. Programming methodologies for the WPRAM, BSP, and related models are therefore of crucial importance. We are currently investigating the design of scalable algorithms, using case studies from linear programming, linear algebra and computational geometry. Of particular interest is the provision of scalable data structures with predictable performance, which can be available to programmers as abstract data types. We have already implemented scalable shared linked lists and queues, and we are currently studying the implementation of binary trees.

Our current experience is that WPRAM programming is not particularly easy, since the inclusion of all realistic costs necessarily involves a significant degree of complexity. Ultimately we view the WPRAM as a target onto which higher-level programming models can be mapped. A major challenge here is to enable performance to be realistically predicted at the higher levels of abstraction. The method described here, using the strong PRAM, is a first step in this direction.

Definition and measurement of scalability is an important independent issue which is the subject of intensive current research. A suitable scalability metric should be well-defined, capable of both theoretical prediction and experimental measurement. We are continuing to investigate practical scalability metrics, with the aim that suitable measurement tools be developed to support algorithm designers.

5.8 Acknowledgements

Thanks to David May and the Architecture Group at Inmos Ltd, Bristol, UK, for their helpful comments at various stages of this work. J. M. Nash was supported by the Science and Engineering Research Council and Inmos Ltd under a C.A.S.E. studentship.

REFERENCES

A. Aggarwal. On Communication Latency in PRAM Computations. In *Proceedings of the 1st Annual ACM Symposium on Parallel Algorithms and Architectures*, pp 11-21, 1989.

V. Chvatel. *Linear Programming*. W. H. Freeman and company, 1983.

D. Culler, R. Karp, D. Patterson, A. Sahay and K. Schauser, E. Santos, R. Subramonian and T. vonEicken. LogP: Towards a Realistic Model of Parallel Computation. In *Proceedings of the 4th ACM SIGPLAN Symposium on Principles and Practice of Parallel Programming, PPOPP*, San Diego, California, volume 2, pp 19-22. ACM Press, May 1993.

S. Fortune and J. Wyllie. Parallelism in Random Access Machines. In *Proceedings of the 10th Annual Symposium on Theory of Computing*, pp 114-118, 1978.

A. V. Gerbessiotis and L. G. Valiant. Direct Bulk-Synchronous Parallel Algorithms. In *3rd Scandinavian Workshop on Algorithm Theory*, Lecture Notes in Computer Science, volume 621, pp 1-18. Springer-Verlag, 1992.

P. B. Gibbons. A More Practical PRAM Model. In *SPAA'89 Proceedings of the 1989 Symposium on Parallel Algorithms and Architectures*, 1989.

A. Gottlieb and R. Grishman and C. P. Kruskal and K. M. McAuliffe and L. Rudolph and M. Snir. The NYU Ultracomputer - Designing an MIMD Shared Memory Computer. *IEEE Transactions on Computers*, **C-32**: 175-189, 1985.

T Heywood and C Leopold. Models of Parallelism. In *Abstract Machine Models for Highly Parallel Computers*, J. R Davy and P. M Dew (eds), Oxford University Press, 1994.

A. Jones. *Hashing in Virtual Address Translation*. Technical report, INMOS Ltd, Aztec West, Bristol, UK, 1992.

C. P. Kruskal and L. Rudolph and M. Snir. A Complexity Theory of Efficient Parallel Algorithms. *Theoretical Computer Science* **71**: 95-132, 1990.

V. Kumar and V. N. Rao. Scalable Parallel Formulations of Depth First Search. In *Parallel Algorithms for Machine Intelligence and Vision*. Springer-Verlag, 1990.

K. Mehlhorn and U. Vishkin. Randomized and Deterministic Simulations of PRAMs by Parallel Machines with Restricted Granularity of Parallel Memories. *Acta Informatica* **21**: 339-374, 1984.

J. M. Nash. Scalable Parallel Algorithm Design using the XPRAM Model. Ph.D. Thesis, University of Leeds, 1993.

D. Pountain and D. May. *A Tutorial Introduction to Occam Programming*. BSP Professional Books, 1987.

V. N. Rao and V. Kumar. *Analysis of Scalability of Parallel Algorithms* University of Texas at Austin, 1989 (unpublished).

L. G. Valiant. Optimality of a Two-Phase Strategy for Routing in Interconnection Networks. *IEEE Transactions on Computers* **C-32**(9): 861-863, 1983.

L. G. Valiant. A Bridging Model for Parallel Computation. *Communications of the ACM*, **33**(8): 103-111, 1990.

L. G. Valiant. General Purpose Parallel Architectures. in *Handbook of Theoretical Computer Science*, volume A, pages 945-971, North Holland, 1990.

L. G. Valiant. *A Combining Mechanism for Parallel Computers*, TR-24-92. Technical report, Aiken Computation Laboratory, Harvard University, Cambridge, MA 02138, November 1992.

G. V. Wilson. Using Opportunistic Combining Networks to Reduce Contention in Multicomputers. In *Proceedings PARLE'92*, pp 651-666. Springer-Verlag, Berlin 1992.

6
CTDNET III - AN EAGER REDUCTION MODEL WITH LAZINESS FEATURES

Padam Kumar and J. P. Gupta

Department of Electronics and Computer Engineering
University of Roorkee
Roorkee 247 667 (India)

S.C. Winter

School of Computer Science & Information System Engineering
University of Westminster
115 New Cavendish Street, London W1M 8JS (UK)

Abstract

A message-passing multiprocessor model for computation based on functional languages has been suggested. The model follows an applicative (eager) style of reduction, giving it an edge in exploiting parallelism over those models following normal order. However, applicative strategies are prone to being unsafe and are incapable of handling recursion. To take care of this, a concept of partial task has been introduced. The sluggishness of partial tasks in reduction is utilized to achieve a controlled version of recursion. The model has certain other features which make it selectively lazy so that the reduction tends to be more need-based and hence more safe.

Keywords: Functional architectures, lazy evaluation, multiprocessing, reduction machines, supercombinator reduction

6.1 Introduction

Functional languages (Peyton Jones 1987) seem to provide a natural approach towards the tricky problem of programming general-purpose multiprocessors. Compared to many other languages offering concurrent structures, they score better for the following reasons:

1. Parallel activity need not be specified by the programmer for the runtime synchronization between various processes;
2. Functional programs can be represented as graphs and evaluation may proceed through graph transformation, which is inherently a distributed parallel activity;

3. Due to freedom from side effects, various sub-expressions in a program can be evaluated in any order, even in parallel, without any fear of changing the meaning of the program.

In the present paper we define an abstract model of computation for functional programs, named CTDNet-III, which is suitable for a message passing multiprocessor. The model adopts an applicative or eager style of reduction where all arguments are reduced to normal form before being used by a function (a normal form is one containing no eligible redexes). Although this strategy helps in utilizing parallelism more effectively, laziness has its own virtues: it helps in avoiding wasteful activity and makes the computation safe, that is, it is guaranteed to terminate unless the program itself is undefined. To take advantage of laziness, the eager model can switch over to lazy reduction wherever it can benefit, such as in the evaluation of conditionals, shared sub-expressions, and recursion. This selective laziness is achieved by assigning some special properties to a *partial task* (Kumar 1990) – a term used to describe the application of a function to an incomplete set of arguments.

Besides reduction strategy, an important design parameter in a parallel reduction machine is the grain size of a computation unit because it affects the amount of administrative overheads in computation. Two closely related models (earlier versions of CTDNet (Gupta 1989, Winter 1990)) represented programs as λ-graphs where each binary application node is a process. Fine grain size in these models allows them to exploit parallelism very effectively but the overhead of message passing in a direct implementation is rather heavy. Designs of many other abstract machines for functional languages (Johnsson 1983, Keiburtz 1985, Fairbairn 1987) have in general favoured the idea of coarse grain redexes. The developments reported here are an attempt towards realizing a selectively lazy coarse grain version of the CTDNet evaluation model.

The paper is organized as follows: the next section contains structural descriptions of a task, the smallest unit of computation in the model; in Section 6.3, the organization process, which is concerned with framing task packets out of an expression built-up of binary curried applications, is described; Section 6.4 deals with the rules of reduction for a simple applicative model; in Section 6.5, the modifications to the simple model for making it selectively lazy are discussed and the process of taming recursion through them illustrated; finally, Section 6.6 gives conclusions and future projections for the work.

6.2 Task structure

The CTDNet-III model assumes a program to be a set of functions defined in supercombinator form (Hughes 1982). Figure 6.1 describes an abstract

$d ::= f^m x_1 \ldots x_m = e$
$e ::= c \mid x \mid e_1 e_2 \mid p\, e_1 \ldots e_k \mid f e_1 \ldots e_k \mid \text{let } x = e_1 \text{ in } e_2 \mid \text{letrec } x = e_1 \text{ in } e_2$

$$where \begin{cases} c \in C & : \text{Constants} \\ x \in Var & : \text{Variables} \\ p \in Op & : \text{Operators} \\ e \in Exp & : \text{Expressions} \\ f \in F & : \text{Function Identifiers} \\ d \in Def & : \text{Definitions} \end{cases}$$

FIG. 6.1. Syntax for function definitions

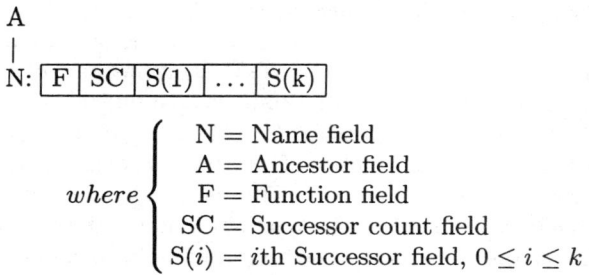

$$where \begin{cases} N = \text{Name field} \\ A = \text{Ancestor field} \\ F = \text{Function field} \\ SC = \text{Successor count field} \\ S(i) = i\text{th Successor field, } 0 \leq i \leq k \end{cases}$$

FIG. 6.2. Structure of a task

syntax for supercombinator definitions where f^m represents a function f of arity m.

Besides the definition set, a program would contain the principal application, the value of which is the value of the program. In the proposed model, the application of a function/operator to its arguments/operands is treated as a unit of computation known as *task*. For reduction, the program expression (the principal application) is organized into a graph of such tasks by the process of *organization*. A task here is similar to the current context stack in the G-machine (Johnsson 1983) or a packet in FLAGSHIP (Watson 1987) or the frame node in the $<\nu\text{-G}>$-machine (Augustsson 1989).

A task-graph is a directed acyclic graph (DAG) where the nodes are tasks and the arcs point to the successor tasks of a node. Each node maintains graph links in the form of *ancestors* and *successors*. Tasks are of variable size because different functions may require different number of arguments. The structure of a task in terms of its various fields is shown in Fig. 6.2.

Let us consider an expression $(f^m e_1 e_2 \ldots e_k)$. Depending on the values of m and k, the expression may be organized into three types of tasks defined below:

1. **complete** (C type): $m = k$, that is, the function arity matches the number of available arguments exactly.
2. **partial** (P type): $m > k$, that is, the function requires more than the number of available arguments. This situation corresponds to partially applied functions (Goldberg 1987).
3. **dummy** (D type): $m < k$, that is, the function arity is less than the number of available arguments. In its F-field, a dummy task holds the complete task (organized from m arguments) while in the successor fields it holds the surplus $(k - m)$ arguments.

The complete task type is further divided into two sub-types called reducible (R type) and waiting (W type) depending on the nature of successors. In a reducible type all the successors are in evaluated form whereas in a waiting type one or more of the successors are in unevaluated form. The information about the task type (R, W, P, D) forms part of the name-field N of a task. The field also carries a count which indicates the number of results awaited from successor tasks (for W type) or the shortage of arguments (for P type) or the number of surplus arguments (for D type). The count is zero for a reducible type. In the graph, a task knows its neighbours by their N-fields, and the type and count information in the field provides it all the necessary knowledge about its neighbours. This structure of the N-field, written as (name, type, count) reduces the need for inter-task communication at run time.

The Ancestor field A is for linkage within the graph. It holds the N-field of an ancestor and an integer s indicating that the task is the sth successor of its ancestor. Sometimes when a task is shared, its A-field will contain a list of A-fields indicating all those who share it. The root task, corresponding to the principal application has a <u>nul</u> value in its A-field.

The F-field contains the function identifier which is to be applied to the arguments in the Successor fields. It could be a user-defined function, a built-in operator or the N-field of another task (in the case of a dummy task).

Successor fields are ordered from left to right. A successor could be an atomic item, a pointer to an unorganized expression (discussed in the next section) or another task. All atomic items are represented as tagged quantities, e. g. the operator '+' would be shown as <u>op</u>+ where <u>op</u> is a tag which identifies the item as an operator.

6.3 Organization

Organization is the name given to the process of making a graph of coarse grained tasks out of the fine grained binary application nodes of an expression tree. It also acts as a pre-condition for reduction because any unorganized expression remains in a dormant state in our model. By postponing

$Org(C) = $ tag C ;tag indicates the type of C such as <u>op</u> (operator),
<u>val</u> (value) etc.

$Org(\hat{\ }x) = \hat{\ }Org(x : e_1)$;$\hat{\ }x$ points to e_1
$= \hat{\ }Org(e_1)$;situation may arise in a let expression

$Org(f^m e_1 \ldots e_k)$
$= [Org(f) \{Org(e_1), \ldots, Org(e_k)\}]$
$= [T_0 \{T_1, \ldots, T_k\}], m = k$;complete task
$= [T_0 \{T_1, \ldots, T_k, \ldots], m > k$;partial task
$= [[T_0 \{T_1, \ldots, T_m\}] \{\underline{ug}\hat{\ }e_{m+1}, \ldots, \underline{ug}\hat{\ }e_k\}], m < k$
   ;dummy task, <u>ug</u> is a tag
   indicating unorganized graph

$Org(IF\ e_1\ e_2\ e_3)$
$= [Org(IF) \{Org(e_1), \underline{ug}\hat{\ }e_2, \underline{ug}\hat{\ }e_3\}]$
$= [\underline{op}IF \{T_1, \underline{ug}\hat{\ }e_2, \underline{ug}\hat{\ }e_3\}]$;then and else arms left
   unorganized

FIG. 6.3. Description of organization process

organization, the evaluation of an expression can be kept in abeyance as long as desired.

A conceptual picture of the organization process is given in Fig. 6.3 where a function Org is specified. Here e stands for an expression, f^m for a function of arity m, C for a constant, and T for a task/atomic value resulting from the application of Org to an expression. In the notation used, a task is enclosed within square brackets, and inside it the argument list is enclosed within braces. A task representation then consists of a function task T_0 applied to a list of argument tasks $T_1, \ldots, T_k$. A partial task has an unfinished list of argument tasks (indicated by a missing brace), for example, [<u>op</u>+ {<u>val</u>5, ...] is a partial task because the operator '+' requires two arguments when only one is available.

It may be seen that the general characteristic of Org is to avoid organization of an argument whenever its need is doubtful, and it is most visible in the organization of a conditional expression where initially only the condition part is organized. Later when need is established, through the result of condition, the relevant part (then or else) only would be organized. On the same lines, whenever $m < k$, only the required m arguments are organized into a complete task. The remaining $k - m$ arguments are held as unorganized successors of a dummy task created for this situation, because nothing is known about their need at this stage.

Example: The organization steps are illustrated below for an expression

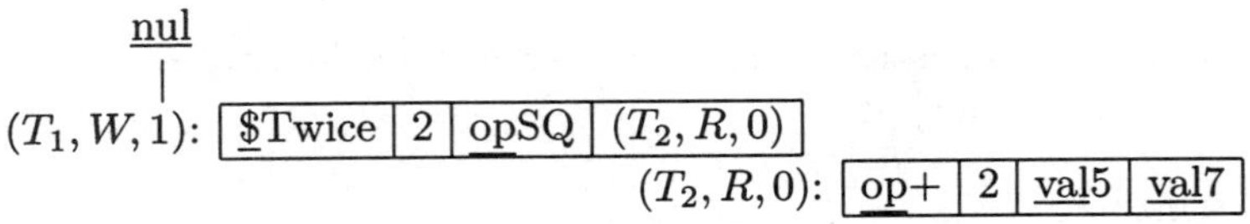

```
        nul
         |
(T₁, W, 1): | $Twice | 2 | opSQ | (T₂, R, 0) |
                              (T₂, R, 0): | op+ | 2 | val5 | val7 |
```

FIG. 6.4. Task-graph for (Twice SQ (+ 5 7))

(Twice SQ (+ 5 7)). (Note: the arity of the user defined function 'Twice' is 2, which is known from the definition part of the program):

$$Org(\text{Twice SQ } (+\ 5\ 7))$$
$$= [Org(\text{Twice}) \{Org(\text{SQ}), Org(+\ 5\ 7)\}]$$
$$= [\$\text{Twice} \{\underline{op}\text{SQ}, [Org(+) \{Org(5), Org(7)\}]\}]$$
$$= [\$\text{Twice} \{\underline{op}\text{SQ}, [\underline{op}+ \{\underline{val}5, \underline{val}7\}]\}]$$

Here $\underline{\$}$ is a tag to indicate a supercombinator function. The task-graph for this expression is shown in Fig. 6.4. The ancestor field (not shown) of task T_2 would contain $((T_1, W, 1); 2)$, indicating that it is the second successor of its ancestor. The name field of T_1 shows that it is a waiting type and is waiting for the result from one successor, while that of T_2 indicates that it is reducible type.

6.4 Eager reduction

6.4.1 *Reduction*

The aim of reduction is to bring a task-graph to normal form, that is, to a form where no reducible tasks are left. In the process whenever a task involving application of a user defined function is encountered, the relevant definition is used for creating an instance of the function (the actual mechanism of instance creation is not the focus of this paper - a scheme of compiling definition bodies into a code of structured director string terms for this purpose has been suggested elsewhere (Kumar 1990). As a task reduces, it communicates the result in the form of a message to its ancestor task(s) and then gets away from the scene. Thus during the tenure of its existence, a task either reduces or handles messages received from other tasks. Figure 5 defines this 'life cycle' of a task.

The reduction strategy is applicative (eager) so as to allow simultaneous action on all reducible tasks at any given time. This leads to a simple set of following conditions for a task to be reducible:

1. the task is complete;
2. all of its successors are in normal form, that is, either an atomic value or a partial type task.

```
task ::= begin
            while task ≠ null do {while alive keep on doing}
            begin
               if reducible then reduce else
               begin
                  * ? M;              {wait for a message M}
                  handle(M)
               end
            end
         end.
```

FIG. 6.5. Life cycle of a task

The two conditions are in the spirit of data flow approach where a node fires on the availability of all its input tokens. A complete task which has some of its successors as further tasks (other than partial type) waits until results in atomic/partial-task form are received from such successors through messages. A partial task has been treated as a normal form and is not allowed to proceed with reduction (later, in the modified reduction this condition is relaxed).

6.4.2 *Message handling*

During reduction, a task has to generate messages for its ancestor. The general structure of a message is a tuple given by

$$M ::= (\text{type, data, link})$$

The 'type' indicates the type of message which helps the recipient to choose its course of action. The 'data' field refers to the contents being sent, and the information in the 'link' field helps in establishing proper links in the dynamic graph. The model uses three types of messages discussed below:
(a) *Result message.* A <u>result</u> message is generated when the result of applying a function is an atomic value. A task receiving this message handles it according to its own type. A waiting or a partial type simply overwrites its appropriate successor field with the data received in the message. A waiting task may become reducible if that was the last result it was waiting for. A dummy task can receive a message from the task in its F-field only because all its successors are in unorganized and hence dormant form. If the data received through the message is a function/operator then the dummy task reorganizes itself according to the arity of the function received. The action is summarized in Fig 6.6, where T_r is the task which receives the message.
(b) *Link message.* The message is generated when the result of applying a function is another task, rather than an atom. The data field in this

$(\underline{\text{result}}, \underline{D}, (N_A; s)) \rightarrow \begin{bmatrix} T_r.\text{type} = \text{W} : \text{overwrite } s\text{th successor by } \underline{D}; \\ \quad \text{decrement wait-count}; \\ \quad \textbf{if } \text{wait-count} = 0 \\ \quad \textbf{then } \text{set type to } R \\ \\ T_r.\text{type} = \text{P} : \text{overwrite } s\text{th successor by } \underline{D} \\ \\ T_r.\text{type} = \text{D} : \textbf{if } \underline{D} \text{ is an operator/function} \\ \quad \textbf{then } \text{reorganize } \textbf{else } \text{error} \end{bmatrix}$

FIG. 6.6. Handling a <u>result</u> message

$(\underline{\text{link}}, T.N, (N_A; s)) \rightarrow \begin{bmatrix} T_r.\text{type} = \text{W: overwrite } s\text{th successor by } N; \\ \quad \textbf{if } T.N.\text{type} = \text{P then} \\ \quad \text{decrement wait-count} \\ \\ T_r.\text{type} = \text{P: overwrite } s\text{th successor by } N \\ \\ T_r.\text{type} = \text{D: } \textbf{if } T.N.\text{type} = \text{W or R then} \\ \quad \text{overwrite F-field by } N; \\ \quad \textbf{if } T.N.\text{type} = \text{P then} \\ \quad \text{supply shortage of } T \\ \quad \text{from own surplus}; \\ \quad \textbf{if } T.N.\text{type} = \text{D then} \\ \quad \text{pass own surplus to } T \text{ and} \\ \quad \text{send link message to ancestor(s)} \end{bmatrix}$

FIG. 6.7. Handling a <u>link</u> message

case carries the N-field of the resulting task. The message is not of much value to a waiting or a partial task, they simply overwrite the appropriate successor. The waiting type will, in addition, decrement its wait-count if the task received is of partial type. This is done because a partial type is assumed to be in normal form. The message is of great value to a dummy task which is always looking for an opportunity to shed its surplus arguments. It gets the opportunity if the task referred to in the message (task T in Fig. 6.7) is a P or D type. For the P case, the dummy task tries to fill the shortage of T from its surplus. For the D case, it passes on all its surplus to T and kills itself (there is no point in keeping two dummies in series). However, before moving out, it links its own ancestor(s) to T in order to maintain proper graph links.

$$(\underline{\text{take}}, n\text{-list}, T_d.A/\text{nc}) \rightarrow \begin{bmatrix} T_r.\text{type} = \text{P: adjust A-field;} \\ \qquad \text{organize arguments in } n\text{-list and} \\ \qquad \text{add to own successor list} \\ \\ T_r.\text{type} = \text{D: adjust A-field;} \\ \qquad \text{add successors in } n\text{-list} \\ \qquad \text{to your own successor list} \end{bmatrix}$$

T_d: The generating dummy; nc: no-change

FIG. 6.8. Handling a <u>take</u> message

(c) *Take message.* As seen above in the handling of a link message, a dummy task may have to pass on its own arguments to another task. It does so through the <u>take</u> message. The data field of this message contains an n-list of successors of the sending dummy, where n is the number of arguments to be passed down. If in the process, the dummy becomes empty of successors it would be killing itself, and so to maintain the graph link it must send its ancestor's name through the link-field of message. However, if it is not going to kill itself it would indicate 'no-change'. The action taken by the recipient of this message is summarized in Fig. 6.8. The message is meant for a partial or dummy task only. It may be noted that a dummy recipient adds the received successors to its list without organizing them. This is in accordance with the policy to postpone organization until need is known.

6.5 Modified reduction

The task reduction mechanism discussed in the previous section is based on a purely applicative reduction strategy. Only those tasks are allowed to reduce whose F-field is an operator or a user defined function and the successors are atomic values/partial tasks. In this section, we consider modifications for dealing with matters such as laziness in the evaluation of conditionals and shared expressions, and recursion.

6.5.1 *Laziness*

Laziness advocates that an expression be evaluated only when needed and then only once. Thus one of its requirements is that the decision to undertake a computation be based on need rather than the availability of data, that is, we must be able to postpone a computation and activate it again when the need is clearly established.

The main instrument for bringing laziness into the model is the process of organization acting as a necessary (but not sufficient) condition for reduction. Lazy evaluation of conditional expressions is achieved through this. In Fig. 6.3 it may be observed that the *then* and *else* parts are kept as inactive (unorganized) successors to the IF-task.* Initially, only the condition part is active, and when a boolean result is received from it, the relevant part (then or else) is organized and linked to the ancestor of the task handling the complete IF-expression. The other part is discarded thus avoiding any wasteful work. To accommodate this, the handling of a <u>result</u> message M by a waiting task (with a little modification to that in Fig. 6.6) may be expressed as follows:

    if F-field = <u>op</u>IF
    then begin
             if M.data = <u>val</u>TRUE **then** temp := organize(S(2))
            else temp := organize(S(3));
            if temp = task **then** send link message to ancestor(s)
            else send result message to ancestor(s)
      end
    else begin   {same as in Fig.6}
        overwrite sth successor by M.data;
        decrement wait-count;
        if wait-count = 0 **then** set type to R
    end;

    Another situation where laziness can be useful is the computation of a shared sub-expression. When a complete task is shared, the applicative strategy will reduce it to normal form and communicate the result to all those sharing it. There is no risk of any repeated computation. A situation of particular interest is the sharing of partial application (Goldberg 1987) of a function, that is, of a partial task in our model. Until now we have been allowing partial tasks to be passed down as normal form arguments to other complete tasks. This approach has the risk of repeating computations enclosed in the partial task, if the function to whom it is passed as argument has several references for it. This is so because the reduction procedure makes copies of an argument task for multiple requirements.

    The above problem can be solved by detecting whether a partial task has a computable part, and if it has one it should be allowed to reduce (with the available arguments) before being passed as argument to another task. It would give rise to an 'incomplete' instance of the function being handled by the partial task. We call it a *residual* definition of the function. This residual definition can be organized and passed as normal form to the ancestor task(s). There is no computation left in it and hence there is no risk of losing laziness in copying it for multiple references.

    As tasks are created at run time also, detection of a computable part in a partial task may be a significant overhead. However, this detection

---

*for convenience of reference, a task is referred by the function in its F-field.

can be done through a compile time computability analysis (Kumar 1990) of all function definitions in the program. In this analysis, a computability number CN is assigned to each definition. A CN of n means that the function requires a *minimum* of the first n arguments so as to have a computable sub-expression. As an example, a function $X defined as

$$\$X \ a \ b \ c = *(+(SQ \ a)b)c$$

has a CN of 1 because the availability of first argument a makes the computation of (SQ a) possible. Thus during organization of an expression into task-graph, a partial task will have computability if

No. of available successors $\geq CN$ of the function in F-field

In view of the above discussion, the conditions of reducibility of a task need to be modified as follows:

1. The task is complete and its F-field contains an operator or a user defined function or a residual definition.
2. All the successors are in normal form (atomic value or partial type tasks with no computability or residual definition).

OR

1. The task is partial having computability, and all its successors are in normal form.

To keep message-passing simple, the partial types may be further classified into computable and non-computable at the time of organization and the idea of CN acts as a handy tool for this classification.

6.5.2 *Recursion*

Many function definitions have a local variable defined recursively through a *letrec* block. One way of dealing with this is through the fixed point combinator Y treated as a built-in operator (Peyton Jones 1987). However, applicative order cannot handle recursion because in its eagerness to compute everything, it gets entangled in the non-terminating task of reducing the application $(Y \ H)$ to normal form. Thus the reduction strategy needs modification so as to be able to postpone the reduction of $(Y \ H)$ until an appropriate time.

Revesz (1987) has suggested a controlled recursion through an operator Y' which fires only once. The reduction rule for Y' is

$$Y' \ H \longrightarrow H \ (Y \ H)$$

where Y is a disabled combinator which cannot initiate further reduction. The blocked reduction is restored through a modified β-reduction rule which states that during a β-reduction an occurrence of Y be changed to Y'.

The main idea behind taming the Y operator is that it be allowed to fire only when there is need (normal order reduction). To achieve this in our eager model, we propose that Y, as an operator, has no fixed arity of its own; rather, it should acquire the arity of the function to which it is applied. Thus, if H is a function of two parameters then the arity of Y is two and hence the expression $(Y\ H)$ would be organized as a partial task (with no computability) whereas $(Y\ H\ A)$ would be a complete one. The reduction rule for $(Y\ H\ A)$ may then be given as

$$Y\ H\ A \longrightarrow H\ (Y\ H)\ A$$

It can be seen that in the modified rule, the presence of the argument A on the left-hand side simply acts as a 'catalytic agent' which does not take part in actual reduction but without which the reduction cannot proceed further. On the right-hand side, the application $(Y\ H)$ is inhibited because Y is short of one argument. In general, if H requires r arguments then the expression $(Y\ H\ e_1 \ldots e_k)$ is a complete task (and hence reducible) only if $k = r - 1$. In other words, a Y-task is organized as partial unless the required number of arguments are available.

In our notation, a complete Y-task is of the form $[Y^r\{H, T_1, \ldots, T_{r-1}\}]$ where Y^r is a Y of arity r. After a reduction this task becomes

$$[H\ \{\overbrace{[Y^r\{H, \ldots\}]}^{seed},\ T_1, \ldots, T_{r-1}\}]$$

The reduction results in a complete task having H in the F-field position. The first successor of this complete task (marked by an upper brace) will always be a partial Y^r-task having only one argument, viz. H. This partial task contains the 'seed' of recursion from which a fresh cycle of recursion can be initiated whenever required (by providing it with the necessary number of arguments).

6.6 Conclusions

Supercombinator reduction has been advocated (Peyton Jones 1987) as a means of coarse-grain processing leading to lesser communication overheads. In accordance, a task reduction model has been proposed where a reducible task represents either a supercombinator redex or a redex involving a built-in operator (similar to computation units in some other machines (Keiburtz 1985, Augustsson 1989)).

Program expressions in CTDNet-III have been represented as taskgraphs where nodes are tasks and the edges link the nodes through ancestor and successor fields of the task structure. Information about task-type forms part of the name field making the tasks more knowledgeable about their neighbours, thereby cutting down the need for unnecessary inter-task communication.

The process of framing task-graphs from program expressions, called organization, acting as a pre-condition for task reduction, serves to delay the computation of a sub-expression if useful. The strategy is expected to help in avoiding unnecessary computations and thus make the model more safe. Its need in the handling of conditional expressions is obvious.

The concept of a partial task, introduced to represent a function applied to too few arguments, is gainfully employed in avoiding recomputation of a shared sub-expression. A compile time computability analysis of supercombinator definitions has been suggested which can help in deciding whether to allow reduction of a shared partial application or not. The 'inertia' of a partial task in reducing on its own has been further utilized in realizing a controlled recursion through Y combinator in a model which is basically applicative in its reduction strategy. The Y combinator has been made a built-in operator with flexible arity for this purpose. dummy tasks have been introduced as pseudo tasks which simply act as place-holders for keeping surplus arguments in a dormant form. The combined use of dummy and partial tasks makes the model selectively lazy.

Task structure has been designed with a view to keep the message passing simple and less frequent. Work on modelling and designing a simulation scheme is in progress so that the actual amount of message passing overhead and other performance factors may be assessed. Further, it appears that the properties of partial task may be utilized to implement a controlled version of mutual recursion also. This is being explored. Another direction of future work is towards adding data structure handling in the model.

6.7 Acknowledgements

The support of British Council and UGC (India) for the work, in terms of supporting the exchange visits of research scientists between the University of Westminster, London, and the University of Roorkee, Roorkee, is gratefully acknowledged.

REFERENCES

Augustsson, L. and Johnsson, T. (1989). Parallel graph reduction with the $<\nu\text{-G}>$ machine., *Proc. IFIP Conf. on Functional Programming & Computer Architecture*, London, pp 202-13.

Fairbairn, J. and Wray, S. (1987). TIM: a simple lazy abstract machine to execute supercombinators. *Proc. 3rd Conf. on Functional Programming & Computer Architecture*, Oregon, USA, *LNCS* **274**, pp 34-43, Springer-Verlag.

Goldberg, B. (1987). Detecting sharing of partial applications in functional programs., *Proc. 3rd Conf. on Functional Programming & Computer Architecture*, Oregon, USA, *LNCS* **274**, pp 408-25, Springer-Verlag.

Gupta, J.P., Winter, S.C. and Wilson, D.R. (1989). CTDNet - A mechanism for the concurrent execution of lambda graphs. *IEEE Trans. Software Eng.* **15**, 1357-67.

Hughes, R.J.M. (1982). Supercombinators: a new implementation method for applicative languages. *Proc. ACM Symposium on Lisp & Functional Programming*, Pittsburgh, USA, pp 1-10.

Johnsson, T. (1983). The G-machine: an abstract machine for graph reduction. *Proc. Declarative Programming Workshop*, University College London, UK, pp 1-19.

Keiburtz, R.B. (1985). The G-machine: a fast graph reduction evaluator. *Proc. IFIP Conf. on Functional Programming & Computer Architecture*, Nancy, *LNCS*, **201**, pp 400-13, Springer-Verlag.

Kumar, P. (1990). *Design of a functional computation model for multiprocessor architecture*, PhD Thesis, University of Roorkee, India.

Peyton Jones, S.L. (1987). *The implementation of functional programming languages*, PHI series in Computer Science, PHI(UK).

Revesz, G.E. (1987). Rule-based semantics for an extended λ-calculus. *Proc. 3rd Workshop on Mathematical Foundations of Programming Language Semantics*, New Orleans, Louisiana, USA, *LNCS*, **298**, pp 43-56, Springer-Verlag.

Watson, I., Sargeant, J., Watson, P. and Woods, V. (1987). Flagship computational models and machine architectures. *ICL Tech. J.*, **5**, 555-74.

Winter, S.C. (1990). *A distributed reduction architecture for real time computing*, PhD Thesis, The Polytechnic of Central London, UK.

Winter, S.C., Wilson, D.R. and Neale, D.F. (1990). Real time functional programming systems. *Proc. 16th Euromicro Symposium on microprocessing and microprogramming*, Amsterdam, 491-8.

7

EXPLOITING PARALLELISM IN FUNCTIONAL LANGUAGES: A 'PARADIGM-ORIENTED' APPROACH

Fethi A. Rabhi

Department of Computer Science
University of Hull, Hull HU6 7RX, UK
far@dcs.hull.ac.uk

Abstract

Deriving parallelism automatically from functional programs is simple in theory but very few practical implementations have been realized. Programs may contain too little or too much parallelism causing a degradation in performance. Such parallelism could be more efficiently controlled if *parallel algorithmic structures* (or skeletons) are used in the design of algorithms. A structure captures the behaviour of a *parallel programming paradigm* and acts as a template in the design of an algorithm. This paper presents some important parallel programming paradigms and defines a structure for each of these paradigms. The iterative transformation paradigm (or geometric parallelism) is discussed in detail and a framework under which programs can be developed and transformed into efficient and portable implementations is presented.

7.1 The 'paradigm-oriented' approach

In recent years, there has been a steady improvement in the design of high-performance parallel computers. However, writing parallel programs is still a complex and expensive task which requires a detailed knowledge of the underlying architecture. It is now argued that functional languages could play an important role in the development of parallel applications. Their *implicit parallelism* eliminates the need to explicitly decompose a program into concurrent tasks, and to provide the necessary communication and synchronization between these tasks.

Following this principle, a typical parallelizing compiler for functional languages would have the structure displayed in Fig. 7.1. Given an arbitrary functional program, the first phase analyses this program to detect parallelism, and encodes decisions such as evaluation order, partitioning of data, load balancing, and granularity control in the form of *annotations* (Burton 1987). There are two forms of inherent parallelism: *horizontal* and

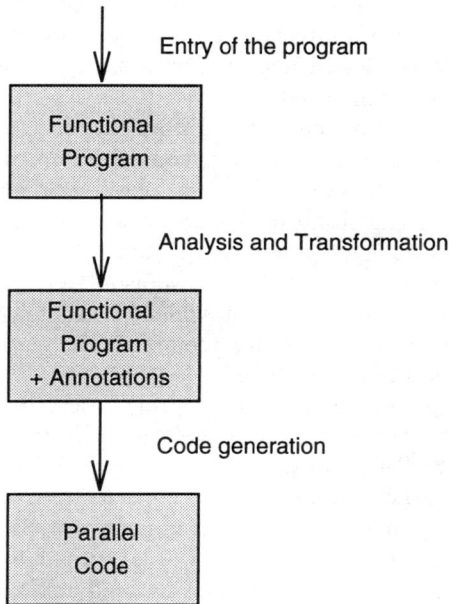

FIG. 7.1. Implementation based on implicit parallelism

vertical parallelism (Kelly 1989). Horizontal parallelism evaluates arguments of a function in parallel, requiring a strictness analysis to determine situations where it is safe to do so. Vertical parallelism on the other hand often relies on lazy evaluation for concurrent evaluation of aggregate structures. The second phase in the compiler generates code from the annotated program, according to some abstract parallel evaluation model (e.g. graph reduction).

Several implementations have been developed around this scheme (e.g. (Augustsson and Johnsson 1989, Burn 1989, Peyton Jones *et al.* 1989)). However, their success has been limited due to the following reasons (Rabhi and Manson 1991a):

- the compiler fails to detect parallelism, or there is no parallelism because of the way the program is written;
- the compiler generates too much parallelism and the resources of the machine are swamped by an overwhelming number of tasks;
- the compiler produces parallel code with an unpredictable behaviour, causing problems such as uneven load balancing, fine granularity, and

excessive non-local memory references.

These problems can be partially solved by allowing the programmer to enter annotations directly, such as in para-functional programming (Hudak 1991). This is somewhat similar to explicit parallel programming with all the difficulties involved.

Another approach inspired from the idea of *algorithmic skeletons* (Cole 1989) suggests that efficiency could be improved, providing that the implementation 'knows' about the parallel programming paradigm used as opposed to the situation where parallelism arises in a rather 'accidental' way. This has been confirmed through experiments involving Divide-and-Conquer problems (Rabhi and Manson 1991a). A skeleton can be thought of as a higher-order function which, when supplied with the functions representing the different parameters of the problem, yields the *executable specification* of this problem.

The 'paradigm-oriented' approach described in this paper consists of taking the parameters of a skeleton and using them to synthesize an efficient parallel program. This approach is half-way between implicit and explicit parallelism and can be considered as an alternative in the control of parallelism in functional languages. An effective implementation becomes more feasible because a program is always an instance of a known paradigm whose behaviour can be predicted. This approach also provides the opportunity for portability across different architectures. The resulting implementation is illustrated in Fig. 7.2.

Parameters are entered using templates, that we will call *structures*, and each structure represents a parallel paradigm. A higher-order function (the skeleton) can be applied to the parameters of the structure producing the executable specification, which could be used for testing and verification. The structure can also be used to generate the annotated program, then the parallel code. Code generation is made efficient by using the inherent knowledge about the paradigm being used.

This paper reviews some important paradigms for parallel algorithms and their associated structures, paying special attention to the iterative transformation paradigm. The paper is not concerned with *implementation* techniques as each structure needs to be further investigated before it can be efficiently executed.

7.2 Parallel programming paradigms

A paradigm, such as Divide-and-Conquer or dynamic programming, is a high-level methodology that helps in the design of efficient algorithms. In general, programming paradigms encapsulate useful information about the different stages in the computation and the patterns of data references. In a parallel programming paradigm, several computational stages can execute

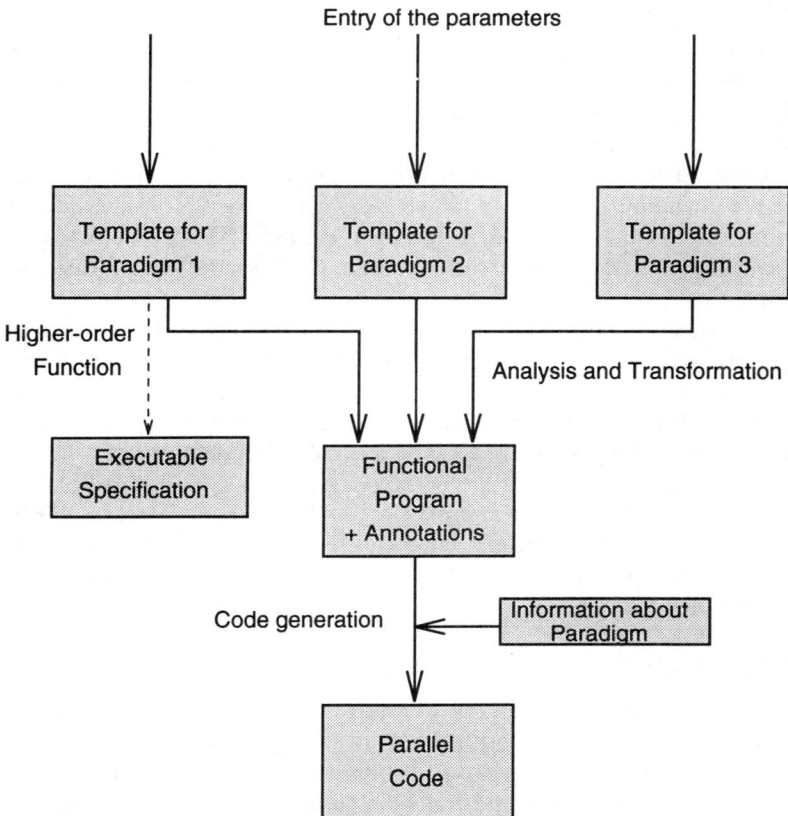

FIG. 7.2. Structure of a 'paradigm-oriented' implementation

simultaneously, and data references generally correspond to *communication patterns*.

Parallel algorithms are often characterized by the regularity of their communication patterns which can form chains, trees, meshes, or hypercubes. An algorithm is said to be *static* if the number of stages is known in advance, and *dynamic* if the concurrent computations are generated at run-time. We use the term *distributed algorithm* if the computational stages only communicate by message-passing and *centralized algorithm* if the execution depends on shared global values.

In this paper, we identify four important paradigms:

- Recursively Partitioned
- Process Networks

- Distributed Independent
- Iterative Transformation

There are other classifications for parallel algorithms, some of which are more detailed (Kung 1989, Burkhart *et al.* 1993), others are closer to existing architectures (Quinn 1987). The classification presented here is independent from any machine model, although one particular machine will be more suited to a class of parallel algorithms than another. As an example, a shared memory architecture is more adapted to centralized algorithms that a distributed memory one; or a SIMD-type of machine would be better suited to some static systolic and iterative transformation algorithms.

The rest of this paper defines a structure for each of these paradigms and presents some examples. From these structures, only (sequential) executable specifications are derived but we will whenever possible provide insights about some implementation issues. Skeletons for the first three paradigms exist in the literature so this paper concentrates mainly on the fourth paradigm.

The language used in the examples is Haskell (Hudak 1992), a newly proposed functional programming language standard. A summary of the Haskell operators and functions used in this paper can be found in the appendix to this chapter.

7.3 Recursively Partitioned Algorithms

These algorithms solve a problem by dividing it into subordinate problems, which are themselves recursively solved by dividing them further. The final result is obtained by recursively combining the solutions of the subproblems. Parallelism arises naturally in such problems as each of the subproblems can be solved independently from the others. The parallelism is also dynamic in nature, as the tasks are generated at run-time and the communication pattern formed by these tasks is a tree.

Divide-and-Conquer is a category where all the sub-problems need to be solved to compute the solution, thus exploiting *conservative parallelism*. Divide-and-Conquer is important to many applications such as sorting, the Fast Fourier Transform, and computing convex hulls(Kung 1989).

Most existing classifications include Divide-and-Conquer as a paradigm but fail to mention its general form which includes algorithms which work by building a partial solution, then trying in parallel all possible improvements to this solution. When no further improvements can be made, a partial solution is discarded. Sometimes, the first solution obtained is accepted so all the other attempts have to be interrupted. These algorithms are called Generate-and-Solve in (Finkel 1987). Examples include the N-queens problem and combinatorial search problems(Finkel and Manber 1987). Unlike Divide-and-Conquer algorithms, sub-problems are solved without knowing if their results will be useful or not, thus exploiting *speculative parallelism*.

7.3.1 Parallel algorithmic structure

To specify a recursively partitioned algorithm, the following components need to be defined:

- `ind prob`: returns `True` if the problem `prob` is indivisible
- `solve prob`: solves an indivisible instance of a problem `prob`
- `divide prob`: divides the problem `prob` into sub-problems
- `combine prob sols`: combines the solutions `sols`, possibly using information from the original problem `prob`

We assume that a problem is defined using some suitable type. As a convention, the functions which produce executable specifications have a '$' sign appended to their name. Recursively partitioned algorithms are represented in the form of a higher-function rp$.

```
rp$ ind solve divide combine prob
    | ind prob   = solve prob
    | otherwise = combine prob
                    (map (rp$ ind solve divide combine)
                     (divide prob))
```

In any parallel implementation, the map function is concurrently applied to the list of sub-problems thus exploiting horizontal parallelism.

7.3.2 Example

Quicksort is a well known example of a Divide-and-Conquer problem. The divide stage selects the head of the list as a pivot and splits the rest of the list into a lower and an upper part. The combine phase builds a list with the sorted lower part, the pivot, and the sorted upper part.

The specification of the problem would be:

```
ind l    = if (length l < 2) then True else False
solve l = l
divide (pivot:rest) = [ [x | x <- rest , x <= pivot],
                        [x | x <- rest , x >  pivot] ]
combine (pivot:_) [b,c] = b++[pivot]++c
qsort list = rp$ ind solve divide combine list
```

The Recursively Partitioned structure as defined above can be inefficient if the combine stage needs some information derived during the divide stage. In the previous example, the pivot is not directly available to the combine function and has to be extracted again from the original problem. This would be time-wasting if other methods of selecting the pivot are used, as the following example shows:

```
divide l  = [ [x | x <- rest , x <= pivot],
              [x | x <- rest , x >  pivot] ]
          where (pivot,rest) = computePivot l

combine l [b,c] = b++[pivot]++c
          where (pivot,_) = computePivot l
```

In this example, the call (`computePivot l`) is computed twice.

7.4 Process networks

These algorithms are characterized by a division of the computation into *stages*, with the data *flowing* through the stages. Parallelism is evident as stages can operate concurrently, depending on the availability of input data. Process networks are *uniform* if the same operation is repeated in every stage or *multifunction* if stages implement different functions. They also can be static or dynamic, and can have any communication pattern.

7.4.1 Parallel algorithmic structure

The parameters of the problem just consist of the list of functions and a description of how they are interconnected. To improve the readability of programs, the Caliban language provides some additional notation (Kelly 1989). A graphical representation of static process networks would also be a suitable alternative.

To derive the executable specification, process networks can be modelled using the concept of a *stream*, which is a list constructed *lazily*, thus exploiting vertical parallelism. A stage corresponds to mapping a function on one or several streams of input values, producing one or several streams of output values. This is only valid for deterministic algorithms, where the ordering of communicating actions can be determined statically. There are approaches that model indeterministic general communication capabilities using *bags* (Darlington *et al.* 1991).

An executable specification for any static process network can be obtained using the higher-order function defined in (Kelly 1991). Dynamic problems are very difficult to capture, particularly those presenting an irregular communication pattern such as computing the prime numbers using Eratosthenes' Sieve.

7.4.2 An example

A list of functions `fs` is sufficient to define a pipeline that repeatedly applies the composition of these functions to a stream of objects. The corresponding executable specification can be obtained using the following higher-order function:

```
pipe$ fs = (foldr (.) id fs)
```

The (.) operator is the composition function and `id` is the identity function. Other regular pipelined structures are described in (Cole 1990).

As an example, all possible bit patterns of a word of length n can be computed using a pipeline of n processes, each process periodically adds a 0 or a 1 to a partially-built pattern before passing it to the next process. Processes have periods ranging from $1, 2, 4$ to 2^{n-1}.

```
not 0 = 1
not 1 = 0

bits p i v (x:xs) | (i==1)    = (v:x):(bits p p (not v) xs)
                  | otherwise = (v:x):(bits p (i-1) v xs)

bitpatterns n = take (2^n) (pipe$ (ps n) nils)
                where ps 0 = []
                      ps (i+1) = (bits (2^i) (2^i) 0):(ps i)
                      nils = []:nils
```

This is a dynamic pipeline because the number of stages is not known in advance. An example is illustrated in Fig. 7.3.

7.5 Distributed independent algorithms

These algorithms solve a problem by dividing it into a number of *independent* computations, and then combining the results of these computations. They are similar to the Recursively Partitioned Algorithms but with two differences. The first one is that the control of the operations is centralized, that is, there is a master process which distributes the work to slave processes. The second one is that the number of independent computations can be specified by the programmer or determined by the implementation based upon the number of available slaves. Distributed independent algorithms are generally called *processor farms*.

7.5.1 *Parallel algorithmic structure*

The most important parameter is the number of slaves **n**. The other parameters of the problem consist of a `divide` function that specifies how the work is divided amongst the slaves, a function `f` that is executed by the slaves, and a function `combine` that combines the results on the host.

The corresponding executable specification is obtained through the following higher-order function:

```
di$ f n divide combine p = combine (map f (divide n p))
```

7.5.2 *Example*

In this example, random numbers are generated using **n** independent computations, each computing one sub-sequence of random numbers. The

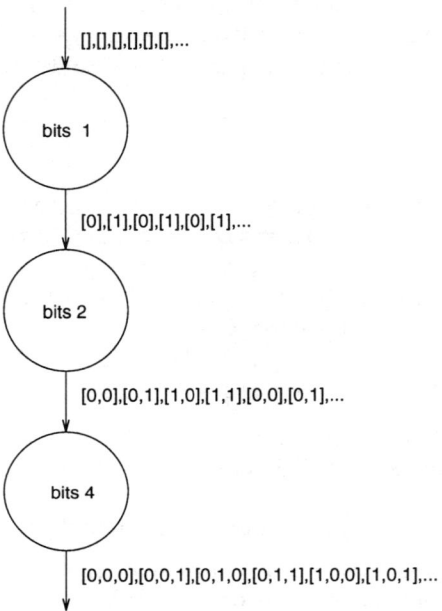

FIG. 7.3. Generating 3-bit patterns using a pipeline of processes

central control gathers all the sub-sequences generated to form the final sequence. Given a suitable choice of the constants a and m, a sub-sequence of random number $X_k (k \geq 1)$ is computed according to the following rule (Knuth 1973):

$$X_{k+n} = \left(a^n X_k + \frac{a^n - 1}{a - 1}\right) \bmod m$$

The corresponding Haskell program generates a list of n initial seeds, then maps a function **gen** that generates an infinite sub-sequence of random numbers using the seeds. The central control function **collect** collects the results (cf. Fig. 7.4).

```
gen a m n seed = seed:(gen a m n r)
        where r = (a^n * seed + ( (a^n -1) 'div' (a-1)))
              'mod' m
initseeds a m n x1 = take n (gen a m 1 x1)

randnumbs n seed = di$ (gen a m n) n divide collect seed
        where divide n firstseed  = initseeds a m n firstseed
```

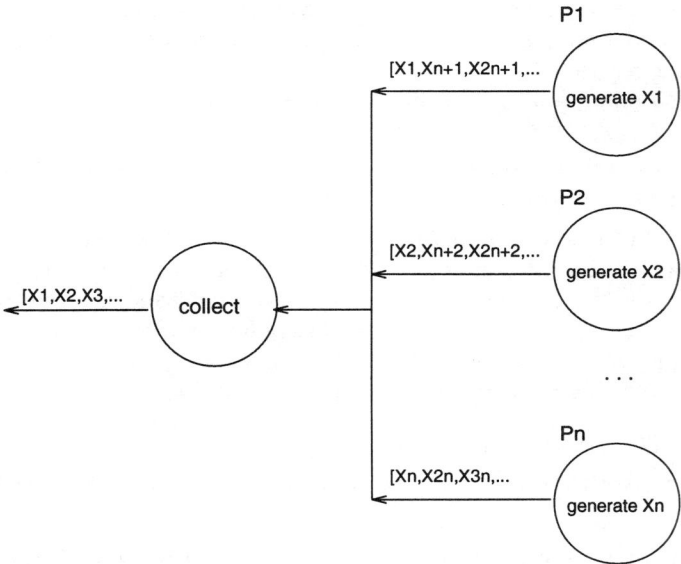

FIG. 7.4. Generating random numbers using n distributed independent computations

```
collect rnbs =  (map head rnbs)
                ++ (collect (map tail rnbs))
```

In this example, the value of the parameter n does not affect the result. The implementation can determine a value depending on the number of processors available. This problem can also be expressed as a process network.

7.6 Iterative transformation algorithms

All previous structures have been presented in some form or another in the literature. The paper now concentrates on this specific class of algorithms called iterative transformation algorithms. Special cases of the iterative transformation paradigm are also called iterative combination algorithms (Cole 1989), iterative relaxation algorithms (Finkel 1987), or Compute-Aggregate-Broadcast algorithms (Nelson and Snyder 1987). Hey (Hey 1989) refers to them as *geometric parallelism*. These algorithms are typically used in numerical analysis, image processing, parallel branch and bound and graph problems.

Such an algorithm operates on a set of objects, each containing some local data. There can be global values that are shared by all the objects. The objects are transformed through several *iteration* steps. During an iteration step, one or several of the following operations can be performed:

128   EXPLOITING PARALLELISM IN FUNCTIONAL LANGUAGES

- local operation: each object performs a computation using local data, or data contained in other objects. Global data may be used during a local operation;
- combine operation: groups of objects are combined to form another set of objects;
- global operation: global data is computed, possibly using local data contained in all the objects.

An algorithm is static if there is no combine operation, and dynamic if the number of objects changes at run-time. In a static algorithm, parallelism arises when applying a local operation simultaneously to every object. In a dynamic algorithm, additional parallelism arises when a group of objects can be combined independently from other groups. In the rest of this section, we restrict ourselves to static algorithms.

7.6.1 Parallel algorithmic structure

To design an algorithm, the following components have to be specified by the user:

- The local and global information : the state held by each object and the global data are respectively represented as tuples $(s_1, s_2, \ldots, s_p)$ and $(g_1, g_2, \ldots, g_q)$.
- The communication pattern and communication primitives for these objects: each object is uniquely identified within a coordinate system and a communication primitive c_i allows an object to select the list of coordinates of a group of objects it needs to communicate with.
- A *transformation* to be applied at each iteration step: this transformation takes a set of states and a tuple of global values, and returns a new set of states and a new tuple of global values.
- The starting and ending conditions : the initial state of each object, the initial global values and the termination condition.

As the number of objects does not change, we assume that the size of the set is defined as a special constant *wsetsize* whose value can be accessed anywhere in the specification. To ease the process of defining transformations, we allow such a transformation to be defined as:

$$transf((s_1, s_2, \ldots, s_p), (g_1, g_2, \ldots, g_q)) = (s', g')$$

where s' represents the new state (in each object) and g' represents the new global data. These expressions may contain:

- references to state variables (in s' only): this corresponds to an object accessing the value of a local variable;
- references to the self index *coord* (in s' only): this corresponds to an object accessing the value of its coordinates;

ITERATIVE TRANSFORMATION ALGORITHMS

- references to global variables: in s', this corresponds to a *broadcast* operation;
- references to *external expression lists*: this is expanded below.

An external expression list is of the form $exp@dest$ where exp is an arbitrary expression and $dest$ is a list of neighbour coordinates. It means that the expression exp is computed by each of the objects whose coordinates are in $dest$ and the cumulated results are returned as a list. An expression computed remotely may refer to state variables before or after they are modified in the iteration cycle. The external expression list notation described above is a *pure syntactic construct*. Its semantics are given by its sequential equivalent (cf. Section 7.6.3.1).

The initial state and global variable are defined using a parameterless transformation *init*. Given a transformation *transf* and a termination condition *terminate*, the entire problem can be expressed using the function $it\$$ (defined later) which produces the executable specification:

$$problem = it\$ \; terminate \; transf \; init$$

To avoid communication with the objects when testing for termination, the function terminate only takes global values as arguments (see next example).

7.6.2 Examples

The following examples shows how to define specifications for various SIT algorithms.

7.6.2.1 Example 1: Solving Laplace's equation on a square
This example uses the Jacobi iterative method to solve Laplace's equation on a square. Given a matrix of discrete values, a new value is computed after each iteration. This value is the average of the 4-neighbours' values. The algorithm stops when the difference between the old value and the new one at every point is less than some tolerance. This algorithm is static because the number of objects does not change. It is centralized because there is a need to test all the differences before the next iteration.

The objects are arranged into an $n \times n$ grid, n being considered as the size of the set. Each object is uniquely identified by its row-column coordinates (i, j) where $1 \leq i, j \leq n$. The state of each object consists of one variable only, which is the value of the corresponding point. We choose to keep the maximum difference between two successive values in all the grid as a global variable.

Each object that is not on the border needs values from its North, East, West, and South neighbours, so a function **news** as well as a predicate function **border** are defined:

```
wsetsize = n

news (i,j) = [ (i,j-1),(i+1,j),(i-1,j),(i,j+1) ]

border (i,j) = or [ (i==1),(j==1),(i==n),(j==n)]
```

Communication is also needed to collect all differences between two successive values in order to compute the global value. Therefore a function `interior` which returns the coordinates of all the objects (except those on the border) is defined:

```
interior = [ (i,j) | i <- [2..n-1] , j <- [2..n-1] ]
```

Next, the transformation that should be applied between two iteration steps is defined:

```
laplacetransf (v , g) = ( v' , g' )
            where
               v' = if not (border coord)
                    then (foldr1 (+) v@(news coord)) / 4
                    else v
               g' = maximum abs(v-v')@interior
```

where the function `foldr1` adds all the elements in the list `v@(news coord)` and the function `maximum` returns the maximum element in a list (see their definition in the appendix to this chapter).

The last component to be defined is the initial state and global value, and the termination condition:

```
tolerance = ... constant ...

init = ( ... defined for object coord ... , (tolerance + 1))
terminate g = (g < tolerance)
```

All these parameters are now supplied to the `sit$` function to complete the specification:

```
laplace = sit$ terminate laplacetransf init
```

7.6.2.2 *Example 2: Topological sort* The topological sort algorithm presented here has been adapted from (Nelson and Snyder 1987). The input is a directed acyclic graph of n nodes G represented by a boolean adjacency matrix g where g!(i,j) is True if there is an arc from node i to node j. The algorithm works in two steps: the first one computes a level number for each node and in the second step sorts the nodes by increasing level numbers. The level number of a node is 0 if it has no incoming arc, otherwise it is equal to the longest path from an 0-level node.

Computing the level for each node is done through $(log_2 n - 1)$ successive iterations, using a matrix l where l!(i,i) is the current longest path to node i. For simplicity, we assume that n is a power of 2. Each object in the algorithm is associated with an entry in the matrix l. During one iteration, each object of coordinates (i,j) performs the following:

- If the object is on the diagonal, at the mth iteration the new level entry is equal to the largest of the entries in the column j whose corresponding G^k entry equal to true plus 2^m. If there are no such entries, the old value is retained. G^k is the graph of paths of length k, and is computed using a boolean matrix multiplication from the original boolean adjacency matrix.
- If the object is not on the diagonal, its value is the one which has just been computed by the object on the diagonal.

Therefore, the global values in this algorithm are the index m and the matrix G^k. The parameters of this problem can be defined as follows:

```
truecolumn (i,j) gk = [ (l,j) | l <- [1..n] , gk!(l,j)]
-- communication primitive, selects column
-- values where G^k = TRUE

diagonal (i,j) = [ (i,i) ]
-- to select a value from the object on the diagonal

ondiagonal (i,j) = (i == j)
--to test whether the object is on the diagonal

boolmatmul m1 m2
    = array ((1,1),(n,n))
      [ (i,j) := (f i j) | i<-[1..n],j<-[1..n]]
        where
          f x y = or [(m1!(x,k))&&(m2!(k,y)) | k <- [1..n]])
          ((1,1),(n,m)) = bounds m1
--boolean matrix multiplication



tsorttransf ( l , (m,gk)) = (l' , (m',gk'))
   where
      (m',gk') = (m+1 , boolmatmul gk gk)
      l' = if (ondiagonal coord)
           then maximum (l:others)
           else head(l'@(diagonal coord))
      others = map (\x -> 2^m + x) l@(truecolumn coord gk))

init = ( 0 , (0 , initboolgraph))
```

```
terminate (m,_) = (m == (logBase 2 n))

computelevels = it$ terminate tsorttransf init
```

After the last iteration, the level of a node i is stored in the entry l(i,i), so the last step in the algorithm sorts the nodes into an increasing level number order (e.g. using quicksort).

This is an example of a *multi-paradigm* algorithm because the matrix multiplication boolmatmul could itself be expressed using the Divide-and-Conquer (Cole 1989) or systolic paradigm (Nelson and Snyder 1987), and the function qsort could also be expressed using the Divide-and-Conquer paradigm (Section 7.3.2).

Implementing such multi-paradigm algorithms is a challenge. The system has to decide for every parallel sub-part if it is to be executed in sequence or parallel depending on its individual characteristics. Some interaction with the user might be necessary, so that he can provide more information about his program, or give priority for parallel execution to the paradigm that could result in better efficiency.

7.6.3 *Generating the executable specification*

7.6.3.1 *General principles*
Once the specification is completed, a sequential functional program can be derived from it. This is important for two reasons:

- the semantics of the parallel problem are given by its sequential version;
- the sequential version helps in identifying many errors that are difficult to detect in a parallel environment.

The sequential program is modelled as to generate a succession of sets and global variables and finishes when the termination condition is satisfied. Assuming that each object contains a state $(s_1, s_2, \ldots, s_p)$, the first requirement is to simulate the set of objects using a functional aggregate type. We assume that the following operations on this type are available:

- *access set coord*: returns the state of the object in *set* whose coordinates are *coord*;
- *setmapcoord f set*: maps a function f to every state in the set with its coordinates in the set;
- *setmk f*: creates a set by applying the function f to every possible coordinate.

Producing the executable specification is achieved through a program transformation called the $\mathcal{S}$ transformation. It only consists of removing the 'syntactic sugar' (see Section 7.6.1) by evaluating external expression lists within the appropriate context. These lists are only found in the

ITERATIVE TRANSFORMATION ALGORITHMS

iterative transformation so each transformation $transf$ is converted into a local version $ltransf_s$ and a global version $gtransf_s$. The local version corresponds to the function executed during a local cycle while the global function indicates how global data is modified during an iteration cycle. The $\mathcal{S}$ program transformation operates as follows:

$$transf(s,g) = (s',g')$$
$$where$$
$$s' = \ldots e_1 @dest \ldots$$
$$g' = e_2$$

$\Rightarrow_{\mathcal{S}}$

$$ltransf_s(wset,g)(wset',g') \; s \; coord = s'$$
$$where$$
$$s' = \ldots map \; (\lambda \; coord \rightarrow let \; s \; = access \; wset \; coord$$
$$\qquad\qquad\qquad\qquad\qquad let \; s' = access \; wset' \; coord$$
$$\qquad\qquad\qquad\qquad\qquad in \; e_1)$$
$$\qquad dest \ldots$$

$$gtransf_s \; wset \; wset' \; g = g'$$
$$where$$
$$g' = e_2$$

The local definition ($let \; s' = access \; wset' \; coord$) can be omitted if the expression e_1 does not refer to s'. Similarly, the definition ($let \; s = access \; wset \; coord$) is not introduced if e_1 does not refer to s. If s and s' are tuples, references to their variables are made by pattern-matching in the usual way. A similar transformation is applied to external expression lists found in the expression e_2.

The initial transformation $init = (s,g)$ is also converted into its $init_s$ version:

$$init_s = (setmk \; wsetsize \; (\lambda coord \rightarrow s) \; , \; g)$$

where $wsetsize$ is the constant that represents the size of the set (see Section 7.6.1). It is now possible to define the function $sit\$$ which produces the executable specification:

$$sit\$ \; terminate \; (ltransf_s, gtransf_s) \; init_s = sit' \; init_s$$
$$where \; sit'(set,g) \; | \; (terminate \; g) = (set,g)$$
$$\qquad\qquad\qquad | \; otherwise \quad\; = sit' \; (set',g')$$
$$\qquad\qquad\qquad where$$
$$\qquad\qquad\qquad\quad set' = setmapcoord \; (ltransf_s \; (set,g) \; (set',g')) \; set$$
$$\qquad\qquad\qquad\quad g' \;\;\; = gtransf_s \; set \; set' \; g$$

7.6.3.2 Example: The executable specification of Laplace's problem

In the case of solving Laplace's equation on a square (see Section 7.6.2.1), the set of objects can be implemented using a Haskell $n \times n$ two-dimensional array. Therefore, manipulating the set is done through the following array operations:

```
access (i,j) m = m!(i,j)

setmk n f = array ((1,1),(n,n))
            [ (i,j) := (f i j) | i<-[1..n],j<-[1..n]]

setmapcoord f m = setmk n (\x y -> f (m!(x,y)) (x,y) )
                  where ((1,1),(n,n')) = bounds m
```

The iterative transformation `laplacetransf` can now be converted into the following functions:

```
llaplacetransfs (wset, g) (wset', g') s coord = s'
  where
    s' = if(border coord)
         then s
         else  (foldr1 (+)
                       (map (\coord ->
                                    let s = (access wset coord)
                                    in s)
                            (news coord)))
              / 4

glaplacetransfs wset wset' g = g'
  where
    g' = maximum(map (\coord ->
                            let s = access wset coord
                                s' = access wset' coord
                            in abs(s-s'))
                     interior)
```

7.6.4 Parallel implementation

The issues related to the parallel implementation of the structure for iterative transformation algorithms are discussed in (Rabhi and Schwarz 1993). In short, to each object is associated a virtual process, containing the state of this object. A virtual process runs several tasks concurrently:

- the main task, implementing the list of transformations that are applied to the state during each cycle;
- several *update tasks* that send *update messages* to other virtual processes. An update task is associated with every external expression

ITERATIVE TRANSFORMATION ALGORITHMS

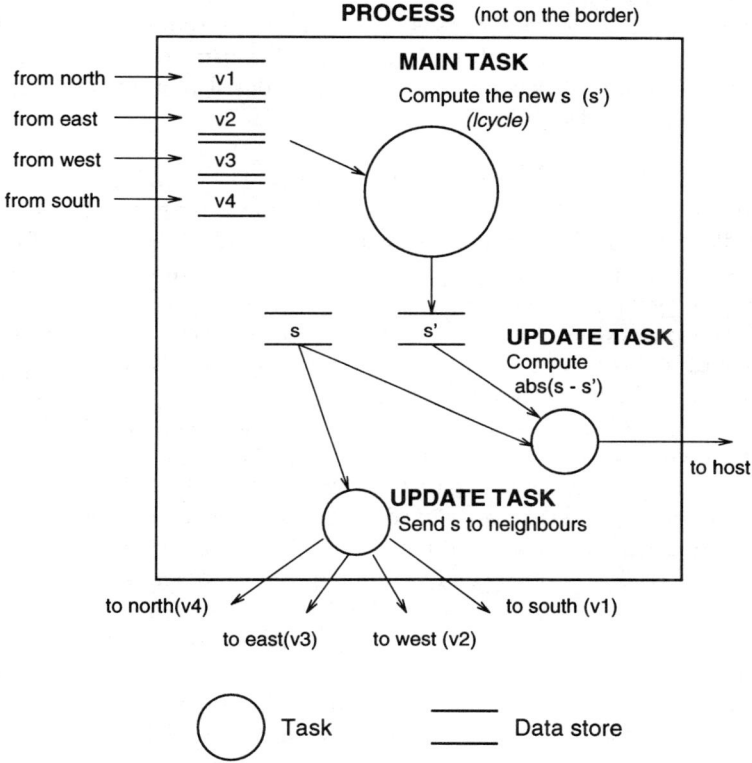

FIG. 7.5. Structure of a virtual process

list of the form $e@(c_j\ coord)$ which means that expression e is computed and sent to the appropriate process. Destination addresses are determined by computing the inverse of c_j.

For example, a virtual process that is not on the border of the grid in Laplace's problem would have the structure illustrated in Fig. 7.5.

This implementation is based on *distributed graph reduction* (Bevan et al. 1989) so every value corresponds to a node in the graph. Any node being evaluated by a task is *blocked* until it is overwritten by the result. Any task attempting to read the value of a blocked node is *suspended* and reawakened only when the evaluation of the node is completed.

The graph reduction mechanism guarantees synchronization between the tasks. For example, there might be a situation where the main task attempts to read the value of an input node whose value has not arrived yet. The main task is blocked and moved to a *suspended tasks pool*. It is reactivated when the value becomes available.

Finally, several virtual processes can be grouped and executed by a single physical processor, but care is taken to ensure that states and other variables are uniquely identified. Communication between two virtual processes located on the same processor will just consist of a memory transfer between two variables.

7.7 Conclusion and future work

This paper advocates a 'paradigm-oriented' approach for the development of parallel functional programs. Dealing with well-identified forms of parallelism eases the design, implementation, and performance analysis of programs, and enhances portability. All the benefits associated with functional languages are retained: higher-order capabilities, data abstraction, absence of side-effects and opportunity for formal analysis and transformation. The drawback of our approach is some loss of efficiency because automatic partitioning and scheduling is generally not as effective as a direct implementation with an explicit parallel language.

This paper focuses on a particular paradigm, called iterative transformation, where data is partitioned amongst objects that transform this data in an iterative manner. Having identified a framework under which the parameters of a problem can be entered, the paper shows how to derive an executable specification by combining these parameters within a higher-order function. Iterative transformation algorithms could also be expressed as process networks because each data object could be thought of as a process communicating with other objects by message-passing. However, we believe that these algorithms require a separate parallel structure which exhibits the synchronization of the transformations and shows the global operations on the set. In addition, the specification of a problem is much clearer than the corresponding process network. This paper also sketches a parallel implementation of this paradigm based on graph reduction.

More research work is needed to improve the structures presented in this paper, and to allow for more general classes of parallel algorithms to be expressed. A structure must clearly indicate the paradigm used, allowing for the extraction of the important characteristics needed by the implementation. Defining structures could be helped by providing additional notation, and an exhaustive library of functions and data structures, depending on the paradigm used. An example is the iterative transformation structure where the description of most used communication patterns and global set transformations could be part of the system.

Having specified a problem using the appropriate structure, the next step is to transform the problem specification into a program targeted for a particular parallel architecture. Several implementation techniques have already been developed for particular applications (Fox *et al.* 1988) and it is hoped that most of them could be adapted to more general cases.

REFERENCES

Augustsson L. and Johnsson T. (1989). Parallel Graph Reduction with the $<\nu, G>$-machine, In *Functional Programming Languages and Computer Architecture*, ACM Press, pp. 202-213.

Bevan D.I. et al. (1989). Design principles of a distributed memory architecture for parallel graph reduction, *The Computer Journal*, **32(5)**, pp. 461-469.

Burn G.L. (1989). Overview of a Parallel Reduction Machine Project II, In Proc. *PARLE '89*, Odijk E. et al. (eds), LNCS 365, pp. 385-369.

Burkhart H. et al. (1993). BACS : Basel Algorithm Classification Scheme, Technical Report 93-3, University of Basel.

Burton F.W. (1987). Functional Programming for Concurrent and Distributed Computing, *The Computer Journal*, **30(5)**, pp. 437-450.

Cole M. (1989). *Algorithmic skeletons : A structured approach to the management of parallel computation*, Pitman, London.

Cole M. (1990). Towards Fully Local Multicomputer Implementations of Functional Programs, Report CSC 90/R7, Department of Computing Science, University of Glasgow.

Darlington J. et al. (1991). Structured Parallel Functional Programming, In Proc. *Workshop on Parallel Implementation of Functional Languages*, Technical report, Department of Electronics and Computer Science, University of Southampton.

Finkel R.A. (1987). Large grain parallelism - Three case studies, In *The Characteristics of Parallel Algorithms*, Jamieson L.H. et al. (eds.), The MIT Press, pp. 21-63.

Finkel R.A. and Manber U. (1987). DIB - A Distributed Implementation of Backtracking, *ACM Transactions on Programming Languages and Systems*, **9(2)**, pp. 235-256.

Fox G.C. et al. (1988). *Solving Problems on Concurrent Processors*, Prentice Hall International.

Hey A.J.G. (1989). Experiments in MIMD parallelism, In Proc. *PARLE' 89*, Odijk E. et al. (eds), LNCS 365, pp. 28-42.

Hudak P. et al. (1992). Report on the Programming Language Haskell, *SIGPLAN Notices*, **27(5)**.

Hudak P. (1991). Para-Functional Programming in Haskell, In *Parallel Functional Languages and Compilers*, Szymanski B.K. (ed.), ACM Press, pp. 159-196.

Kelly P. (1989). *Functional Programming for Loosely-coupled Multiprocessors*, Pitman, London.

Kelly P. (1991). Parallel Functional Programming in Hope+ using Skeletons, Department of Computing, Imperial College, London.

Knuth D.A. (1973). *The Art of Computer Programming*, Volume 2, Addison Wesley.

Kung H.T. (1989). Computational models for parallel computers, In *Scientific Applications of Multiprocessors*, Elliott R.J. and Hoare C.A.R. (eds), Prentice Hall International, pp. 1-15.

Nelson P.A. and Snyder L. (1987). Programming Paradigms for Non-shared Memory Parallel Computers, In *The Characteristics of Parallel Algorithms*, Jamieson L.H. *et al.* (eds), The MIT Press, pp. 3-20.

Quinn M.J. (1987). *Designing Efficient Algorithms for Parallel Computers*, McGraw Hill International.

Peyton-Jones S.L. *et al.* (1989). High Performance Parallel Graph Reduction, In Proc. *PARLE' 89*, Odijk E. *et al.* (eds), LNCS 365, pp. 193-206.

Rabhi F.A. and Manson G.A. (1991). Divide-and-Conquer and Parallel Graph Reduction, *Parallel Computing*, **17**, pp. 189-205.

Rabhi F.A. and Manson G.A. (1991). Experiments with a transputer-based parallel graph reduction machine, *Concurrency : Practice and Experience*, **3 (4)**, pp. 413-422.

Rabhi F.A. and Schwarz J. (1993). 'Paradigm-Oriented' Design of Parallel Iterative Programs Using Functional Languages, In *Proc. Applications of Supercomputers in Engineering*, Bath.

Appendix

This appendix contains the definition of the Haskell functions used in this paper.

```
abs a = if a<0 then (-a) else a
max a b = if a>b then a else b      True && True = True
                                    _    && _    = False
[] ++ ys = ys                       head (x:_) = x
(x:xs) ++ ys = x : (xs ++ ys)       tail (_:xs) = xs

map f [] = []                       take _ []        = []
map f (x:xs) = (f x):(map f xs)     take 0 _         = []
                                    take (n+1)(x:xs) = x:take n xs
foldr f z [] = z
foldr f z (x:y)=f x (foldr f z y)
                                    a ^ 0 = 1
foldr1 f [x] = x                    a ^ (n+1) = a * (a ^ n)
foldr1 f (x:y)=f x (foldr1 f y)
                                    id x  = x
length = foldr (\n _ -> n+1) 0
f . g = \x -> f (g x)               maximum = foldr1 max
```

8

BUILDING PARALLEL APPLICATIONS WITHOUT PROGRAMMING

J. Darlington and H.W. To

*Department of Computing, Imperial College,
London SW7 2BZ, UK
email: {jd,hwt}@doc.ic.ac.uk*

Abstract

To counter the problems of the parallel programming, we propose an approach based on using a fixed range of computation patterns, which we call skeletons. This approach provides a path to efficient but correct programs by separating the issues of behaviour and meaning, and through the exploitation of the knowledge we have about the fixed patterns of computation and communication of the skeletons.

The method can be applied to specific application domains by observing that recurring patterns of data and control structures appear in these domains. Often, for any one domain, there may be many different sets of patterns reflecting abstract and concrete structures. Equivalences can be derived between these sets, through transformations, to enable problem expression at an application's level, but efficient execution at the machine level. We demonstrate this through an example in solid modelling.

8.1 Introduction

Even though there has been a rise in the number of commercially available parallel machines, there are still too few real applications which use them. The main factor in this is software. Programming parallel machines is considered to be more difficult than is the case for sequential machines. The diversity of parallel machine architectures and their associated computational models make **predictability of performance** and **portability** of software difficult. These problems did not arise so critically for sequential machines because there exists a single computational model, the von Neumann model, to which all sequential machines adhere.

The von Neumann model of computation provides a simple relationship between language constructs and their implementation. Issues such as memory allocation are resolved by the compiler with no performance implications, allowing the programmer to concentrate on higher level aspects of programming. Thus the programmer can predict the performance of their program without worrying about the run-time resource allocation. This is not true for parallel programming where mapping a program to a multiprocessor machine is a complex task involving decisions about task allocation, scheduling of competing processes' communication patterns, etc. Usually the only way to achieve the desired performance is through explicit control, which adds to the complexity and non-portability of the program.

The universality of the von Neumann model guarantees portability of sequential programs at the language level, with no danger of unforeseen degradation in performance. With the explicit task allocation of parallel programs there is rarely any portability. Even when portability is possible, through the use of high-level programming languages, which can be compiled to other machines, there is no guarantee that performance will be maintained.

Given the success of the von Neumann model for sequential computing the conventional solution to the parallel software dilemma is to attempt to produce a 'parallel von Neumann' model. The search is then for a uniform abstract machine model for parallel computing. The rôle of such an abstract machine model is to provide an abstraction of the execution behaviour of the target machines. Programming languages are then defined on these models, with programs written to organize the machine's activities to produce correct answers and behave efficiently. The need to express both correctness and efficiency in the same framework requires that there needs to be a close correlation between the abstract and concrete machine models. So even if such a model is found, and there are several candidates (Fortune and Willey 1978, Valiant 1990), programming will still remain a fairly low level and laborious activity. Critically, in the conventional programming language framework, the meaning and behaviour of an application is still required to be expressed via a single program. It is not possible to specify one without effecting the other.

Another approach is to exploit the **implicit parallelism** of functional languages. These languages provide a much higher degree of abstraction than their imperative cousins. Using this implicit parallelism would alleviate the need to explicitly decompose the problem into concurrent tasks, and to state the control synchronization and communication between tasks. Unfortunately, automatic exploitation of this has not yet succeeded. This is because the problems of automatic resource allocation have still to be overcome. Compliers can generate too much fine grain parallelism which overwhelms resources and is too costly to spawn. It is difficult to automatically produce load balanced code given the general nature of the problem. Even though the parallelism is implicit, it can be difficult to detect and is dependent on the author of a program expressing it a way which reveals the natural parallelism.

However, functional languages do have some important properties:
- there is a clean separation of behaviour from meaning (Church-Rosser property);
- correctness-preserving transformations can be applied to functional programs;
- it is possible to write higher-order functions, which provide a powerful abstraction mechanism.

In the next section we consider another solution which exploits the advantages of functional languages, but in a new, hopefully more practical, manner.

8.2 Skeletons – a radical but practical solution

Both previous solutions result in the meaning and behaviour of a program being entwined. With explicit parallelism, through a uniform abstract machine, the two are expressed in the same language. With implicit parallelism, the behaviour is inferred from the programming language with no user control. Successful parallel programming requires that both meaning and behaviour are expressed. However, the two are orthogonal, programs should be written to be correct, then tuned to be efficient. This implies a separation of meaning and behaviour.

To achieve the separation we propose the use of a range of algorithmic forms, known as **skeletons**, which abstract the useful computational structures from applications. Applications are then constructed as instantiations of these algorithmic forms. The approach follows that taken by Cole (Cole 1989), for imperative languages, and Backus's idea (Backus 1978) of programming functional languages through a fixed set of operators.

The skeletons are expressed in a functional language as higher-order functions. The skeleton's declarative meaning is then established by this function definition. This meaning is independent of any implementation issues and hence any behavioural constraints. It also provides a sequential prototype which is portable across different parallel machines. Behaviour is defined by the implementation of the skeleton on a particular architecture. So the behaviour of a skeleton can vary from machine to machine, as one would expect. The only parallelism in a program arises from the use of skeletons. All other functions are executed sequentially. Thus, the parallel behaviour of a program is given by the behaviour of its constituent skeletons. All aspects of a skeleton's behaviour, such as process placement and interconnectivity, are documented with the skeleton's implementation, along with optional behaviours which can be specified at compile or run time. There is no reason why each skeleton could not be implemented onto every type of architecture. However, different skeletons are more naturally implemented on particular groups of architectures. It would seem wise to implement skeletons only on those architectures that were suitable. Thus the range of skeletons make up the programmer's target abstract machine model, but one that is expressed at a much higher level and is capable of accommodating a much greater variety of machines and behaviours more easily than for a single abstract machine model.

Transformations are involved at all levels of building applications using skeletons. Using transformation it is possible to derive equivalences between skeletons. We have a fixed set of skeletons, so the possible equivalences can be computed once and stored for future use. These relationships between skeletons provide a route for portability. An application naturally expressed in one skeleton may find that there are no implementations on a particular machine. Choosing the right equivalences will map the application onto the desired machine. Once that particular skeleton-machine pair has been selected then further transformations can be used to optimize the skeleton for the particular machine. An example of a common transformation would be to vary the grain size through partial evaluation (Darlington 1988). Through transformation we have effectively preserved meaning while providing alternative implementations and behaviours.

For each skeleton-machine pair we can construct an associated performance model. The performance models are formulae which predict the performance and whose variables are machine and problem parameters. These models are used to predict the performance of instantiations of skeletons on particular machines, and once a particular machine has been selected to optimize the combination.

The ultimate goal of this work is to replace the inventive process of producing solutions by creating programs from scratch with the development of programs through selection and instantiation of a fixed range of alternatives. Thus there is an explicit identification of all the decisions that have to be addressed to produce an efficient mapping rather than having these decisions made implicitly by writing a program with the desired properties. By recording all the decisions for replay we have a facility for supporting retargeting and portability of applications.

8.3 Some skeletons

8.3.1 *Initial skeletons*

To highlight and explore some of the issues of skeletons, we introduce an initial, but incomplete, set of skeletons. The meaning of each skeleton is expressed in Haskell (Hudak *et al* 1992). For the definitions we use some standard primitive functions which can be found in (Bird and Wadler 1989).

Pipelining or linear process parallelism is captured by the PIPE skeleton. A list of functions are composed together and applied to a list of inputs. Parallelism is achieved by creating a different process for each function and streaming the problem list through the list of processes.

$$\text{PIPE} :: [\alpha \rightarrow \alpha] \rightarrow (\alpha \rightarrow \alpha)$$

PIPE = foldr1 (.)

Simple data parallelism is expressed by the FARM skeleton. A function is applied to each element of a list, each element of which is thought of as a task. The function also takes an environment, which represents data which is common to all tasks. Parallelism is achieved by distributing the tasks on different processors.

FARM :: $(\alpha \rightarrow \beta \rightarrow \gamma) \rightarrow \alpha \rightarrow ([\beta] \rightarrow [\gamma])$
FARM f env = map (f env)

So far the skeletons reflect computational patterns which are still quite close to the machine level. It is proposed that there are more abstract skeletons which are closer to the application level. One such skeleton is RaMP, reduce and map over pairs. It describes systems where we have two pools of objects. Each object from one pool can interact with every object in the other pool. We are interested in the combined results of these interactions for each object in the first pool. The pools are represented as lists. It is not intended that there should be a direct, efficient parallel implementation of RaMP, rather it provides a way to express certain problems naturally, but a naïve implementation of RaMP would be inefficient. Efficient implementations of RaMP can be achieved through transformations to less abstract, but more efficient skeletons.

RaMP :: $(\alpha \rightarrow \beta \rightarrow \gamma) \rightarrow (\gamma \rightarrow \gamma \rightarrow \gamma) \rightarrow [\alpha] \rightarrow [\beta] \rightarrow [\gamma]$
RaMP f g as bs= map h as
    where h x = foldr1 g (map (f x) bs)

Other skeletons have been identified and defined. They reflect other common algorithmic forms and computation patterns. Examples of these are the divide and conquer paradigm, DC, and parallel map, PMAP. Others like dynamic message passing architecture, DMPA, express specific types of architecture. A fuller description and examples of their use can be found in (Darlington *et al* 1993).

8.3.2 *Transformations between skeletons*

The ability to apply transformations can be used to split the programming process into two phases. The program is first expressed at a natural level then transformations are applied to produce an efficient version targeted towards a particular architecture. Applying this process to skeletons means that we can, similarly, operate with a fixed repertoire of pre-derived equivalences rather than having to invent transformations on each occasion.

It is envisaged that application level skeletons will be transformed to lower level skeletons or compositions of them. This provides our route not only to efficiency, but portability. Porting from one architecture to another will be achieved by re-expressing the skeleton in terms of other skeletons.

As an example, reconsider RaMP which we know to be an inefficient, but application-oriented skeleton. It can be expressed as a PIPE.

RaMP f g as bs
= (map snd . PIPE (map map (map g' bs)) .
  map (pair unitg)) as
  where g' b (a, c) = (a, g (f a b) c)

The formal transformation of RaMP to PIPE can be found in (Kelly 1989). Informally, we can see that each stage of the pipe is constructed from an element of the second list, bs. The list passed through the pipe is the first list, as, with each element coupled with an initialized accumulator. Each stage of the pipe maps the interaction function, f, to its element of bs and the as part of the input list. This result is used to update the accumulator field of each item in the list using g. Each stage returns a list of tuples consisting of an element of as and an updated accumulator. There needs to be some pre- and post-processing to build the problem list and to eventually strip away the extra data to leave only a list of final accumulators.

Alternatively, there is a FARM version.

RaMP f g as bs
= FARM h (f, g, bs) as
  where h (f, g, bs) = foldr1 g (map (f x) bs)

The FARM derivation is relatively straightforward. Each slave is allocated an element of as. The processing for each slave involves comparing this with each element of bs and then doing the reduction.

8.3.3 *Implementation*

Implementations of skeletons can take advantage of the extra knowledge one has about a skeleton's behaviour and optimizing opportunities to produce efficient, specialized implementations on particular machines. This contrasts with the situation that pertains when using a general purpose programming language; here one has to extract such information by automatic analysis from arbitrary programs. Using the skeletons defined we can consider some examples. For PIPE there is a logical locality between adjoining stages. This locality information can be used to allocate processes so that the locality is physically realized. FARM provides a more interesting example in that many machines do not have a broadcast network. To prevent the possible bottle-neck at the master, we may employ tree structures. The skeletons provide information that allows us to specialize their implementation. This is particularly important with regard to communication. For example, asynchronous communication is potentially dangerous, but it is also likely to be more efficient. From the structure of certain skeletons we can see when it is safe to perform asynchronous communication, for example in PIPE.

A prototype implementation of the skeletons has been developed. It uses Hope[+] (Perry 1989) as the source language and C as the target language. This initial implementation was carried out on a Meiko Computing

Surface with the CS-Tools library (Meiko 1990) providing the communications layer. The set of skeletons implemented to date are PIPE, DC, FARM, PMAP, and DMPA. The initial implementation had no compiler options. However, a preliminary study and implementation of such options has been carried out (Isaac 1992). For several of the skeletons alternative implementations were identified which exploit knowledge of the application to specialize the behaviour of the skeleton. These were implemented and initial investigations showed improvements over the general versions for selected problems.

8.3.4 An example — ray-tracing

In this section we show the use of skeletons in a particular application. We take as our example a simple ray-tracing problem. It follows closely the example given by Kelly (Kelly 1989). Ray-tracing is a technique used to generate images of scenes. The visual attributes of each pixel are determined by tracing a ray from the viewpoint through the pixel and determining which object is struck first, if any. The intensity and colour of the pixel can then be determined from the properties of the object. Taking the rays and objects as two pools of items we can see how the problem can be naturally expressed using RaMP. The rays and objects interact by intersecting and we are interested in the earliest intersection for each ray.

RayTrace :: Int → Int → Point → [Object] → [Impact]
RayTrace wd ht viewpoint scene
    = RaMP TestForImpact Earlier (GenerateRays wd ht viewpoint)

TestForImpact :: Ray → Object → Impact
-- Returns the impact of a ray with an object

Earlier :: Impact → Impact → Impact
-- Given two impacts it returns the one closest to the view point

GenerateRays :: Int → Int → Point → [Ray]
-- Returns a list of rays for a given screen size and viewpoint

Typical parallel implementations of raytracers employ farm or pipeline parallelism. Using the equivalences between RaMP, FARM, and PIPE we find that we arrive at the same parallel implementations. However, a 'eureka' step of expressing the problem in terms of its possible forms of parallelism was not necessary. The jump from natural specification to a parallel program was one of selection and instantiation.

RaMP wd ht viewpoint scene
    = (map snd . PIPE(map map (g' scene)) . map (pair NoImpact))
    (GenerateRays wd ht viewpoint)
    where g' b (a, c) = (a, Earlier (TestForImpact a b) c)

```
RaMP wd ht viewpoint scene
  = FARM h (TestForImpact, Earlier, scene)
    (GenerateRays wd ht viewpoint)
    where h (f, g, bs) x = foldr1 g (map (fx) bs)
```

8.4 Application-specific skeletons

The skeletons previously discussed are general-purpose. The skeleton methodology can be further refined by observing that many specific application domains have characteristic data and control structures. The concept of skeletons can be taken closer to the application level by abstracting these structures and capturing them as skeletons. Even from a sequential programming viewpoint this has great benefits. Such a system would enable application specialists to develop applications quickly and efficiently without being burdened with low-level programming. Once the structures have been identified and given a meaning via a definition then its parallel behaviour can be considered. To facilitate very high-level domain specific operations, it may be necessary to contemplate unorthodox mappings to machines.

The next section presents a case study of application-level skeletons.

8.4.1 Solid modelling

Solid modelling is the term used to describe the methods used for representing three-dimensional objects within a computer. Often these representation are only approximations, hence the term modelling. Associated with these modelling techniques are functions used for manipulating the models and for gathering information from these models. Typically users wish to view the models at a high level of abstraction. One representation of models at such a level is Constructive Solid Geometry (CSG) trees. A scene is constructed from the composition of primitive solids. They can be composed using standard set operations. Each primitive solid is defined not only by its size, but also its orientation and position in space. An example is given in Fig. 8.1, but without the orientation and position information.

This can be expressed functionally in Haskell as:

```
data PrimitiveSolid = Block Int Int | Sphere Int | Cylinder Int Int |
                     Torus Int Int

type PrimitiveInstance = (PrimitiveSolid, CoordinateSystem3D, Material)

type CoordinateSystem3D = (Position3D, Orientation3D)

type Position3D = (Int, Int, Int)
```

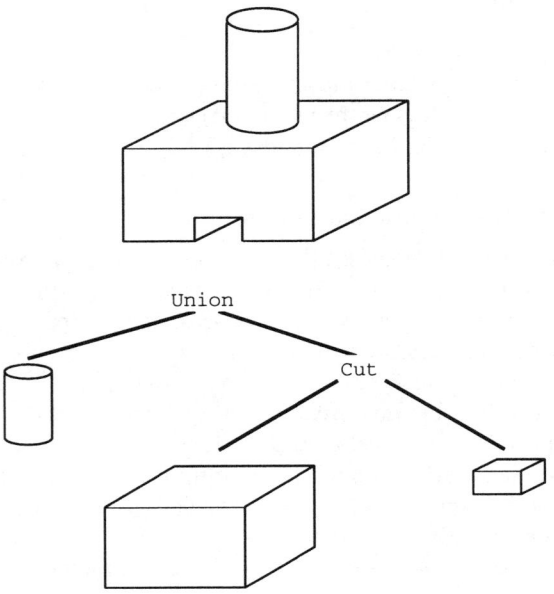

FIG. 8.1. Example of a CSG tree.

type Orientation3D = (Int, Int, Int)

type Material = MATERIAL

data CsgTREE = NullSolid | Primitive PrimitiveInstance |
        Composite CsgTREE SetOp CsgTREE

data SetOp = Join | Cut | Intersect

Two classes of operations are performed over CsgTREEs; transformations and reductions. These can be expressed as higher-order functions, transformCSG and reduceCSG. Intuitively, transformCSG allows us to change the primitives in the tree whilst keeping the structure of the tree unchanged. This provides the ability to make global changes to the model in a succinct way. To extract information from the tree reduceCSG is provided. Thus questions can be posed at a model rather than primitive level.

transformCSG :: (PrimitiveInstance → PrimitiveInstance)
        → CsgTREE → CsgTREE
transformCSG primF NullSolid
    = NullSolid
transformCSG primF (Primitive primInst)
    = Primitive (primF primInst)

transformCSG primF (Composite s1 op s2)
    = Composite (transformCSG primF s1) op
      (transformCSG primF s2)

reduceCSG :: $\alpha \rightarrow$ (PrimitiveInstance $\rightarrow \alpha$) $\rightarrow$ (SetOp $\rightarrow \alpha \rightarrow \alpha \rightarrow \alpha$)
    $\rightarrow$ CsgTREE $\rightarrow \alpha$
reduceCSG base primF compF NullSolid = base
reduceCSG base primF compF (Primitive primInst) = primF primInst
reduceCSG base primF compF (Composite s1 op s2)
    = compF op r1 r2
      where r1 = reduceCSG base primF compF s1
            r2 = reduceCSG base primF compF s2

One would expect functions of these shapes, since they correspond to map and fold over trees. The next two examples show the use of these skeletons in expressing typical manipulations upon models. In the first example we write a function which translates a model by a given vector.

translateCGS :: Vector3D $\rightarrow$ CsgTREE $\rightarrow$ CsgTREE
translateCSG v = transformCSG (translatePrim v)

translatePrim :: Vector3D $\rightarrow$ PrimitiveInstance $\rightarrow$ primitiveInstance
translatePrim v (prim, cs, mat) = (prim, cs', mat)
                where ((x,y,z), or) = cs
                      (dx, dy,dz) = v
                      cs' = ((x + dx, y + dy, z + dz), or)

A more interesting and computationally intensive problem is to decide if a given point lies inside the model. It would also be desirable to know the type of material the point lies in.

data Class = Out | In MATERIAL

testPointCSG :: Position3D $\rightarrow$ Class
testPointCSG p = reduceCSG Out (testPointPrim p) combinePointClass

combinePointClass :: SetOp $\rightarrow$ Class $\rightarrow$ Class $\rightarrow$ Class
combinePointClass Join class1 class2
        = class1 , if (class1 = In mat)
        = class2 , otherwise
combinePointClass Cut class1 class2
        = class1 , if (class1 = In mat) and (class2 = Out)
        = Out   , otherwise
combinePointClass Intersect class1 class2
        = class1 ,if (class1 = In mat1) and (class2 = In mat2)
        = Out   , otherwise

```
testpointPrim :: Position3D → CsgTREE → Class
testpointPrim p (prim, cs, mat)
     = In mat , if ( inPrimitive p' prim)
     = Out    , otherwise
       where p' = pointToGivenCS3D cs p
```

CSG trees are a very natural way of representing objects. Unfortunately they are not very efficient to process. Other representations are more efficient to process, but are less natural to use. One such method is octrees. It is one of a class of methods known as spatial decomposition. Here the bounding volume of the scene is divided into regular blocks and an object is constructed by marking a block when it contains part of the object. With octrees the space of interest must be a rectangular box. This box is divided into eight regular boxes, octants. An octant is marked white if it contains only space, black if it contains only material. If it is mixed then it is marked grey, and the octant is recursively subdivided to some resolution. This information is held in a tree with white and black leaves and grey nodes. Each node has eight children and the depth indicates the resolution. A two-dimensional example is shown in Fig. 8.2.

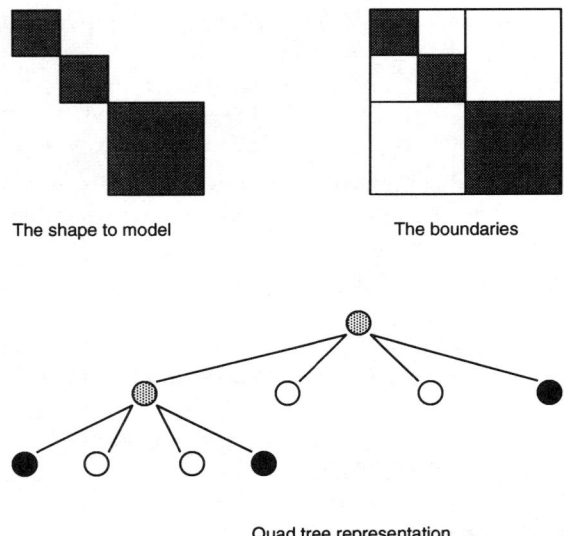

FIG. 8.2. Example of a quadtree; a 2-D octree

It is a simple matter to define octrees in Haskell.

type Octree = (Position3D, Size, CubeSubdivisionTree)

data CubeSubdivisionTree = Cube Class | Subcubes [CubeSubdivisionTree]

APPLICATION-SPECIFIC SKELETONS

data Class = Out | In material

Again, we need to process over the model so we must provide functions to manipulate octrees, viz.

reduceOctree :: (position3D → size → α) → ([α] → α) → Octree → α
reduceOctree solve combine (p, s, Cube Class)
       = solve ps Class
reduceOctree solve combine (p, s, SubCubes octants)
       = (foldr1 combine . map F) octants'
        where F = reduceOctree solve combine
          octants' = zip (generateSubPositionSize p s) octants

We have presented two methods of solid modelling. CSG trees provide a natural way of representing objects, but are inefficient to process. Octrees are low level and unnatural to use, but are efficient to process. Functional languages allows us to combine the best of both approaches. Transformation allows us to derive equivalences between CSG tree functions and equivalent octree functions. The user could therefore work entirely at the conceptual level of CSG trees, while all the processing is performed at the octree level.

To generate equivalent functions we need to assume that a function csgToOct, which transforms any CsgTREE to an Octree, exists. One is defined in (Papachrysantou 1992). Then for any function f over CsgTREE's we require a function g over Octree's which conforms to the transitive diagram of Fig. 8.3. Usually, either α is equal to β or α is equal to Octree and β is equal to CsgTREE. Given csgToOct and a function f then g can

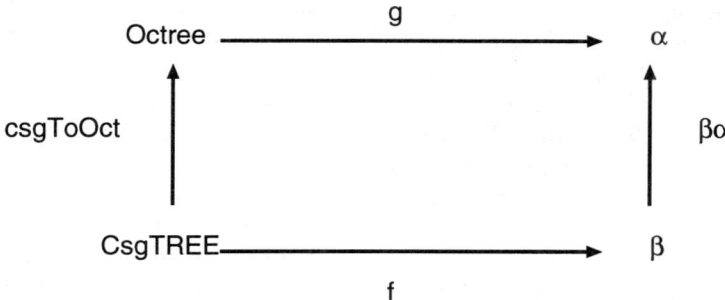

FIG. 8.3. Relationship between CsgTREE and Octree.

be derived through transformations. As an example consider reduceCSG. If combineF obeys certain properties then the following equivalence holds.

reduceCSG nullVal primitiveF combineF csgtree
    = reduceOctree cubeF combineFF (csgToOct csgtree)

```
       where cubeF p s Out = nullVal
             cubeF p s (In mat) = primitiveF (makePrimitiveBlock p s mat)
             combineFF = combineF Join
```

This example is explored in greater detail in (Papachrysantou 1992) and the technique for transformation is described in (Harrison 1992). Since both the abstract and the concrete types are known in advance, the equivalences can be developed once and then stored for future use.

Given these equivalences the intention is that although the user would specify a computation in terms of CSG trees all execution would take place on the octree representation.

Possible implementations of the octree skeletons are guided by its tree structure. For message-passing architectures we may consider dynamic divide-and-conquer to traverse the tree, if the tree is balanced or nearly balanced. For sparse trees we may consider a master-slave implementation with the master walking the tree and handing leaves to slaves; this may also be the case if the tree is large, so to reduce communication only the leaves of the tree need to be passed. A method for processing octrees with SIMD architectures is studied in (Papachrysantou 1992).

8.5 Conclusions

Writing parallel programs is difficult. There are many issues; resource allocation, communications protocol, and so on, that require explicit machine dependent programming to ensure efficiency. This, of course, has an adverse effect on portability, predictability of performance for any ports, and complexity. The skeletons approach addresses these problems by identifying the common computation patterns in problems. By isolating these as the building blocks of parallel programming, we can build efficient implementations for each block. Portability is provided by various implementation on different machines and by transformations between skeletons. Performance can be predicted through the use of performance models attached to each skeleton. The use of skeletons allows for the separation of behaviour and meaning. Programming for complexity is reduced by allowing the programmer to first address the correctness or meaning of a program, and then to tackle the behaviour to ensure efficiency.

This approach can be applied at a higher level to application domains. The recurring data and control structures for applications can be abstracted and incorporated into skeletons. The applications builder can then work at a familiar level, and need not address the unfamiliar low-level details of parallel machines. We have seen this applied to solid modelling and investigations are continuing into other fields, such as numerical problems and databases.

The aim of the work is to provide programming through the selection and instantiation of skeletons aimed at the desired application area.

Further work needs to be done to investigate the information needed to specify behaviour of a skeleton completely, and a notation for expressing it. So far we have not discussed the composition of skeletons. For efficiency, the behaviour of interacting skeletons must be compatible. We must consider how these behaviours can be optimized and whether this process can be automated. These research areas are being actively pursued. Work is also continuing with the specification of performance models for skeletons.

8.6 Acknowledgements

We would like to thank our colleagues in the Advanced Languages and Architectures Section at Imperial College for their assistance and ideas. Much of the work in Subsection 8.4.1 is based on (Papachrysantou 1992). The work reported here was initially developed in the UK SERC/DTI funded project 'The Exploitation of Parallel Hardware using Functional Languages and Program Transformation' and used equipment funded under the SERC's Parallel Equipment Initiative. This work is being continued under a SERC Research Scholarship to the second author.

REFERENCES

Backus, J. (1978). Can Programming Be Liberated from the von Neumann Style? A Functional Style and its Algebra of Programs. *CACM*, **21(8)**, 613-41.

Bird, R. and Wadler, P. (1988). *Introduction to Functional Programming*. Prentice Hall.

Cole, M. (1989). *Algorithmic Skeletons: Structured Management of Parallel Computation*. Pitman/MIT Press.

Darlington, J. and Pull, H. M. (1988). A Program Development Methodology Based on a Unified Approach to Execution and Transformation. *Partial Evaluation and Mixed Computation*. North-Holland.

Darlington, J., Field, A.J., Harrison, P. G., Kelly P. H. J., Sharp, D. W. N. and Wu, Q. (1993). Parallel Programming Using Skeleton Functions. *Parallel Languages And Architectures, Europe:Parle '93*

Fortune, S. and Willey, J. (1978). Parallelism In Random Access Machines. 10th *ACM Symposium on Theory of Computing STOC*.

Harrison, P. G. and Khoshnevisan, H. (1992). The Mechanical Transformation Of Data Types. *The Computer Journal*, **35(2)**.

Hudak, P., Peyton Jones S. L., Wadler, P. I., Boutel, B., Fairburn, J., Fasel. J., Guzmán, M., Hammond, K., Hughes, J., Johnsson, T., Kieburtz, R., Nikhil, R. S., Partain W. and Peterson J. (May 1992). Report on the Functional Programming Language Haskell. *SIGPLAN Notices*, **27(5)**.

Isaac, C. A. (1992). *Structured Implementations of Functional Skeletons*. MSc Project Report, Dept. of Computing, Imperial College.

Kelly, P. H. J. (1989). *Functional Programming for Loosely-coupled Microprocessors*. Pitman/MIT Press.

Meiko Ltd. (1990). *CS Tools for SunOS*. 83-009A00-02.02.

Papachrysantou, G. (1992). *High Level Forms for Computation in Solid Modelling*. MSc Project Report, Dept. of Computing, Imperial College.

Perry, N. (1989). $Hope^+$. Internal document IC/FPR/LANG/2.5.1/7, Dept. of Computing, Imperial College.

Valiant, L. G. (1990). General Purpose Parallel Architectures. *Handbook of Theoretical Computer Science*. North-Holland.

9
CATEGORICAL DATA TYPES

D.B. Skillicorn*

*Department of Computing and Information Science
Queen's University, Kingston, Canada
skill@qucis.queensu.ca*

Abstract

An ideal abstract model for parallel computation must carefully balance requirements for effective software engineering with requirements for efficient implementation. Models based on sets of fixed communication/computation patterns satisfy these requirements but, in general, the sets of patterns are chosen arbitrarily. Categorical data types are a way of building such models while automatically generating operations, equations, and a guarantee of completeness. We illustrate this construction, and its usefulness for practical problems, by building the type of chemical molecules and showing how molecular properties can be computed in parallel.

9.1 Introduction

An ideal abstract model of parallel computation is constrained by both its upward and downward relationships. Upwards, a model must allow programs to be constructed from specifications in a reasonable way and must hide unnecessary details of target architectures. Downwards, a model must allow effective implementations to be built. The central problem in choosing abstract models for parallel computation is to find the correct balance between these two sets of constraints.

The relationship between specification and model seems to require the following properties:

- *Architecture independence.* Since the range of possible target architectures changes frequently and there is no style of architecture that is likely to remain the best as technology changes, long lifetimes for software can only be achieved if they are insulated from architecture considerations. Thus an ideal model cannot include any assumptions about how abstract operations are achieved; for example, explicit

---

*This work was supported by the Natural Sciences and Engineering Research Council of Canada.

message passing or assumptions about memory organization must be ruled out. We can summarize architecture independence by saying that it means that program source does not need to be changed when a program is ported to a new architecture.

- *Development method.* We have been casual about the way sequential software is developed in all but the most safety-conscious and potentially litigious situations. In a parallel world, correctness is not an optional issue since incorrect programs will tend not to run at all. While it is just conceivable that a program development strategy based on verification could be used, it is much more natural to consider the calculational or derivational style of program development as the way to build parallel software. This can be based either on refinement or equational transformation. It requires that an abstract model have a sufficiently strong semantics to make derivations possible.

- *Intellectual simplicity.* Humans need to be able to keep in mind what a piece of software they are working on does, at least in some weak sense. It is hard to maintain a mental model of software with many thousands of threads, multiplexed onto some smaller number of processors, in which the detail of what is going on is important. Thus an abstract model must allow programmers not to think about substantial parts of the computation. While some of the management of this abstraction can doubtless be hidden by a compiler, it strongly suggests that regularity is important.

All of the properties listed above can be achieved by making the model sufficiently abstract. However, the requirements of interacting with an implementation exert a pull in the other direction, towards models that are very concrete. In particular, we can identify the following properties for this relationship:

- *Congruence.* If software is to be built in calculationally, it must be possible to make decisions along the way with some understanding of their implications on the final cost of the computation. The model must expose some of the details of the implementation, without violating either architecture independence or intellectual simplicity. Thus the model must permit a set of cost measures that can be computed from a small number of parameters. A cost calculus, that is, cost measures that respect the structure of the transformation or refinement system, would be even better.

- *Efficient implementation.* An even stronger property than congruence is that the model should not itself introduce any non-intrinsic inefficiency into an implementation. Thus a comparison between the cost of a computation within the model and an abstract cost (using say Boolean circuits or the PRAM) should not reveal any hidden cost differences.

- *No preferred granularity*. The appropriate grain size to execute on each processor of a parallel system is growing quickly as processors become deeply pipelined or superscalar. A model in which the appropriate grain size is bounded above will eventually fail to utilize the processors on which it runs.

These requirements are challenging and it may not be possible to satisfy them all in general. Nevertheless, a clear picture of the ideal enables us to assess both goals and progress.

A consideration of these six criteria suggests two ways in which a good model might be found. The first is to take a model that is very abstract and then try to make it efficiently implementable. This is the approach taken with Unity/PCN (Chandy 1988), higher-order functional programming (Peyton-Jones 1987), and Concurrent Rewriting Logic (Maude) (Meseguer 1991). The other approach is to take a model that has a limited set of communication and computation patterns and try to make it an effective platform on which to build software. This is the approach taken by models that are called 'skeleton-based' in a functional setting (Cole 1989, Darlington 1993, Danelutto 1992) and 'data-parallel' in an imperative setting (Sabot 1989, Blelloch 1990). It is also the approach that we advocate in this paper.

9.2 Categorical data types

The major drawback of models based on sets of fixed patterns is the choice of patterns. These have typically been chosen based on some perception of usefulness, but without any consideration to whether the resulting set is either complete or non-redundant. Building datatypes using the categorical data type (CDT) construction of Malcolm (Malcolm 1990) avoids this drawback while producing models in which parallelism is perfectly natural.

Rather than give the details of the construction, we will illustrate by building a slightly unusual datatype – that of chemical molecules. The construction will automatically generate interesting parallel operations that can be applied to molecules, as well as a transformation system for parallel molecular computation programs. An elementary knowledge of category theory is assumed.

We begin with a category of sets and computable functions between them. One particular object, A, of this category will be our starting point – it is the set of chemical elements with which we wish to work. Other objects in the category are the set of integers, the set of reals, and so on. An example of a useful function is the function *atomic weight*, which is an arrow from A to the object *set_of_naturals*.

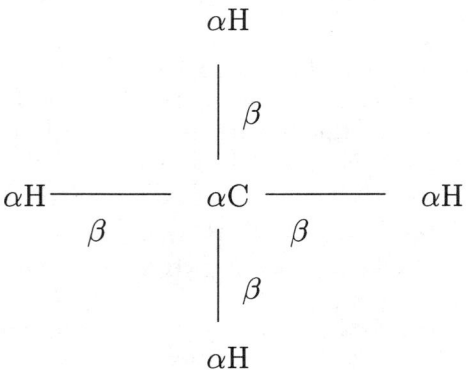

FIG. 9.1. A methane molecule

We define three constructors for the new type M (for Molecule) that we are going to build.

$$\alpha : A \to M$$
$$\beta : M \times M \times R^3 \to M$$
$$\gamma : M \times R^3 \to M$$

Intuitively, α takes an atom and makes it into a molecule, consisting of that single atom (essentially a type coercion). The constructor β takes two molecules and a vector giving the relative orientation of two of their component atoms and creates a new molecule, modelling a new bond between the two molecules. The constructor γ takes a molecule and a vector and creates a new bond between two of its component atoms, thus allowing cycles within the bond structure of molecules. We require that β and γ are mutually associative and β is commutative in M – intuitively this means that it doesn't make a difference in which order molecules are connected together, provided the resulting bonds are the same. This requirement can be expressed as a set of equations on β and γ and carry over into the algebras that we will shortly introduce. The atoms used for each of these constructions are "remembered" by the physical arrangement of the resulting molecule; that is, we use a straight line to represent the application of the both the β and γ constructors.

Instances of the constructed type correspond closely to the stick diagrams that are normally used to draw molecules, except that they contain information about the orientation and length of each bond. A methane molecule is shown in Fig. 9.1.

Now consider the endofunctor whose action is the "inverse" of the constructors, that is, which takes an object and maps it to its component objects. This functor T_A should therefore map the object M to its components like this:

$$T_A M = A + M \times M \times R^3 + M \times R^3$$

and is thus the functor

$$T_A = K_A + Id \times Id \times K_{R^3} + Id \times K_{R^3}$$

(where K_X is the constant X functor that maps every object to X and every arrow to the identity arrow on X). The properties we wish for T_A occur when it is a polynomial functor – if it had turned out not to be, we would have had to go back and choose different constructors.

Now define a T_A algebra to be the pair and arrow between them:

$$T_A X \to X$$

(written $(T_A X, X)$ when the arrow is obvious and (TX, X) when the base type is also obvious) and consider the category T_A-Alg whose objects are T_A-algebras, and whose arrows are T_A-algebra homomorphisms, that is, pairs of arrows in our underlying category that respect T_A-algebra structure. This is shown in Fig. 9.2. This category is where all of the interesting parallel computation takes place. If T is polynomial, then we define the T_A-algebra

$$TM \to M$$

to be the initial object in the category T_A-Alg, which we know must exist because the functor T_A is polynomial. There is a unique arrow from this algebra to any other T_A-algebra. Furthermore, there is an isomorphism

$$TM \leftrightarrow M$$

or

$$A + M \times M \times R^3 + M \times R^3 \leftrightarrow TM$$

Intuitively, this means that a constructed type is isomorphic to the pieces from which it was built, in their proper relationships.

The initiality of $TM \to M$ in the category of T_A-algebras means that there is a unique arrow from it to any other T_A-algebra. We call these unique arrows *catamorphisms*; they are homomorphisms on the constructed type.

Let us now consider what T_A-algebras are for the molecule type we have constructed. A T_A-algebra is an arrow as shown in Fig. 9.3. Such an

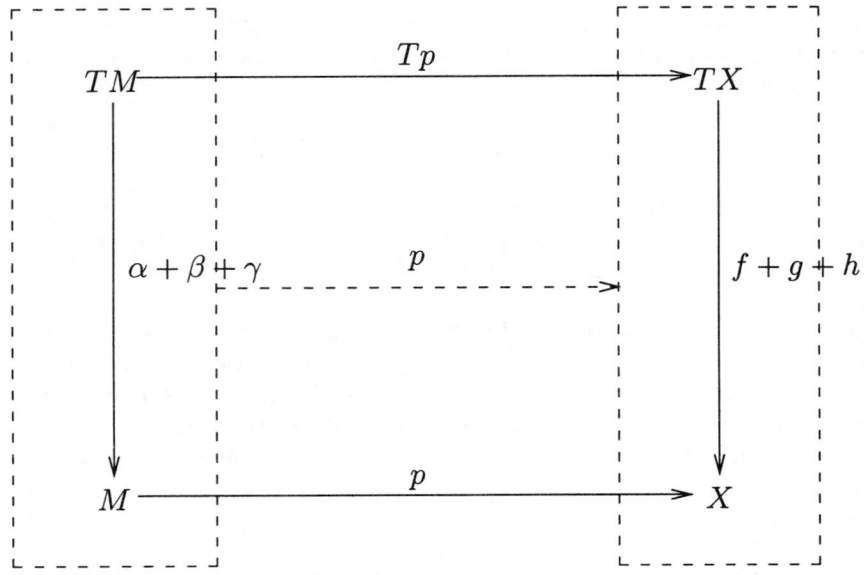

FIG. 9.2. Objects and an arrow in T_A-Alg

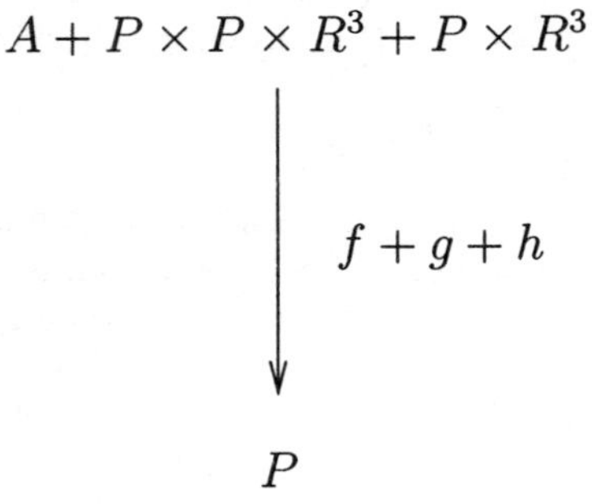

$$A + P \times P \times R^3 + P \times R^3$$
$$\downarrow f + g + h$$
$$P$$

FIG. 9.3. A T_A-algebra

algebra exists precisely when there are three functions

$$f : A \to P$$
$$g : P \times P \times R^3 \to P$$
$$h : P \times R^3 \to P$$

CATEGORICAL DATA TYPES   161

that is, when P is some *property* that is

1. computable for each individual atom (and hence there must have been an appropriate arrow in the underlying category with which we began);
2. recursively dependent on the properties of component molecules together with their relative orientation and distance;
3. dependent only on the addition of new bonds in the molecule.

Thus catamorphisms include all functions that compute molecular properties that are recursive and independent of the order in which the molecule was assembled.

Some examples of catamorphisms are:

- **Volume.** Given the volume of each individual atom it is possible to compute the volume of molecules from the volumes of the components and their relative positions.
- **Energy.** Given an energy value for each individual atom, the energy of a molecule depends on the energy of its component molecules and the orientation and distance of the bond that joins them. The energy also depends on any internal bonds that have been created using the γ constructor.
- **Convex hull.** Given the convex hull of the individual atoms, the convex hull of the molecule can be computed from the convex hulls of the components and their orientation and distance.

It is easy to see that molecule catamorphisms can be evaluated in parallel, and we will make this intuition more precise in the next few paragraphs. Consider Fig. 9.4, which shows the arrows involved in a catamorphism in more detail. The lower horizontal arrow is the catamorphism that we wish to compute. One way to compute it is to go around the other three sides of the diagram. The first step is really pattern matching; that is, we examine the molecule to which we are applying the catamorphism and decide which constructor was used to build it. The second step is to follow the upper horizontal arrow, which computes the catamorphism on base cases or recursively on components of the molecule. The third step is to use the resulting values to compute the final result of the catamorphism.

We have built a new type based only on the original object A. These molecules should, strictly speaking, be written as M_A because we can use exactly the same construction for any of the other objects in our underlying category. For example, if we start with the object *Int* of integers, then we can define constructors:

$$\alpha : \text{Int} \to M_{Int}$$
$$\beta : M_{Int} \times M_{Int} \times R^3 \to M_{Int}$$
$$\gamma : M_{Int} \times R^3 \to M_{Int}$$

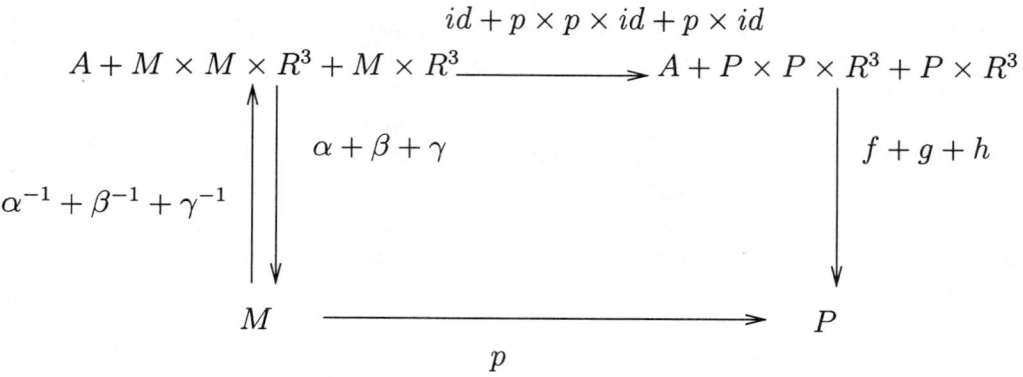

$$p.\alpha = f.id$$
$$p.\beta = g.(p \times p \times id)$$
$$p.\gamma = h.(p \times id)$$

FIG. 9.4. Catamorphism recursion structure

and a functor T_{Int}. As before, we can build a category T_{Int}-Alg in which an algebra of "molecules" whose nodes are integers, rather than atoms, is initial. In fact, all of the types constructed in this way have the same *structure* but different *content*; that is, what is at the nodes of the "molecule" will differ, but the oriented graph structure remains the same.

All of the T_X-algebra categories are related in the following way. If the original underlying category has an initial object $\emptyset$, then the category $T_\emptyset$-Alg has $(TM_\emptyset, M_\emptyset)$ as its initial object. All of the other T_X-Alg categories are subcategories within it. Arrows in the underlying category lift to arrows in $T_\emptyset$-Alg called *generalized maps*. Given an arrow $h : X \rightarrow Y$ in the underlying category, the lifted arrow $h* : (TM_X, M_X) \rightarrow (TM_Y, M_Y)$ is an algebra homomorphism that takes a "molecule" whose nodes are of type X and maps it to an identically structured "molecule" whose nodes are of type Y using function h. Clearly maps are parallel operations. This is shown in Fig. 9.5. The maps "paste together" the initial objects of each subcategory in a way that reflects the arrow structure of the original underlying category. Thus we can define a functor from the underlying category to $T_\emptyset$-Alg mapping object X to $(T_X M_X, M_X)$ and arrows h to $h*$.

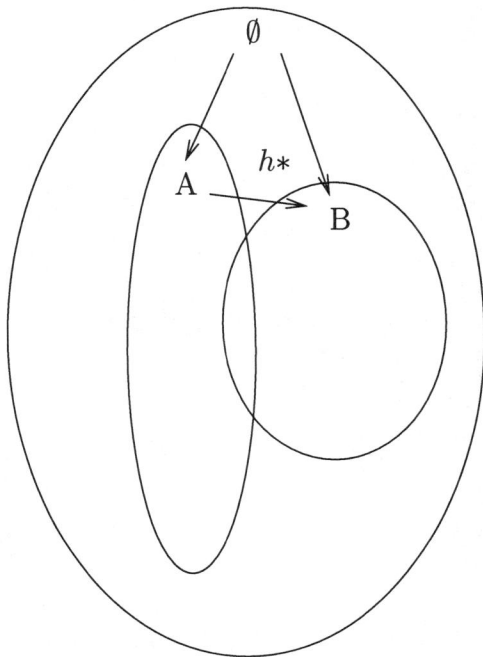

FIG. 9.5. Structure of the category $T_\emptyset$-Alg (where $\emptyset$ is the algebra $(TM_\emptyset, M_\emptyset)$, A is the algebra (TM_A, M_A), B is the algebra (TM_B, M_B), and h is an arrow from A to B in the underlying category)

The (polymorphic) type M we have built is a separable type, that is the functor T_A can be separated into two pieces, one of whose codomains is the original object A and the other of whose codomain does not contain an A. Separable types have the following property:

Property Every catamorphism on a separable type can be expressed as the composition of a generalized map and a generalized reduction.

A generalized reduction is the catamorphism from an initial algebra to a T_A-algebra (TA, A) in which the function $f : A \to A$ is an identity. Its effect is to alter the structure without changing the content, and it generalizes the associative reduction on lists. Reductions are also parallel operations, because a reduction is some function of reductions of its component parts, and these can be evaluated in parallel.

The property above can be seen clearly in some of the catamorphisms

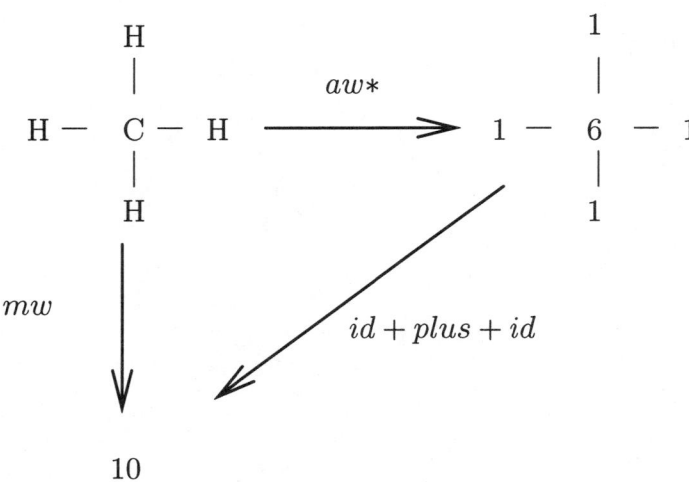

FIG. 9.6. Computing the molecular weight of a molecule

given as examples above. The energy of the molecule can be computed more or less directly as we described above. However, it can also be computed by mapping the molecule to a "molecule" whose nodes are energy values; and then applying a recursive function to this "molecule" to compute the final value of the energy. The first step, the generalized map, is solely concerned with manipulating the content of the structure; the second step manipulates only the structure.

Another example is the function from molecules to natural numbers that computes the molecular weight (mw). It is an extension of the function aw that computes atomic weights, an arrow in the underlying category from atoms to natural numbers. Figure 9.6 shows the way in which mw decomposes into a generalized map (the lifting of aw) composed with a generalized reduction.

Another useful benefit of the categorical construction is the property known as *promotion*. Promotion means that there is a canonical form for every catamorphism and that compositions of catamorphisms with homomorphisms can be reduced to a single catamorphism. Suppose that there is a catamorphism
$$p : (TM, M) \to (TX, X)$$
and a T_A-algebra homomorphism (that is, a function that respects the algebra structure)
$$q : (TX, X) \to (TY, Y)$$
Then there is a catamorphism

$$r : (TM, M) \to (TY, Y)$$

by the initiality of (TM, M), and the following equation holds:

$$r = q.p$$

Many other equations arise as the result of diagrams in the category T_A-Alg. These equations can be used for transformational software development of programs that compute with the constructed type.

9.3 Properties of the construction

We can summarize the properties of the categorical data type construction as follows. Many of these properties should now be obvious from the type of chemical molecule that we have constructed.

- The construction is polymorphic over the types in the underlying category (although this was only used in a peripheral way in the chemical example, it is crucial for more conventional types).
- Based only on the chosen set of constructors, a set of (second order) functions on the constructed type is generated automatically by the construction. These operations are a generalized map and a generalized reduction, both of which are inherently parallel for many interesting types. They form a set of basic skeletons, and they can be composed to make further, more abstract skeletons to capture more complex operations.
- Many interesting functions on the constructed type can be expressed as catamorphisms; examining catamorphisms is therefore a good way to find new and interesting functions.
- A set of equations is automatically generated; these can be used for equational transformational software development.
- The second order functions take other functions as arguments. These arguments can themselves be skeletons, so that nested skeletons are possible.

Two further properties hold that are not obvious from the example above:

- The set of equations forms a *complete* transformation system in the sense that the syntactic form of any homomorphism on the constructed type can be transformed into any other equivalent syntactic form within the system.
- The communication properties of the constructors are directly reflected in the locality of the communication properties of the second order operations; so that for some types, implementations that require only local communication on a range of parallel architectures are possible; when this occurs, cost calculi that apply to a

wide range a target architectures can be constructed (Skillicorn 1990, Skillicorn 1992).

The construction of fixed pattern models of parallel computation using categorical data types has all the advantages of such models with respect to the six criteria with which we began; but it also removes any *ad hoc* choices of patterns to use, while at the same time providing a framework for software development.

The type of chemical molecule is directly useful for parallel computation in molecular applications. It also generalizes naturally to a data type of *graph* (the objects become anonymous and the distance and orientation can be dispensed with). Categorical data types for lists (join, cons, and snoc) (Bird 1987), bags, finite sets, trees (Gibbons 1991), and arrays (Banger 1994) have also been built. In each case, we get a similar fixed set of parallel operations with known communication and computation patterns. Most of the work has been done with lists, for which many derivations are known (Bird 1987, Bird 1989, Bird 1989a), efficient architecture independent implementations have been built (Skillicorn 1990, Skillicorn 1992a), and there is a cost calculus (Skillicorn 1992).

The categorical data type construction is an important general technique for building fixed pattern abstract models for parallel computation. We have suggested why, in both abstract and practical terms, this should be so.

REFERENCES

C.R. Banger. Construction of Multidimensional Arrays as Categorical Data Types, PhD thesis, Queen's University, Kingston, Canada, 1994.

R.S. Bird. An introduction to the theory of lists. In M. Broy, editor, *Logic of Programming and Calculi of Discrete Design*, pages 3–42. Springer-Verlag, 1987.

R.S. Bird. Algebraic identities for program calculation. *The Computer Journal*, 32(2):122–126, February 1989.

R.S. Bird, J. Gibbons, and G. Jones. Formal derivation of a pattern matching algorithm. *Science of Computer Programming*, 12:93–104, 1989.

G.E. Blelloch. *Vector Models for Data-Parallel Computing*. MIT Press, 1990.

K.M. Chandy and J. Misra. *Parallel Program Design: A Foundation*. Addison-Wesley, 1988.

M. Cole. *Algorithmic Skeletons: Structured Management of Parallel Computation*. Research Monographs in Parallel and Distributed Computing. Pitman, 1989.

M. Danelutto, R. di Meglio, S. Orlando, S. Pelagatti, and M. Vanneschi. A methodology for the development and the support of massively parallel programs. *Future Generation Computer Systems*, 1992. Also appears as "The P^3L language: an introduction", Hewlett-Packard Report HPL-PSC-91-29, December 1991.

J. Darlington, A.J. Field, P.G. Harrison, P.H.J. Kelly, Q. Wu, and R.L. While. Parallel programming using skeleton functions. In *PARLE93, Parallel Architectures and Languages Europe*, June 1993.

J. Gibbons. *Algebras for Tree Algorithms*. D.Phil. thesis, Programming Research Group, University of Oxford, 1991.

G. Malcolm. *Algebraic Data Types and Program Transformation*. PhD thesis, Rijksuniversiteit Groningen, September 1990.

J. Meseguer and T. Winkler. Parallel programming in Maude. In J.P. Banâtre and D. Le Métayer, editors, *Research Directions in High-Level Parallel Programming Languages*, pages 253–293. Springer Lecture Notes in Computer Science 574, June 1991.

S. Peyton-Jones. *Implementation of Functional Programming Languages*. Prentice-Hall, 1987.

G. Sabot. *The Paralation Model: Architecture-Independent Parallel Programming*. MIT Press, 1989.

D.B. Skillicorn. Architecture-independent parallel computation. *IEEE*

Computer, 23(12):38–51, December 1990.

D.B. Skillicorn and W. Cai. A cost calculus for parallel functional programming. Submitted for publication. Also appears as Department of Computer Science Technical Report 92-329, Queen's University, Kingston, Canada, 1992.

D.B. Skillicorn and W. Cai. Equational code generation: Implementing Categorical Data Types for Data Parallelism, Proceedings of TENCON'94, Singapore, 1994.

10

RETRAN: A RECURRENT PARADIGM FOR DATA-PARALLEL COMPUTING

A.V. Shafarenko

*Department of Electronic and Electrical Engineering
University of Surrey, GU2 5XH, UK*

Abstract

An applicative paradigm of array processing based on recurrence relations and a data-parallel overloading of constants is presented. It is shown that the principle of anti-currying with function-based eager arrays generates a language with simple and intuitive notation yet complete for data-parallel applications. The evolution of data structures (which corresponds to program execution in an imperative language) is captured by the notion of recurrently defined stream.

The quest for an apposite parallel-computing paradigm that the dormant "broad-minded" user will eventually accept (even though it could hardly be called Fortran by any honest one) has been going on resultlessly for some years now (Pancake 1992). At the top end of the abstraction range the expectations associated with functional programming, although not yet found abortive, have not so far been gratified with any demonstrably scalable model which one could target towards massively concurrent machines (Peyton Jones 1992), which was also mentioned at the recent SERC/DTI Workshop on Declarative Systems Architectures in London. At the bottom end there are quite a few message-passing models in which hardware features are very far from being hidden from the programmer. Here high performance is arguably achieved – sadly at the expense of portability and flexibility, which makes the respective programming paradigms hard to use in the scientific computing context.

Another class of methods is based on high-level machine-independent executable specifications restricted to certain patterns, usually in terms of "standard" high-order functions, or paradigms, such as those defined in (Rabhi 1992), of which the executing agent may have special implementation-related knowledge that is not inferable from their formal definition. In a broader context these are called skeletons (Cole 1989, Backus 1978, Bird 1989).

This paper offers an account of one such paradigm, which cannot (or should not) be expressed in a standard functional language because it introduces not only certain high-order functions (which could be expressed in an existing programming language) but restricted mechanisms of combining these into further functions in such a way that certain important properties could still be inferred and hardware-specific annotation avoided. On the other hand it uses a more intuitive set of notions for a present-day scientific programmer, which enable not only declarative (equational) but also quasi-imperative reasoning about the resultant problem specification. Data parallelism is achieved by mapping all primary functions onto all possible index spaces by default (the *anti-currying* rule) and making them strict in those spaces only. The so overloaded set of functional constants then generates a complete nonscalar static-rank/dynamic-shape typed calculus with the standard Hindley-Milner type system.

The rest of the paper is organized as follows. Section 1 reveals the main obstacles to the static inference of data-parallelism in a functional program. Section 2 introduces an array algebra which ensures complete inferability of data-parallelism in a simple and generic way. Section 3 concentrates on the problem of evolving data structures in numerical applications and how this could be achieved by exploiting the notions of stream and partial data-parallel definition (assignment). Finally, Section 4 briefly discusses the issues of visual (languageless) design and persistency with reference to data-parallel streams and exposes some future work.

The author is grateful to the Nuffield Foundation for their financial support, to Professors Steve Schuman and Chris Jesshope of Surrey University for stimulating discussions and constructive suggestions and to Dr V. Muchnick for commenting on the original draft.

10.1 Data parallelism is difficult to deduce even in a functional program

Let us first make a small experiment. Consider a simple numerical method for a one-dimensional diffusion equation:

$$\frac{\partial u}{\partial t} = \frac{\partial^2 u}{\partial x^2}$$

An $O(\tau, h^2)$ finite-difference scheme will look thus:

$$u_i^{[n+1]} = u_i^{[n]} + \tau \frac{u_{i+1}^{[n]} - 2u_i^{[n]} + u_{i-1}^{[n]}}{h^2}$$

With the boundary conditions and the initial data given by

$$u_0^{[n]} = a;\ u_d^{[n]} = b;\ u_i^{[0]} = v_i,$$

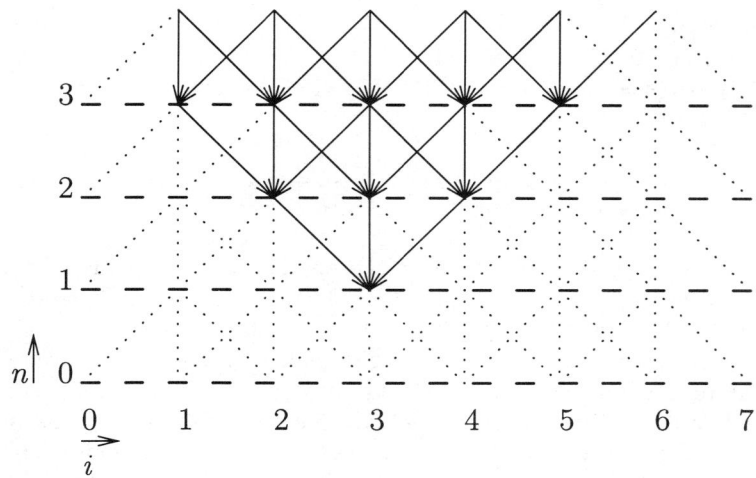

FIG. 10.1. Data-flow graph of the the function u

this scheme can be rendered into Miranda* by slightly changing the syntax:

```
u (0,i)    = v(i)
u (k+1,i)  = a, if i=0
           = b, if i=d
           = u(k,i) + coef*(u(k,i-1)-2*u(k,i)+u(k,i+1)),
                                                  otherwise
```

If one types this program in to a Miranda system, defines (say) $d = 7$, and then enters

[u(20,i)| i<-[0..d]]

to compute the state of the diffusion system after the first 20 time steps, the result will not appear immediately, but when it does it will be... an error message saying that there is not enough memory for the program data, although one might have thought the Miranda system had only some two hundred numbers to crunch!

As it turns out, the small example above causes a grave inference failure. Indeed, let us have a look at this function dataflow graph (Fig. 10.1). The figure shows the dependencies between different points in space/time according to the specification. Every edge depicts a recursive invocation

---

*Miranda is a trademark of Research Software Ltd

```
u 0   = v
u n+1 = (array (0,d) ((0:=a):(d:=b):
        [i:=(u(n)!(i-1)-2*u(n)!i+u(n)!(i-1))*coef
        | i<- [1..d-1]]!)
```

FIG. 10.2. Diffusion equation in Haskell. It is assumed that a, b, d, coef, and function v are defined outside

of u with the lower end of it placed at the spatio-temporal point to be computed and the upper end going to the point it is immediately dependent on. Since the system cannot deduce that a function will have to be computed at the same value of its arguments many times (and that there will be a finite number of those values), the evaluation of say $u(1,3)$ will be triggered as many times as it is required by further time steps (this is highlighted in Fig. 10.1). It should be evident now that the number of re-evaluations is equal to the number of edges in the highlighted subgraph. For n-th step in time and a vertex located a long way from the spatial borders the computational complexity is of the order 3^n, rather than n as in a straightforward imperative implementation, where the function u is represented by a one-dimensional mutable array. It is not surprising therefore that since the invocation pattern is not statically recognized, this functional specification simply doesn't terminate withing reasonable time with n equal to just a few tens — and the countless recursive calls in order to reach the bottom fill up the available memory.

Note that still a data-parallel call [u(20,i)| i<-[0..d]] is possible given the Miranda specification above, which is at least in this sense easily inferable. However complex the spatial access pattern could be, this would not change the fact that all of the components u(20,i) can be computed in parallel; automatic prevention of re-evaluation, however, would not be possible wherever a more complex access to the space is being made.

To deal with effects such as this one, functional languages introduce an entity which would otherwise be redundant: the array. From a mathematician's point of view this makes little sense, since an array is just a function of one or more integer variables that has a compact domain. It is this compact domain that makes such a function representable by a data structure, in which elements once evaluated ("rewritten" in terms of graph reduction) are never evaluated again. This is indicated in Fig. 10.1 by the dashed horizontal lines. It should be evident now that, once arrays are introduced along the horizontal lines, vertex (3,3), for example, will be rewritten only once no matter that 3 vertices, (4,3), (4,2), and (4,4), will require the result. Which, of course, brings us back to the normal linear estimate of complexity. The respective program in Haskell (Hudak and Wadler 1988) is given in Fig. 10.2 Two important aspects of the code should be realized. Firstly, the array constructor **array** is in fact a function taking a list into an array. Formally, since the list has to be finite (the function is strict in the respective argument) it could be viewed as a parallel data structure with all the list elements evaluated concurrently. However, there is no way in an ordinary functional language to encode a parallel composition of a list – all one can do is to add one element at a time using a version of CONS. Secondly, although the array in the Haskell program is strict in all indices supplied with the list, it is nonstrict in values coming with these indices so that even if parallel evaluation of the list *were* possible, the semantics of Haskell would forbid it since there would generally be no guarantee of termination.

We therefore need a solution that is based on data structures to use overwriting as a safeguard of efficiency, but we need these to be strict and manipulable in data-parallel ways.

10.2 Indexed objects in RETRAN

Introduction of arrays in functional programming languages has already a few precedents. SISAL (Livermore 1992), a very pragmatic language for which there exists an amazingly fast implementation on a Cray machine, does not suggest arrays have anything to do with functions. It offers an array "constructor", which has to be given a wild mixture of type specifications, nonscalar indexing expressions, and element values. This obviously is based on its authors' wrong belief that 'type inference in the presence of the overloaded operators is not a well-defined procedure' (F-code (Muchnick *et al.* 1993) for one has a much more complex overloading and typing system yet is statically polymorphic, to say nothing about Haskell.)

SISAL arrays are eager and therefore perfectly data-parallel, however operands have to conform in shape (while not having to conform in type), or one of them could be a scalar. Needless to say that already user-defined overloaded operators acting on different-rank arrays in any systematic way are not possible.

Bird's work (Bird 1987) on axiomatization of lists (later made into multidimensional array theory (Bird 1988)) and the practical implementation of the array calculus in (Hains and Mullin 1993) help deriving data-parallel programs from specifications; however, despite the claim of Skillicorn's paper (Skillicorn 1990) that a constructive calculus provides architecture neutrality, it is presumptuous of geometric construction of arrays (any function of an array is constructive-parallel as long as it can be expressed in terms of application to adjacent subarrays). Already the fact that a reduction can be defined by a *commutative* operation and therefore applied to any *set partition* pattern rather than a subarray concatenation is completely outside of the constructive approach, which diminishes the possible parallelism of a constructive specification. Also, constructivists introduce array shape as a component of type, but have to admit that it is impossible to deduce array conformity (which is part of type match) or lack thereof in the calculus (Hains and Mullin 1993).

In our treatment of arrays we return to the fundamental proposition that an array is a function of an integer, limited-variation argument(s). Standard intuition is that it is also a *total*, nonrecursive function (which explains why the data-structure representation looks so natural – it is a record of this function application to the whole of its domain). We shall designate an array made out of a function f as

$$\text{array}(b_1, b_2, \ldots, b_n)f$$

where the bounds $b_1, b_2, \ldots, b_n$ are each either a pair of expressions of the same index type $d : u$ with d defining the lower and u the upper bound of the respective index of the array, or just one expression of an index type defining the upper bound, with the lower bound defaulting to the lower bound of the type. The function should match the arguments in type (but not necessarily in number). The type of the result consists of a component denoting the output type of f augmented with the number n of array dimensions. We assume that all index types are enclosed in a superclass Ind and that an upgrading coercion of any member type to some type $GenInd$ belonging to this class exists. Thus only rank information is a component of the array type.

For example

`r x = array (5,'a':'c') \i j -> j, if i>x then j else 'd'`

The signature of this function is

`f::Int->2#Char`

indicating that it receives an integer and returns a two-dimensional array of type `Char`

The arguments to a function can be curried (fewer items can be supplied as the arguments to a function application), although currying subscripts is not allowed (this explains why the rank is fixed).

An alternative way to define a nonscalar function is by using a nonscalar right-hand side. In either case, the compiler will infer the rank unambiguously.

In data-parallel applications two objects of different ranks (e.g., a matrix and a scalar) can be the operands to an element-wise binary operator. In such a case the lower-rank object is given extra dummy subscripts, and the operator is applied throughout the resulting space. To take this into account, introduce the *anti-currying* principle for subscripts:

$$(\forall r \in S_n) r[x_1, x_2, \ldots, x_n, x_{n+1}, \ldots] \equiv r[x_1, x_2, \ldots, x_n]$$

where S_n is the class of expressions that have exactly n subscripts and the brackets denote indexing. Put simply, it is allowed to add any number of dummy subscripts to the subscript list of a function application. Note that now by "rank" we shall always mean the minimum rank. Any function will assume certain minimum ranks of all its arguments. It is those argument ranks that appear in the inferred signature.

10.2.1 Recast.

In order to enable normal "geometric" manipulation of indexed objects, the mechanism of remapping is introduced. RETRAN offers a generic remapping abstraction with the following syntax:

recast	→	*function* [*expression-list* { <- *pattern-list* }]
function	→	*expression*
pattern-list	→	*pattern* {, *pattern*}
expression-list	→	*expression* {, *expression*}
pattern	→	*var* | *var bound* | *dummy*
bound	→	*bar range*
range	→	*expression* {: *expression* }
bar	→	|
dummy	→	_

[12pt]
This is somewhat reminiscent of the classical typed λ-abstraction except that a pattern is not necessarily a "bare" variable (it can be conditioned by the range, or else it could be dummy, which means the result will not depend on the respective subscript). Besides, a recast defines the mapping of more than one subscript at once and allows anti-currying.

A few semantical restrictions are required to guarantee domain inferability. A variable occurring in a pattern without range specification has to be used in the expression-list in the following contexts:

1. when it alone constitutes an expression.

2. as a linear combination $expr_1 * var \pm expr_2$, where either $expr_1$ with the asterisk or $\pm expr_2$ can be omitted. Neither of these expressions should depend on any pattern variable.

3. when it alone or within a linear combination constitutes the value of a non-dummy subscript of any function.

Only one instance of a variable has to satisfy any one of the above constraints: all other instances are unrestrained. A ranged variable can be used in the expression list arbitrarily.

For example, m[a,a<-a] denotes the diagonal of matrix m, m[b,a<-a,b] transposes the matrix, and m[a+5, b+5 <-a|9,b|9] defines a 10×10 submatrix of m whose upper-left corner is located at element (5,5) of the matrix m. Recasts can be as complex as this one:

`v[a[j,i,2*i] <-i,j]`

which denotes a matrix built from the elements of vector V using some elements of a three-dimensional object A as indices. Note that all pattern variables are only visible within the recast brackets.

Finally, it is possible to omit the pattern list together with the arrow, which makes the recast into an element selection operator.

10.2.2 Data-parallel operators

RETRAN generalizes arithmetic operators over indexed objects. For any unary element-wise operator $\odot$ (such as $-$) and any indexed entity f, the meaning of $\odot f$ is:

$$\odot f \to \lambda[x_1, \ldots, x_n]. \odot f[x_1, \ldots, x_n]$$

where n is the number of free subscripts (i.e. the rank) of the expression f (Note that there must not be any unbound arguments to f). The domain of the result coincides with that of the operand. For example, for any matrix m the expression -m denotes the matrix differing from m only in the sign of the elements.

For any binary element-wise operator $\otimes$ the semantics on indexed objects is defined thus:

$$f \otimes g \to \lambda[x_1, \ldots, x_n]. f[x_1, \ldots, x_n] \otimes g[x_1, \ldots, x_n]$$

Here n is the highest rank of the operands. The operand with the lower rank is anti-curried. The range of each subscript of the result is determined by the intersection of the respective ranges (in the sense of the $GenInd$ type if the types do not match), with any dummy subscript assumed to have the maximum range of its type.

For example, the following code

```
r::real->2#real
r z = array (n,m) \i j -> (i+j)*z

p::3#real
p = array (2,m,2*n) \i j k -> j^2/cos(0.1*k)-i

q::1#real
q=r[i,3<-i](1)

x = -r(3) + p[c,b,a<-a,b,c]/r[a,a+1<-a](7)-q[x<-_,x]*q
```

defines x to be a 3-index object with the following components and domain:

$$x_{ijk} = -3(i+j) + \frac{\frac{j^2}{\cos 0.1i} - k}{14i + 7} - (i+3) * (j+3)$$
$$i = 0 \ldots \min(n, m-1)$$
$$j = 0 \ldots \min(m, n)$$
$$k = 0 \ldots 2$$

Note that the declaration for x uses r as a matrix in the first term, but also slices a subdiagonal off it and anti-curries it to rank 3 for the division operator. The last term encodes what APL calls the outer product.

It should be noted that data-parallel notation used for x not only adequately represents index parallelism, but also ensures automatic inference of the signature and domain transformation. The range of all subscripts will be inferred as above.

To end this section let us consider the cases of function application and composition. A nonscalar function can be applied to nonscalar arguments; due to currying we can always assume there is only one argument to any function. Consider the following signature:

$$\texttt{f} :: n \# \texttt{x} \to \ldots \to m \# \texttt{y},$$

which means function F depends on a nonscalar argument. What happens if we apply it to an expression with the signature like this:

$$\texttt{e} :: k \# \texttt{x} \ ?$$

If $k = n$ the application is given by the f definition. If $k < n$, the argument is anti-curried to rank n. If $k > n$ anti-currying is applied to function f in the following way: the list of subscripts of f is expanded with $k - n$ last subscripts of e, with the signature of the result becoming as follows:

$$\texttt{f e} :: \ldots \to (m + k - n) \# \texttt{y}.$$

The result can be described with the following λ-formula:

$$\texttt{f e} = \lambda[x_1, \ldots, x_{m+k-n}].\texttt{f}[x_1, \ldots, x_m](\texttt{e}[x_1, \ldots, x_n \texttt{<-} x_1, \ldots, x_k]).$$

The range of the subscripts occurring twice on the right of the dot separator is equal to the intersection of the respective ranges.

This has a few pleasant implications. First, we can now apply basic scalar functions such as sin, exp, log, and so on, to nonscalar objects, for example sin(X) means the element-wise application of sin to X. Moreover, one can safely regard arithmetic operators mentioned above as ordinary scalar functions of one or two (generally nonscalar) arguments in infix notation. An important instance of such an operator is if-then-else which has the infix ...?...!... form in RETRAN. Using a clever application of this operator we can encode, for instance, the following quite complex three-dimensional object B:

B=a[z<-_,_,z]?b[x<-x,_,_]!c[y<-_,y,_]

that has the following components:

$$B_{i,j,k} = \text{if } a_i \text{ then } b_j \text{ else } c_k \ .$$

Secondly, it appears one can get heterogeneous data parallelism for free. For example, applying the following function G (where nat is the vector of natural numbers):

g(x,n) = odd(nat[i<-i|n])? sin(x) ! cos(x)

to a vector computes the sine or the cosine of the components depending on the subscript. One can envisage a heterogeneous function applied across a higher-dimensional object with anti-currying, too.

Finally let us have a look at the function composition. Since the list of subscripts is not a data structure in RETRAN, no function can yield such a list and function composition does not apply to subscripts. It is therefore a normal dyadic operator acting on arguments, for which data-parallel subscript conventions apply. The declaration a=f.g can therefore be rewritten as follows:

a x = f (g x)

where the semantics of nonscalar application defined above apply.

10.2.3 Reductions

In order to represent standard reductions such as $\sum$, $\prod$, and so on, RETRAN defines a generic function fold (which is similar to the one in Miranda):

fold:: (x->x->x)->x->1#x->x

which reduces a vector (3rd argument) to a scalar using an associative binary operator(1st argument) and its identity element (2nd argument). The difference from Miranda/Haskell fold is in that the last argument is a one-subscript object rather than a list; anti-currying and argument recast will provide all spatial versions of reduction. Note that fold cannot be applied to a scalar but can be applied to a dummy dimension of a nonscalar object, in which case it just removes that dummy dimension.

For example, let us define sum as usual

sum = fold ((+), 0)

Now, matrix multiplication can be defined like this:

```
mmul (a,b) = sum(a[i,j<-j,i]*b[i,j<-i,_,j])
```

One more example: the sum of all elements of a matrix m is written as `sum(sum(m))`.

10.2.4 Geometric composition

Recast exemplifies a unary "geometric" operator on an indexed object. Although possible in conventional functional ways, it is more convenient to have a binary geometric operator (which we shall call geometric composition, or composition for short) as a pre-defined operator.

Its signature is this:

`n#x->n#x->m#x`

where $m = max(n, k)$. Informally, it is an array concatenation operator acting in dimension k on two objects with rank n (anti-currying loosens this as well). Syntax-wise, it is a pseudo-trinary operator (the middle operand must be a decimal number), which is written as follows:

`a~4~b`

to denote concatenation of a and b in dimension 4. The semantics of it is as follows*. In the dimensions unaffected by concatenation the standard minimization of range is performed as if it were a binary element-wise operator. In the marked dimension, the concatenation of the index ranges is done, with a dummy subscript treated as one having range 1, that is, from 0 to 0.

Formally the semantics of A~k~B is given by the following λ-expression:

$$\lambda[x_1, \ldots, x_m].\text{if } x_k < r_k^a \text{ then } a[x_1, \ldots, x_k, \ldots, x_m]$$
$$\text{else } b[x_1, \ldots, x_k - r_k^a, \ldots, x_m]$$

where r_k^a is the upper bound of the kth dimension of A. The domain of the result is this:

$$\left(\min(r_1^a, r_1^b), \ldots, \min(r_{k-1}^a, r_{k-1}^b), r_k^a + r_k^b, \min(r_{k+1}^a, r_{k+1}^b), \ldots, \min(r_m^a, r_m^b)\right)$$

Note that in the minima the upper bound of any dummy subscript should be assumed to be infinite nevertheless.

Examples:

`m1~2~m2` is the row-wise and `m1~1~m2` is the column-wise concatenation of matrices m1 and m2. Composition `v1~2~v2`, where the operands are vectors, produces a rectangular matrix $2 \times k$, where $k = \min(r_1^{v1}, r_1^{v2})$.

---

*For the sake of simplicity all indices are assumed to be of nonnegative-integer type, but this is not a constraint in RETRAN

stream-env	→	*label* **stream** *output-tuple* [*input-tuple*] *stream-def* [*arg-tuple*]
label	→	*identifier*:
output-tuple	→	(*var*{, *var*}) [**check** *condition*]
input-tuple	→	(*var*{, *var*}) [**hold** *condition*]
stream-def	→ |	**begin** *def-list* **end** **init** *init-list* **update** *adj-list* **end**
def-list	→	*def*{ ; *def*}
def	→	*function-def* | *head* <- *initial* , *step*
head	→	*var* | *var index-pattern*
initial	→	*expression*
step	→	*expression*
init-list	→	*init-exp* { ; *init-exp*}
init-def	→	*head* = *initial*
adj-list	→	*adj-def*{ ; *adj-def*}
adj-def	→	*function-def* | *geom-expr* := *expr*
arg-tuple	→	(*var*{, *var*})

FIG. 10.3. The syntax of stream environment

10.3 Streams and evolution

A definitive feature of a functional language, which also explains why people find it difficult to switch to one, is that it defines no evolution of data. This, however, should not be misleading: lazy evaluation provides an analogue to imperative evolution without breaking referential transparency. This analogue is the notion of stream, or lazy list, found in Miranda, Haskell and some other lazy functional languages. The main idea of RETRAN is to capitalize on the very clean notion of recurrence relation which underpins the semantics of a stream. This notion is common for computational mathematics (which expresses almost everything in terms of iteration rather than recursion) and lazy functional programming. Our pursuit is inferability and implicit specification; as will be seen below that is achieved with streams in a most easy and natural way.

We define a stream generator in RETRAN as is shown in Fig. 10.3.

Its intention is to generates a series of output tuples which contain instances of the respective variables.

The *function-def* is a normal function definition as described above. Its scope will be limited to the stream environment. The other alternative is to define the recurrence relation between two consecutive states of the stream.

Consider the total recurrence first. The head looks like the left-hand side of a function definition possibly with the index pattern but never with arguments. The variable on the left-hand side occurs in the step expression on the right-hand side; these two occurrences refer to the current and next state of the recurrent relation. The *initial* specifies the initial data for the relation. It can depend either directly or indirectly only on the variables from the optional *arg-tuple*. The domain of the next state of the head variable does *not* have to coincide with the domain of the current state. All that is required is for all stages of the same variable to have the same rank.

For example, this is a stream of natural numbers starting at k:

```
stream (m)
  begin
    n<-k, n+1
  end
(k)
```

The partial recurrence relation is unique to RETRAN and owes its existence to the nonscalar character of potentially all RETRAN objects. It is sensible to allow the programmer to use the natural default value for any element of the next stage of a nonscalar recurrently-defined object: namely, the value of the respective element of the previous stage. The relation can therefore be partial, that is, only some elements of a nonscalar stream object have to be re-defined.

The *geom-expr* is an arbitrary expression that is built using the stream variables and recast, choice and geometrical composition combinators, with the interior of any recast and the condition part of any choice being ordinary expressions dependent on the current stage and the arguments without restrictions. The domains of all stream objects participating in any partial recurrence are fixed and determined by the initial state. Any one stream variable can either be defined by a partial or total recurrence relation. Inside a partial recurrence relation definition one variable can occur in more than one update; the value of repeatedly updated elements is non-deterministic. As regards the domain transformation, the operator := behaves like a normal dyadic element-wise operator, namely it allows anti-currying of the right-hand side (anti-currying of the left-hand side is meaningless since it causes massive non-determinism) and takes the intersection of the respective ranges.

Example:

```
stream(v)
```

```
        init
            a=0;
            b=0;
            v= array (n) \k -> k;
        update
            a ~1~ v[i+3] <- i|n-6] ~1~ b := f(v,a,b);
                                         v[1] := r*(v[1])
        end
(f,n,r)
```

This stream generator updates the concatenation of a scalar a, a central part of vector v, and another scalar b with the array value yielded by function f at every stage of evolution. Independently, element number 1 of the vector is multiplied by some scalar r.

The output tuple may include fewer recurrently defined variables than there are inside the stream environment. That allows the programmer to make what is called projections in the database terminology.

10.3.1 *Filtering*

If the input tuple is supplied then the stream construct becomes a filter. The input tuple specifies the stream variables to be used. These are "zipped" (see Miranda function zip) so that all the variables occurring in the tuple are re-instantiated simultaneously. Every time a new stage is being computed by the stream construct, the variables in the input tuple are instantiated to their next stage, too. The output stream is controlled with the check condition which determines whether or not the output tuple must be re-instantiated. The re-instantiation of the input tuple is done every recurrence step unless the hold condition is satisfied, in which case the input tuple is "frozen" until the hold condition becomes false. Here is a simple example of a stream construct that inputs an element from each of the two input streams and outputs the lesser one first and the greater one second:

```
stream (c) (a,b) hold cond
  begin
    cond <- True, not(cond);
    c = (a<b & cond) | (a>b & not(cond)) ? a ! b
  end
```

Note that the input streams may well be nonscalar in which case the ordering will be done for all pairs of corresponding elments.

This is a filter that accumulates a running sum of elements from its input stream:

sum: stream (b) (a)

INSTANTIATION AND SCHEMAS

```
    begin
      b <- a, b+a
    end
```

RETRAN defines a few standard stream generators that help handle recurrence relations on arrays:

```
select: stream (x') check k=n (x) hold k=n
        begin
          k <- 0, k+1;
          x' = x
        end
(n)

a2s: stream (x',stop)
     begin
       k <- 0, k+1;
       x' = x[k];
       stop = (k==rangeX-1)
     end
(x)

s2a: stream (a) check stop (s,stop) hold stop
     begin
       a <- s; a~1~s
     end
```

The **select** outputs an infinite sequence of objects, each being equal to the nth stage of stream x. The generator **a2s** disassembles an array into a steam of objects along the first dimension, with the variable **end** indicating the array layers have all been selected. The standard function **dom** yields the domain of its single argument. The filter **s2a** restores the object from such a stream and outputs it as an stream of identical objects.

10.4 Instantiation and schemas

A schema is the only program unit of RETRAN. It is a directed acyclic graph (dag) whose vertices are instances of the stream environments supplied by the programmer and the edges are defined by the equivalence relation between the stream variables in the input and output tuples at the vertices. Unbound input variables define the dag inputs and any output variables (besides those not bound to input variables) can be marked as the dag outputs.

Note that a schema is a stream environment in its own right. By collecting the *arg-tuple* arguments, dag inputs, and dug outputs into three tuples and adding a label we create further building blocks which can be instantiated within new schemas. That is the RETRAN concept of encapsulation.

If the schema dag has inputs the schema is called a *transformer*, otherwise it is a *producer*. A producer can be made into a function of its joint *arg-tuple* arguments by selecting certain stages of the dag output and collecting these into a tuple, while gathering all the *arg-tuple* arguments into another tuple. This is done in the schema header. Alternatively, the dag inputs can be associated with some input files and its outputs with some output files or channels. Such a transformer plays the role of a "main program". It executes until any one of the input streams is exhausted and then terminates.

A top-level schema is a persistent entity in RETRAN. Due to applicativity it can be viewed as a relation between files and be manipulated by a relational set of database tools. This is a different story though.

A producer-based function can be integrated into any further stream environment by importing it through an *arg-tuple* — just like any other library function.

We have too little room here to concentrate on the top level of RETRAN; also, it still has a bit too much fluidity as we continue our research. All we are prepared to do now is indicate the ways schemas can be created and give some examples.

10.4.1 Stream instantiation

The bottom line is that we wish to be able to develop schemas graphically. In a graphics-based RETRAN system instantiation of vertices will be requested by selecting an appropriate menu entry and the edges will be drawn by identifying the appropriate inputs/outputs with the mouse. In the text form, a schema begins with a header that defines the binding of the loose ends and the joint *arg-tuple*. Then a list of vertices is supplied, whose elements have the following syntax:

instance	$\rightarrow$	*label* [*a-tuple*] *o-tuple* [<- *i-tuple*]
a-tuple	$\rightarrow$	*var mid* (*var*{,*var*})
o-tuple	$\rightarrow$	*var mid* (*var*{,*var*})
i-tuple	$\rightarrow$	*var mid* (*var*{,*var*})

For example, the following schema (header omitted) defines a producer that yields a matrix m' every row of which is the sum of previous rows of a given matrix m:

```
a2s(m)  (stream1, stop)
sum     stream2 <- stream1
s2a     M' <- (stream2, stop)
```

The dag output m' contains identical matrices representing the result. As an aside, this schema will be equally good for any dimensionality; matrices are for illustration only.

10.5 Conclusions

What RETRAN does can be summarized as follows: systematically introduce data-parallelism throughout a Haskell-like language without lists as data-structures; and provide a much more flexible (but again, acyclic, that is, non-recursive) stream environment. The gains are:

1. arrays as parallel data objects;
2. no array copying problem;
3. no garbage collection – inferable life time;
4. no stream deadlock problem;
5. array functions polymorphism with inferable array ranks;
6. inferable process communication structure for distributed; implementations
7. minimum distance from both imperative programming and math specification.

By contrast with the Bird-Meertens school of thought, RETRAN is more extensively data-parallel, since it

1. does not connect arrays with lists in any way;
2. automatically overloads all operators, including user-defined ones, so that they can be applied to arrays;
3. does not require any two arrays to conform, but makes the rank a component of array type;
4. suggests a stream-based solution for array updates in terms of recurrence relations thus avoiding arbitrary recursion; it also supports partial updates (nonscalar l-expressions) which are unique to RETRAN;
5. interprets data parallelism as a function application to independently defined data elements, rather than as a synonym of SIMD; it therefore allows the user to construct arbitrary DP operators.

REFERENCES

J Backus. Can programming be liberated from the von neumann style? a functional style and its algebra of programs. *Communications of the ACM*, **21**(8), 613–641, 1978.

R S Bird. An introduction to the theory of lists. In Manfred Broy, editor, *Logic of Programming and Calculi of Discrete Design (NATO ASI Series F)*, pp 3–42. Springer-Verlag, 1987.

R S Bird. Lectures on constructive functional programming. In Manfred Broy, editor, *Constructive Methods in Computer Science (NATO ASI Series F)*, pp 151–218. Springer-Verlag. 1988.

R S Bird. Algebraic identities for program calculation. *The Computer Journal*, **32**(2), 122–126. 1989.

M I Cole. *Algorithmic Skeletons: Structured Management of Parallel Computation*. Pitman. 1989.

G Hains and L M R Mullin. Parallel functional programming with arrays. *The Computer Journal*, 36(3):238–245. 1993.

P Hudak and P Wadler (eds.). Report on the functional programming language Haskell. Technical Report YALE U./DCS/RR656, Department of Computer Science, Yale University. 1988.

Lawrence Livermore National Laboratory. *SISAL Reference Manual. Language version 2.0*. 1992.

V B Muchnick, A V Shafarenko, and C D Sutton. Data-parallel portable software platform: Principles and implementation. In A Bode, M Reeve, and G Wolf, editors, *Lecture Notes in Computer Science No 694*, pp 161–172. Springer-Verlag. 1993.

C M Pancake. What should we expect from parallel language standards? *The International Journal of Computer Applications*, **6**(1), 112–117. 1992.

S L Peyton Jones. U. K. research in functional programming. *SERC Bulletin*, **4**(11). 1992.

F A Rabhi. A parallel structure for static iterative transformation algorithms. In L Bougé, M Cosnard, Y Robert, and D Trystram, editors, *Parallel Processing: CONPAR 92 — VAPP V*, pp 509–514. Springer Verlag. 1992.

D B Skillicorn. Architecture-independent parallel computation. *IEEE Computer*, **12**(23), 38–51. 1990.

11

WRITING PORTABLE PARALLEL PROGRAMS WITH MATCH AND MOVE

Thomas J. Sheffler

Carnegie Mellon University
Pittsburgh, PA 15213, USA

Abstract

This paper proposes a model of parallel computing based on two operators called **match** and **move**. The **match** operator is used to describe channels of communication between processing sites, while the **move** operator sends data over the channels described by **match**. These two operators subsume the functionality of communication methods of many different abstract parallel models, and generalize many of the ways in which data are rearranged in parallel programs.

Rather than proposing a new theoretical model, this paper demonstrates a *practical* programming model. Programs written in this model have easily derived complexity measures based on the metrics of the EREW Parallel Vector model. However, because the operators may make use of special architectural features available on a given machine, they can guarantee high performance as well. This final claim is supported by discussing data collected from implementations on the CRAY Y-MP and Connection Machine CM-2. Programs implemented using **match** and **move** are competitive with more traditional methods of implementation, while remaining high-level and architecture-independent.

11.1 Introduction

Writing parallel programs that are portable is difficult because parallel computers vary greatly in their capabilities and features. While all parallel computers share the characteristics that they comprise some numbers of processors and an interconnection network for communication, the variants on just these basic features are enormous. One approach to writing a portable parallel program is to use only those features common to many machines. However, this approach has the drawback that only the least common capabilities of each machine will be used and that special features available on a particular machine will go to waste. For example, some machines permit multiple processors to read from the same memory location, but because many machines do not, a truly portable program could not be written using such a feature.

Typically, programmers have been willing to sacrifice portability for high performance. Their programs have been written at a very low level, making use of specialized features of a target machine. In addition to making a program inherently non-portable, this approach to programming is laborious and error prone. Having to consider low-level details along with high-level issues, a programmer can quickly reduce an elegant algorithm into an ugly and confusing program that is difficult to debug.

Program portability is enhanced by raising the level of abstraction at which a programmer views a parallel machine. However, while heightened abstraction simplifies programming, it may also adversely affect performance. Added software layers may incur execution overheads that are simply not acceptable in a supercomputing application. Parallel machines are designed and purchased to run applications at very high speed, and even a beautiful abstract software layer will not find acceptance if it sacrifices high performance.

This paper proposes a way to write portable parallel programs using two operators called **match** and **move**, as proposed by Gary Sabot in his Paralation model of computing (Sabot 1988). Because the main differences between parallel machines are in the design of interconnection networks and processor configurations, the main hindrance to making parallel programs portable is describing data movement operations in a very general manner. Once data is moved to the place where it is needed for computation, most parallel machines provide similar assortments of arithmetic and logical operations. **match** and **move** provide the means to describe data movement in an architecture-independent manner.

The **match** operator is used to describe channels of communication between processing sites, and **move** is used to send data through the paths described by **match**. **match** and **move** clearly distinguish between the description of a communication pattern with **match** and data movement through **move**. Previous approaches have mixed together these ideas, with the result that performance has suffered. For example, the β operation of CM-LISP provided a complex combination of communication and computation, but because its running time was unpredictable it was unusable for real problems (Steele and Hillis).

In addition to helping a programmer to structure programs in terms of the distinct areas of communication patterns and data movement, this division of function assists in achieving high performance. Information about communication patterns and data movement gleaned from instances of **match** and **move** lead to important compile-time and run-time optimizations that guarantee both theoretical efficiency and high performance. Because programs written using **match** and **move** do not dictate policies for implementing low-level operations, the implementation of the operators may use special features present on a particular machine when appropriate. This approach is proven viable through aggressive implementations of the operators on both a massively parallel machine and a vector supercomputer where the test programs achieved *or exceeded* the performance obtained by programs written with conventional means. This result is especially notable considering the higher level of abstraction achieved by programs written with **match** and **move**.

11.2 Parallel machine models

Currently, most parallel algorithms are designed for the theoretical PRAM (Parallel Random Access Machine) models. The PRAM models make explicit the number of processors and allow every processor to perform an arithmetic operation, or to read or write to a global memory in a single time step. This family of models is differentiated by the manner in which memory access collisions are handled. The basic EREW (Exclusive-Read, Exclusive-Write) PRAM model insists that no two processors access the same memory location at the same time, while CR (Concurrent-Read) and CW (Concurrent-Write) variants allow multiple processors to read from or write to the same location. Further variants differentiate the way in which concurrent write operations to the same memory location are resolved. The CRCW-ARB model stores an arbitrary one of the values written to the same location while the CRCW-PLUS model combines values. While the theoretical PRAM models capture some essential properties of parallel machines, they do not provide intellectual leverage to ease the task of writing portable parallel programs. Furthermore, these theoretical models do not reflect the performance of many real machines.

Because programs written for these models assume a specific number of processors and a memory access collision policy, they are generally nonportable. Even though theoretical simulation methods exist that allow CRCW algorithms to run on an EREW machine for instance, these results are of little practical use to programmers. The asymptotic measures of the complexities of these simulations hide important constant factors that must be considered for real applications.

The Parallel Vector models proposed by Guy Blelloch are another family of parallel models (Blelloch 1990). Instead of an explicit number of processors, operations on varying sized vectors have their work evenly distributed over whatever processors are available. In essence, each element of a vector is assigned its own 'virtual processor.' In contrast to the PRAM family, these models are also synchronous, where parallelism arises from operations on entire vectors that may be performed in parallel. This is a style of data-parallel programming where programs have a single thread of control and where parallelism is exploited by operations across large sets of data. The advantage of the data-parallel style over others is that programs naturally scale to larger data sets.

Programs written in this manner are less tightly coupled to a particular machine configuration or topology. However, because the Parallel Vector models also have EREW and CRCW variants, programs written for these models are tied to a particular memory collision policy. Thus, programs written for these models either make use of advanced features of a particular machine and become non-portable, or use only the most general capabilities and possibly sacrifice performance.

Rather than proposing a new theoretical model, this paper demonstrates a *practical* parallel programming model that simplifies the programming problem and ensures architecture independence. It is built on top of the Parallel Vector model using the notions of vectors and virtual processors. In contrast with the Parallel Vector models, this practical approach combines all of the memory access methods under one general mechanism, namely the **match** and **move** operators as proposed in (Sabot 1988). Algorithmic complexity is measured using the metrics of the EREW Parallel Vector model, but actual implementations of **match** and **move** may transparently make use of advanced architectural features on a given machine, guaranteeing high performance for real applications.

11.3 match and move

In the PRAM and Parallel Vector models, when data is sent from one site to another the index of a processor or virtual processor is used, respectively, as a destination address for a communication operator. The **match** operator generalizes the description of an address to the simple notion of a key. In this manner, data are identified in a way that is independent of the number of processors or their interconnections. Keys are further generalized so that they may be composite values, or tuples. Two vectors called the 'source' and 'destination' define a communication pattern. For each element in the source, a path is indicated to each destination element having an equal key. The result returned by the **match** operation, called a 'mapping,' is used with a **move** operation to send data through these paths.

The **match** operation of Fig. 11.1 shows two vectors used to create a mapping. Arrows represent the communication paths of the mapping that are set up between equal keys. Because duplicate keys may exist in both the source and destination vectors, mappings may describe complex patterns. When multiple source sites map to the same destination, variants of **move**, such as **move-add**, **move-max**, and **move-min**, specify combining functions to be applied. Another special combining function, called **move-arb**, selects an arbitrary source value of those sent to the same site. In Fig. 11.1 the mapping is applied to a vector 'sourcedata' with the **move-add** operation. The result vector contains the sums of values from source sites sharing the same key.

Data values mapped to the same destination site may also be gathered into a new vector with the **move-join** operation. With this operation, each

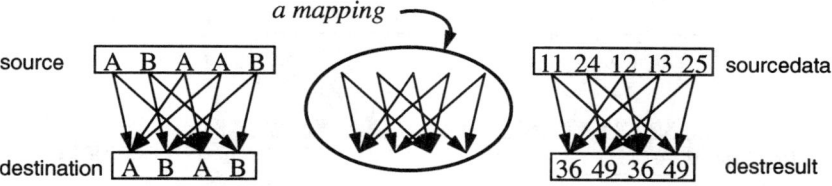

M = **match**(destination, source);
destresult = **move-add**(M, sourcedata);

FIG. 11.1. The **match** operation. The **match** operation constructs a mapping from a source vector to a destination vector. A **move** operation can use this mapping to send data from a source vector to a destination vector with a combining function.

destination site is associated with a contiguous range of sites in a vector containing copies of each of the data values from the source sites in the mapping. An example **move-join** operation is shown in Fig. 11.2. For simplicity, the **move-join** operation is defined to return a vector of tuples. A tuple is an ordered set of values whose member fields are identified by position. These fields contain copies of data from vectors **sourcedata** and **destdata** given as arguments to **move-join**.

The **move-join** operation takes its name from the join operation of the relational data base model (Codd 1970). Another way of viewing **move-join** is as an operation that constructs all of the cross-products between the source and destination data vectors for the groups defined by the keys. This also has the effect of 'flattening' the information encapsulated in a mapping, expressing each path of the mapping explicitly with a tuple in the result vector. The **move-join** operator is a powerful function that is often used to generate new vectors from correspondences between other vectors. Some of its different uses are shown in this paper later.

The family of **match** and **move** operators generalize many of the ways in which data are rearranged in parallel programs. **match** extends the notion of an address from that of a constant site identifier to the more general mechanism of variable keys. Also, in addition to the one-to-many and many-to-one patterns typical for read and write operations, keys allow many-to-many patterns to be described as well. The **move** operators generalize the way in which data values are rearranged in parallel programs to include permutations, combinations and cross-products. This paper illustrates the power of these operators through a number of example algorithms.

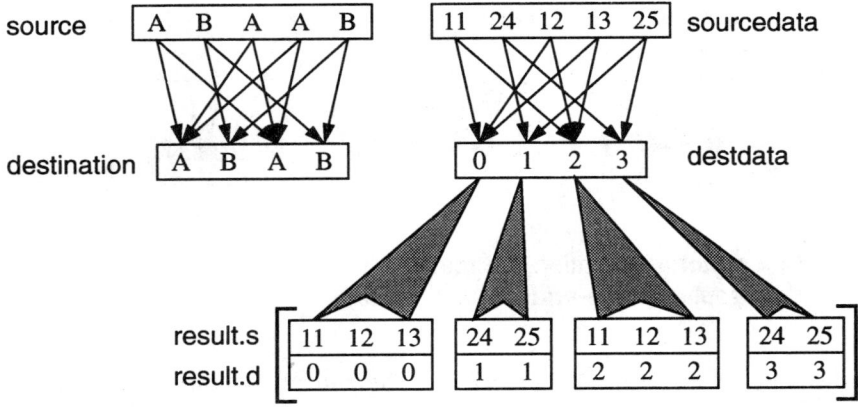

M = **match**(destination, source);
result.<s,d> = **move-join**(M, sourcedata, destdata);

FIG. 11.2. The **move-join** operation. The **move-join** operation provides a combining function that arranges the values colliding at a destination site into adjacent sites in a new vector. The two fields of the result vector relate each site to a pair of sites from the source and destination vectors.

11.3.1 Complexity measures

The complexity of algorithms written using **match** and **move** is expressed using the *step* and *work* complexity measures of the Parallel Scan-Vector model (Blelloch 1990). The *step* complexity measures the total number of vector instructions issued. Another complexity measure, the *work* complexity, measures the total number of vector elements operated on over all instructions.

All of the **move** operations are implemented in a constant number of primitive vector instructions, leading to an $S = O(1)$ step complexity measure for each of them. For a regular combining **move**, the work complexity is $W = O(s+d)$ where s and d are the lengths of the source and destination vector sets. The work complexity of a **move-join** is $W = O(s + d + j)$ where j is the number of sites created in the new vector as the result of the **move-join**.

In general, the match operator is implemented with sorting, and its step and work complexity measures are $S = O(\lg n)$ and $W = O(n \lg n)$ where $n = s + d$, the sum of the lengths of the source and destination. In many cases, these measures over-estimate the complexity of creating a mapping. A large number of optimizations for special instances may be applied to reduce both the algorithmic complexity of creating a mapping, as well as some of the constant factors involved (Sheffler 1992). For example, in many **match** operations involving an index vector, only a simple concurrent-read or concurrent-write is indicated. If the underlying architecture supports these operations, they may be used directly. Another typical optimization employed recognizes when the mapping created is one-to-one, indicating exclusive-read or exclusive-write patterns. These optimizations may be employed in a systematic manner, assuring the programmer using **match** and **move** that programs will not suffer a loss of efficiency.

11.4 A comparison of programming methods

In this approach, a data structure is represented by a collection of tuples. The collection is stored as a vector, and values belonging to the same tuple are located at the same processing 'site.' For example, a simple directed graph is represented as collection of vertices and edges that each have a **tail** and **head** field. While the example shows vertices labeled from 0, general keys could be used as well.

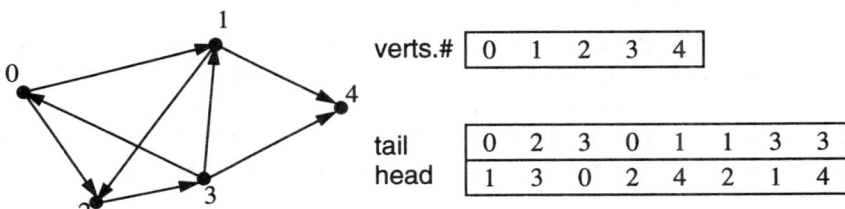

Mappings are used to express relationships between the elements of the data structure. For instance, in a typical parallel connected components algorithm, an important step requires each vertex to find the minimum numbered vertex of its neighbours. Using **match** and **move**, this operation is implemented in two steps.

M = **match**(verts.#, edges.head);
v.minneighbor = **move-min**(M, edges.tail);

The **match** statement creates a mapping relating each edge to the vertices for which it is a head. This mapping is many-to-one, since vertices may have multiple incoming edges. The **move** statement uses the mapping, **M**, to send the **tail** value of each edge to those vertices for which it is an incoming edge. When multiple values are sent to the same vertex, the minimum value is chosen.

There are many ways to implement this operation in the parallel models mentioned earlier. With the combining write of the CRCW-PLUS PRAM model, a single step suffices to find the minimum tail for each group of edges with the same head. If the combining write is not supported, sorting might be used to find the minimum of each group of edges instead. Both of these approaches dictate a detailed implementation strategy, however, rendering the resulting program non-portable. **match** and **move**, on the other hand, describe the desired result of the computation in a more general manner and allow the implementation to choose an efficient method of execution.

To make this point more concrete, this same algorithm is examined as it would be expressed in some other models. In the strict EREW PRAM model, a typical approach would use sorting to find the minimum valued **tail** for each group of edges with the same **head** key. The following sequence of steps describes such an algorithm.

1. Assign each vertex and edge to a separate processor.
2. Sort the edges by **head** value, and then by **tail** as a secondary key.
3. In the sorted order, divide the edges into groups wherever a new **head** value begins.
4. Because of the sorted order, the first member of each group is the minimum **tail** for a head. Send these values to the processor for the **head** specified.

The CRCW-PLUS variant of the PRAM models supports combining write operations. The PLUS operation is generalized to a number of binary associative operators including **min**. Writing the algorithm in this model is very simple, but its implementation is very non-portable because few machines directly support such an operation. The following two steps would suffice.

1. Assign each vertex and edge to a separate processor.
2. In parallel, write the **tail** value to the processor identified by the **head** value using a **min** policy to resolve write collisions.

Finally, in the Parallel Vector model using segmented scans, a different approach might be used. A scan operation computes for each result element, the sum of all preceding elements with respect to a binary associative operator. Segmented vectors are simple vectors that are divided into contiguous sections called 'segments.' A segmented scan is a variant on a simple scan that begins a new running sum at each segment boundary. Thus, a segmented **min-scan** can be used to find the minimum value of each group of edges when the edges have been arranged into segments. The following steps describe an algorithm using segmented scans.

1. Sort the edges by **head** value.
2. Divide the edges into segments so that a new segment begins where the **head** value of an edge differs from the previous one.
3. Use a segmented **min-scan** function to find the minimum of each group of edges with the same head.

The theoretical models of the examples above are useful for complexity analysis, but do little to aid the programmer. For this simple example it is not too difficult to wade through the pseudocode, but for more complicated algorithms it becomes increasingly difficult. Furthermore, these algorithms are very processor-specific and require a detailed understanding of the capabilities offered by a target machine. While a combining **send-min** may be simulated on an EREW PRAM with $t = O(\lg n)$ slowdown, this fact does not provide much information about the real performance of such an algorithm to the programmer. Requiring a programmer to rethink these basic simulations at every turn increases the programming burden and jeopardizes the program's reliability.

11.5 Example algorithms

As described earlier, algorithms described using **match** and **move** are based on simple collections of data objects that use **match** to describe relationships between those objects. A flexible framework for defining data structures is provided through the use of keys that permits an associative model of computing similar to that proposed in (Potter 1988). Keys can provide the functionality of pointers, or can serve to label objects that belong to the same row, column, or other type of class. This section shows how this very general mechanism can be used to implement some important fundamental data types. The expressive power and versatility of the **match** primitives allow most of these algorithms to be described in just a few lines of pseudocode.

11.5.1 *Library functions*

While **match** is the only operator that creates a mapping, it is easy to build functions that compute other common relations. One example is the standard library function **collapse**. This function takes a single vector of keys and produces a mapping to a new vector that has one site for each unique key. An illustration of a mapping produced by **collapse** appears in Fig. 11.3. While **collapse** may be implemented using a constant number of **match** and **move** operations, because its use is so common it is often part of any standard library supporting **match** and **move**.

The **collapse** function is used in many different ways. It forms the basis of a histogramming operation. By applying a vector of the constant **1** to a mapping produced with **collapse**, a histogram of the keys is computed. Duplicate keys may be removed from a set by applying such a mapping to a simple **move-arb** operation. Also, many sparse vector and matrix algorithms can be expressed by combining values with equal keys.

The **collapse** operator is an example of a function producing a mapping that can be used in many different circumstances. The development of libraries of functions that compute familiar relations and produce mappings encourages code reuse. Rather than developing similar algorithms that

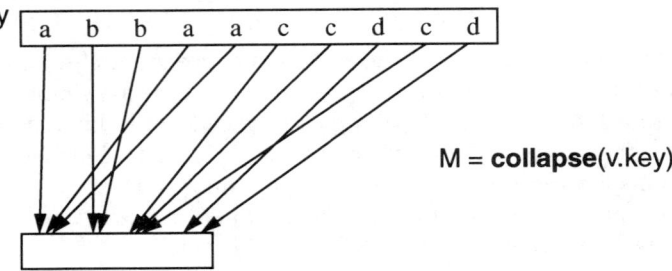

FIG. 11.3. **collapse** gathers equal keys together. This function is generally used to remove duplicate keys or as a precursor to a histogramming operation using a combining **move**.

```
union(s1.key, s2.key) {
    temp.key = append(s1.key, s2.key);                append
    M = collapse(temp.key);                           remove duplicates
    union = move-arb(M, temp.key);
}

intersection(s1.key, s2.key) {
    M = match(s2.key, s1.key);                        pairs match equal elements
    produce a new site for each pair
    temp.<s1,s2> = move-join(M, s2.key, s1.key);
    intersection = temp.s1;
}
```

FIG. 11.4. Basic set functions. The elements of the sets are also the keys used in all of the **match** operations.

exploit common relationships, functions that produce common mappings can be used in many different ways with the **move** operators.

11.5.2 Set operations

A set of n elements is represented by a simple vector of n values, in which there may be no duplicates. In this section, no assumptions are made about the type of the values but there must be an equality function for comparing elements of the two sets. The basic set operations **union** and **intersection** are easily described in just a few lines of code as shown in Fig. 11.4.

The union of two sets is made by first appending the vectors of the two sets and then removing any duplicates that occur. By using **collapse** and **move-arb**, duplicate entries are removed from the appended vector.

The intersection of two sets is most easily implemented by taking advantage of the special properties of the **move-join** operator. Because each

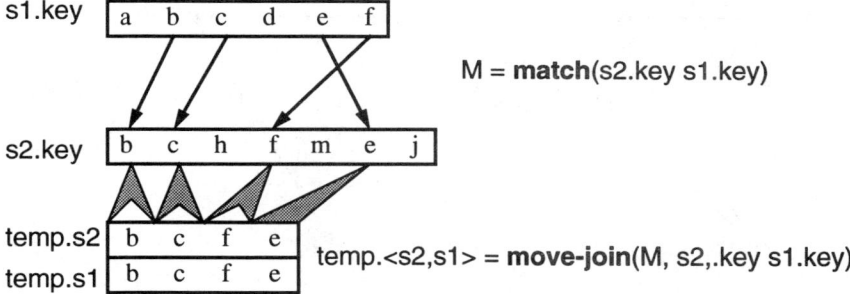

FIG. 11.5. Set intersection: the **move-join** operation ensures that new sites are created only where set elements match.

of the argument sets may have no duplicates, the mapping created by matching from one set to the other is one-to-one, with some elements not mapping to anything. When **move-join** is applied to this type of mapping, exactly one new site is created for each destination site that matches a source site. Figure 11.5 illustrates the simplicity with which **move-join** allows this operation to be implemented.

While the implementation of these operations is nearly trivial, they are important kernels for more complex algorithms. Often, the set elements are the keys of objects and the desired operation is to make a set of objects whose keys have some property. By using **match** on the keys, the mappings created can be used to move the other fields of the objects to their new sites.

11.5.3 Sparse matrix algorithms

A sparse matrix is described by a collection of its non-zero entries. Each is identified by column and row keys and has an associated value. Even though the storage of the row and column keys require more memory per element than a typical dense arrangement, this scheme is more efficient for many sparse matrices of interest. Using this representation, mappings can be created that distribute values along rows or columns, or that sum (reduce) elements along either of these dimensions simply by applying **match** on the appropriate keys. Sparse matrix algorithm construction using **match** is simple and elegant.

11.5.3.1 *Sparse matrix-vector multiplication* The multiplication of a dense vector by a sparse matrix, $x = Ab$, has an appealing graphical representation. The process is illustrated in Fig. 11.6. Each vertex, v_i, is given an

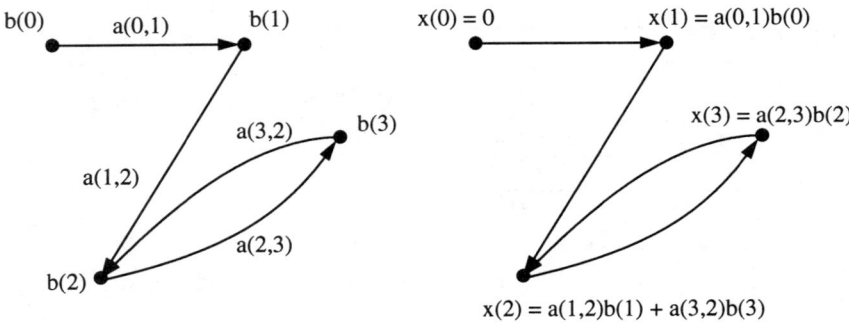

FIG. 11.6. The graphical interpretation of $x = Ab$. The value computed for each vertex is the sum of each of its neighbours being scaled by the edge that connects them.

associated value $b(i)$. At the beginning of the algorithm, two mappings are created. The mapping called **Mtail** relates each vertex to those edges for which it is a tail, and **Min** relates each edge to that vertex for which it is an in-going edge. The matrix-vector product is computed in two phases. First, the value of each vertex is sent to each of its outgoing edges, where it is multiplied by the value of the edge. The result vector, x, is computed by summing the products on the incoming edges of each vertex, The code for function **mvmult** is given in Fig. 11.7.

The more traditional view of matrix-vector multiplication distributes the elements of the vector along the columns of the matrix where the products are computed. Then, the non-zero values along the rows are summed. The two phases of this algorithm correspond directly to this process. Because columns in the matrix correspond to tail vertices in the graph, the first mapping distributes the vector elements over the columns. Similarly, the second mapping gathers elements on the same row so that they may be summed using **move-add**.

11.5.3.2 *Sparse matrix-matrix multiplication* Given two matrices A and B, the matrix product, $C = A \times B$, is defined as follows. An element of matrix A in row i and column j is denoted $A_{i,j}$. Each element of the matrix product C is computed by the summation

$$C_{i,j} = \sum_{1 \leq k \leq N} A_{i,k} B_{k,j}.$$

When A and B are sparse, it is advantageous to avoid storing elements that are zero and computing products where one of the operands is zero. Using

```
mvmult(m.<col,row,val>, v.val) {              matrix-vector multiply
    Mtail = match(m.col, v.#);
    Min = match(v.#, m.row);
    m.copy = move-arb(Mtail, v.val);          distribute values to out-edges
    m.prod = m.copy × m.val;                  calculate all products
    mvmult = move-add(Min, m.prod);           sum along in-edges (rows)
}

mmmult(A.<col,row,val>, B.<col,row,val>) {    matrix-matrix multiply
    PHASE1:
    M1 = match(B.row, A.col);
    pp.<col, Bval, row, Aval>=move-join(M1, B.<col,val>, A.<row,val>);
    pp.val = pp.Bval × pp.Aval;
    PHASE2:
    M2 = collapse(pp.<col,row>);
    new.<col,row> = move-arb(M2, pp.<col,row>);
    new.val = move-add(M2, pp.val);
    mmmult = new.<col,row,val>;
}
```

FIG. 11.7. Basic sparse matrix algorithms.

match, the algorithm consists of two parts. The first phase computes all of the partial products that contribute to the sums, along with their indices in the result matrix, C. The second phase gathers partial products with the same indices in the result matrix and adds them together.

In examining the formula above, each product is of the form $A_{i,k}B_{k,j}$, indicating a 'match' of the columns of matrix A with the rows of B. The partial products created between a column of A that matches a row of B are all of those that result when the elements from a column k of A are distributed over the elements of row k of B. A simple application of **move-join** to the mapping creates sites for all of these partial products. The implementation of **mmmult** is shown in Fig. 11.7.

11.6 Implementation strategies

The **match** operation provides a very general mechanism for specifying communications patterns through source and destination keys. Because paths are identified through key equality, a mapping constructed with **match** associates each site in the source and destination with one and only one *group*. The first step in the implementation of **match** is to discover the unique groups in the source and destination key vectors.

To support the **move** operations, the mapping data structure associates each group with a combining primitive of the underlying hardware. On a machine that supports segmented-scan operations, this may be a segment in a vector. For a machine with a combining write, a single address in memory would suffice. The **match** operator is allowed a great deal of flexibility in how it performs this assignment so that it may employ the most appropriate method for each machine.

The two main methods for creating mappings use either sorting or hashing as a first step to assign keys to groups appropriate for the combining primitives of the underlying hardware. The choice of a method may be based on characteristics of the key vectors, or by the context in which the mapping is created or used with a **move** operation.

11.6.1 *Matching by sorting*

The idea behind constructing a mapping through sorting is that equal keys should be brought near each other by a sorting operation. This type of mapping associates the groups specified by keys with a contiguous segment in a vector. The combining function appropriate for this type of group is the segmented **scan** of the Parallel Vector Models (Blelloch 1990).

The steps involved in building a mapping are illustrated in Fig. 11.8. First, the keys of the source and destination vector are concatenated into one vector called the 'agent' along with pointers back to their sites of origin. These site addresses are called the 'home' pointers for the keys. After sorting the agent elements by their key value, the agents are divided into segments so that each group of equal keys is in one segment. The identification of segments is done in one parallel step by comparing each key to its left neighbour in the sorted agent vector. Finally, the indices of the new sites in the agent vector are sent back to the home site of each key. The agent vector along with the pointers from keys to agent sites becomes the data structure of the mapping.

When creating a mapping with this technique, the complexity measures are dominated by the sort operation. For source and destination vectors of length s and d, respectively, the total number of elements involved is $n = s + d$. Thus, the complexity measures of this type of **match** are $S = O(\lg n)$ and $W = O(n \lg n)$.

A **move** operation using this type of mapping is illustrated in Fig. 11.9. Three steps complete the **move**. First, the source values are sent to the agent. Next, a single **scan** combines the values of the source agents and distributes the sum across the destination elements. The last step sends the sums in the destination agents to their final sites

The operations involved in each of the three steps are unit step ($S = O(1)$) operations in the Parallel Vector model. If the sum of the lengths of the source and destination vectors $n = s + d$, then the work complexity of a combining **move** is $W = O(n)$.

IMPLEMENTATION STRATEGIES

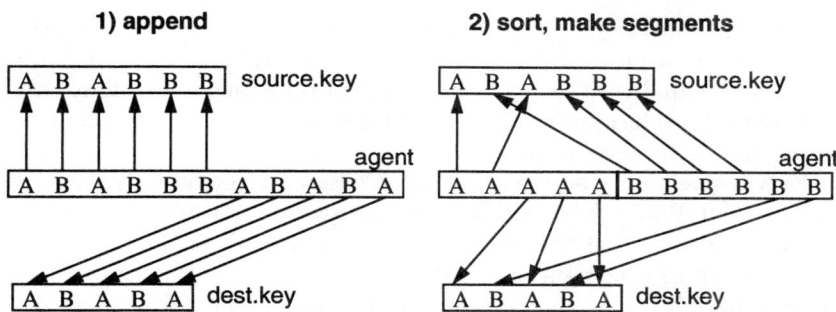

FIG. 11.8. A mapping may be created by using sorting. In the first step, agents are appended from the source and destination vectors into a new vector. After sorting the agents, they are divided into groups by comparing each key to its neighbours.

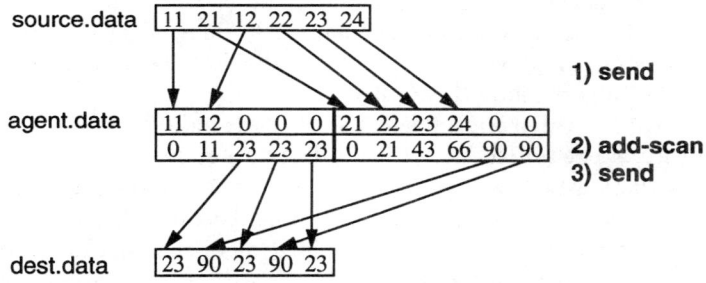

FIG. 11.9. Using a mapping created by sorting to perform a **move-add**. The first step sends the source values to the agent vector, where they are combined with a segmented **add-scan**. The final **send** forwards the results to their destinations.

11.6.1.1 *Implementing* **move-join**   The **move-join** operation is more complicated than the other **move** operations, but uses the same agent vector to perform its work. After moving source and destination values into the agent vector, a full cross-product operation is performed for each segment. This operation may be accomplished in $S = O(1)$ steps in the parallel vector model through a small amount of address arithmetic.

All operations involved in implementing **move-join** are unit step complexity operations in the Parallel Vector model. However, the work complexity of this operation is determined by the number of join sites created, which is highly data dependent. For a **move-join** operation that produces j sites in the result vector, its work complexity is $W = O(j)$.

11.6.2 Matching by Hashing

An alternative method of creating a mapping uses a hashing algorithm to associate each group with a single site in an agent vector. A **move** operation executed with this type of mapping uses a single combining write to combine the values for each source, and a single **get** to copy all of the sums to the destination sites.

We will give only a brief outline of the hashing algorithm here because of space limitations. The parallel hashing algorithm begins by creating a work vector for all of the keys that includes 'home' pointers as before. In parallel, all keys compute the address of an agent site by using a hash function. Initially all agent sites are 'empty', but become 'filled' when they are claimed by a key. When keys find empty agent sites, they are marked 'done' and the agent sites they have probed are marked filled. Other keys continue their probe steps until they find an empty agent site.

After each iteration, there are many sites in the work vector that are marked done, and hence participate in no later iterations. A packing step compresses the remaining active keys into a smaller vector. This packing step has the effect of reducing the amount of work performed on each succeeding iteration, allowing the parallel hashing algorithm to have an expected work complexity of $W = O(n)$ (Sheffler 1992). The parallel step count of the algorithm is $S = O(\lg n)$.

Once the keys are assigned agent sites, these indices are sent back to the 'home' site of each of the source and destination vectors. This completes the construction of this type of mapping. It is interesting to note that because the work complexity for the parallel hashing algorithm is smaller than that of parallel sorting, the hashing algorithm performs less work (fewer total operations) than the sorting algorithm for creating a mapping.

11.6.2.1 Implementing move-join

The implementation of the **move-join** operation for this type of mapping is complicated, and requires the use of a **multiprefix** operator. The **multiprefix** operator is a synchronous version of the fetch-and-add instruction. Many machines do not support the **multiprefix** operation directly since most methods of implementation require special hardware (Ranade *et al* 1988). However, efficient algorithms do exist for the hypercube (Cohn 1988) and for vector processors such as the CRAY Y-MP (Sheffler 1993).

11.6.3 Comparison of sorting and hashing Methods

The two different methods of implementation of **match** are each appropriate in different circumstances, most of which are governed by the context in which they are used in a program. In some algorithms using **match**, such as the **mmmult** algorithm, the keys applied to **match** are pairs of integers. In other cases, they may be even longer bitstrings. Many implementations of parallel sorting are based on a radix sort. As the size of the keys increases, the speed of radix sorting decreases with each added bit. Conversely, the hashing algorithm is immune to increases in key length. For very long keys, the hashing algorithm can perform significantly better than sorting.

The downfall of the hash-based mapping representation is that it requires combining-write operations to implement combining **move** operations. However, to implement the **move-arb** operation, a concurrent write that selects an arbitrary value is sufficient. There are a number of interesting algorithms that only require **move-arb**.

By implementing an algorithm in terms of **match**, a programmer can avoid these considerations until late in the program development cycle. After verifying an application correct, the choice of a particular implementation strategy for different instances of **match** and **move** could be the job of a compiler, or as simple as supplying optional 'hints' that the **match** implementation can use at run time to make decisions about which of the available strategies to use. The latter approach reduces fine-tuning the performance of an application for a particular machine to a very simple process.

11.7 Performance trials

The full suite of **match** and **move** operators has been implemented on both the CRAY Y-MP and the Connection Machine CM-2 using both sorting and hashing-based techniques. On both machines, sparse matrix test programs **mvmult** and **mmmult** implemented with **match** and **move** achieve comparable performance to machine-specific algorithms. The main points observed are that **match** and **move** allow the algorithms to naturally be divided into symbolic *setup* times, and a numerical *evaluation* time. The setup time is that time spent building mappings with **match**, while the evaluation time is that time spent in other computation. In many numerical applications, the same matrix structure is used repeatedly, allowing a high setup time to be amortized over many instances of evaluation. On both the Y-MP and CM-2, algorithms designed to exploit the reuse of setup information are usually hand-coded and highly machine specific. The algorithms written with **match** and **move** are easily adapted to reuse mappings.

Because the implementations on each machine supported both sorting and hashing for implementing **match**, interesting comparisons could be made between these methods. In general, the sorting-based methods traded an increased setup time for a decreased evaluation, while the hashing-based methods spent less time in setup and more time in evaluation. With the algorithms written using **match** and **move** it is easy to explore these tradeoffs, while implementing the sparse-matrix algorithms directly to use either sorting or hashing would be much more difficult. The test data also confirmed that parallel hashing is effective for implementing these sparse matrix kernels, even though hashing is rarely mentioned in numerical supercomputing literature.

Because of space limitations, a full discussion of the performance of each implementation is not possible. The next section discusses the performance achieved by a sparse matrix-vector multiplication routine on the CRAY Y-MP.

11.7.1 Sparse Matrix Vector Multiplication on the Y-MP

Multiplication of a dense vector by a sparse matrix is at the core of many numerical algorithms. This operation appears when solving systems of linear equations by iterative methods such as the Conjugate Gradient method, and in finite element analysis. Many storage schemes have been developed to allow the numerical portion of this algorithm to proceed at near peak speed on vector computers. The Compressed Sparse Row (CSR) storage format is most typically used and arranges the matrix into rows, with the column index of each element stored in a separate vector. This format is very simple and allows the matrix-vector multiply operation to vectorize completely over each row. However, for very sparse matrices, the row lengths can become quite short. Another storage format, called the Jagged Diagonal (JD) format, has been developed for nearly uniform sparse matrices and attempts to overcome the downfalls of the CSR format. It trades an increased preprocessing time for a fast evaluation time. The JD algorithm, however, is highly machine specific.

The times reported in Table 11.1 are broken down into setup, evaluation, and total times. Times measured for the comparisons included only the matrix operation, format conversions were not timed. The order columns gives the number of rows of each matrix and the fraction of non-zero entries is expressed by its density, ρ. The CSR format approach has no setup time associated with it, and performs all of its work during evaluation. The setup time of the JD method is dominated by the sorting of the row lengths and the reordering of the elements into jagged diagonals. In the **match** and **move** approach, the setup time is precisely the time spent building the mappings using **match**. With the sort-based approach, the dominating cost is the sorting step. This is incurred at a fairly substantial cost in time, but allows a fast evaluation step. Because the mappings created involve an index vector, however, the hash-based approach actually performs no iterations.

The results reported show that for very large sparse matrices the JD approach trades a large setup time for a quick evaluation. The hash-based approach performs less of its total work during setup, while the CSR approach suffers from very short row lengths for extremely sparse matrices. Because of the speed of the evaluation phase of the JD approach, its use would be preferable in an application that requires repeated multiplication of the same matrix, while the hash-based approach would be better suited to cases where only one multiplication is performed.

While the JD method has the lowest evaluation time, it also has problems with non-uniform matrices. A different series of trials was timed using matrices from electrical circuit simulation problems distributed with the SPARSE sparse matrix package (Kundert and Sangiovanni-Vincentelli). These matrices are very sparse, with an average of only 7 or 8 elements per row, but have a few very long rows. These rows represent power and ground and are almost completely populated. The results are presented in Table 11.2. In these cases the hash-based approach clearly outperformed both the CSR and JD approaches. This is a known weakness of the JD format, and it has been suggested that such long rows should be handled as a special case (Anderson and Saad 1989). In general, the performance of the hash-based approach is more consistent over matrices of varying structure.

11.8 Conclusion

Programming models shape the way people think about algorithms. Traditional serial programs are built from classical data structures and their algorithms. The basic tools at the disposal of every programmer include methods for sorting, selection, traversing linked-lists and binary trees, and so on. Using these tools, most programmers begin algorithm design by decomposing the problem at hand into constituent pieces that are more easily handled.

Aggregate data movement operations are as fundamental to the description of parallel programs as basic data structures and their algorithms are to serial programs. As conventional data structures provide a framework within which programmers may organize their thoughts, data movement operations shape the way programmers think about parallel algorithms. As suggested by Sabot, the **match** and **move** operations provide a clear distinction between the description of a communication pattern through **match**, and the movement of data through **move**. By distinguishing these two as separate concerns, programmers are guided to address each separately, and avoid considering too many details at once.

The small set of **match** and **move** operators provide the intellectual leverage necessary for programmers to clearly decompose problems, and also encourage code reuse. By providing these in highly optimized implementations, programmers are relieved of the task of re-designing fundamental parallel kernels for each new task. As new parallel architectures are developed, the adoption of a small set of standard architecture-independent operators will allow the porting of algorithms to new hardware with a minimum of effort.

11.9 Acknowledgments

This research was supported by the Defense Advanced Research Project Agency, ARPA Order 7597. Connection Machine time was provided by the

Table 11.1 A breakdown of the setup and evaluation times of three approaches to sparse matrix vector multiplication for matrices of varying size and density on the CRAY Y-MP. For each category, the best times are highlighted. The CSR approach does no preprocessing, while the JD approach trades a large pre-processing time for a very quick evaluation time. The TOTAL time represents the time required to perform one setup and evaluation. When performing only one matrix vector multiply, the hash-based approach excels for very sparse matrices. Surprisingly, for the largest matrix, the rows are long enough that the CSR approach has the lowest total time.

Sparse matrix-vector multiplication (times in ms)					
Order	ρ	CSR	JD	Sort	Hash
		Setup			
20000	0.001		37.23	112.87	**12.09**
15000	0.001		25.01	69.80	**6.03**
10000	0.001		14.85	27.16	**2.67**
5000	0.001		6.67	7.61	**0.81**
2000	0.005		2.97	6.84	**0.94**
1000	0.010		1.52	3.14	**0.36**
100	0.400		0.32	1.60	**0.20**
50	1.000		**0.19**	1.00	0.24
		Evaluation			
20000	0.001	42.16	**6.61**	27.80	39.08
15000	0.001	30.11	**3.84**	20.31	21.42
10000	0.001	19.31	**1.77**	7.97	9.58
5000	0.001	9.37	**0.43**	2.35	2.61
2000	0.005	3.87	**0.37**	1.63	3.35
1000	0.010	1.93	**0.19**	0.91	1.14
100	0.400	0.27	**0.11**	0.42	0.54
50	1.000	0.14	**0.13**	0.32	0.27
		Total			
20000	0.001	**42.16**	43.84	140.67	51.17
15000	0.001	30.11	28.84	90.11	**27.46**
10000	0.001	19.31	16.62	35.13	**12.25**
5000	0.001	9.37	7.11	9.95	**3.43**
2000	0.005	3.87	**3.34**	8.47	4.28
1000	0.010	1.93	1.71	4.05	**1.49**
100	0.400	**0.27**	0.44	2.02	0.74
50	1.000	**0.14**	0.33	1.32	0.51

Table 11.2 A comparison of the setup and evaluation times for some matrices representing electrical circuits on the CRAY Y-MP. For these matrices with a few very full rows, the JD approach suffers a severe performance loss. The **multireduce** instruction used with the hash-based implementation is better suited to this situation.

Title	Order	ρ	CSR	JD	Sort	Hash
Sparse matrix-vector multiplication (times in ms)						
				Setup		
ADVICE2806	2806	0.0030		6.08	8.06	**1.85**
ADVICE3776	3776	0.0019		8.13	8.74	**2.10**
				Evaluation		
ADVICE2806	2806	0.0030	3.60	2.41	3.60	**2.28**
ADVICE3776	3776	0.0019	7.19	3.21	**2.54**	2.72
				Total		
ADVICE2806	2806	0.0030	7.99	8.49	11.65	**4.13**
ADVICE3776	3776	0.0019	7.19	11.34	11.28	**4.82**

Northeast Parallel Architectures Center, and by the Pittsburgh Supercomputing Center, grant number ESC910004P. Cray Y-MP time was provided by the Pittsburgh Supercomputing Center (Grant ASC890018P).

REFERENCES

E. Anderson and Y. Saad. Solving sparse triangular systems on parallel computers. *International Journal of High Speed Computing*, 1,73–96. 1989.

G. E. Blelloch. *Vector models for data-parallel computing.* MIT Press, 1990.

E.F. Codd. A relational model of data for large shared data banks. *CACM*, **13**,377–387. 1970.

E. R. Cohn, The beta operation: A parallel primitive. Technical Report STAN-CS-88-1231, Department of Computer Science, Stanford Universit., 1988.

K. S. Kundert and A. Sangiovanni-Vincentelli, Sparse 1.3, a sparse linear equation solver. Available from EECS Industrial Liaison Program, University of California, Berkeley, CA 94720.

J. L. Potter, Data structures for associative supercomputers. In R. Mills, editor, *Second Symposium On The Frontiers of Massively Parallel Computations*, pp 643–645. IEEE, IEEE Computer Society Press. 1988.

A. G. Ranade, S. N. Bhatt, and S. Lennart Johnsson. The Fluent abstract machine. In J. Allen and F. T. Leighton, editors, *Advanced Research in VLSI, Proceedings of the Fifth MIT Conference*, pp 71–93. 1988.

G. W. Sabot. *The Paralation Model: Architecture-Independent Parallel Programming.* Series in Artificial Intelligence. MIT Press. 1988.

T. J. Sheffler. *Match and Move, an Approach to Data Parallel Computing.* PhD thesis, Carnegie Mellon University. 1992.

Thomas J. Sheffler. Implementing the multiprefix operation on parallel and vector computers. In *Proceedings of the 1993 Fifth Annual ACM Symposium on Parallel Algorithms and Architectures.* 1993.

G. L. Steele Jr. and W. D. Hillis. Connection Machine Lisp: Fine-grained parallel symbolic processing. Technical Report 86.16, Thinking Machines Corporation. 1986.

12

ANDF — SEQUENTIAL TO PARALLEL

Tom Lake[*]

GLOSSA, 59 Alexandra Road, Reading RG1 5PG, UK
Tom.Lake@glossa.co.uk

Abstract

Compiler builders are finding success with wide spectrum intermediate languages. The open system intermediate language ANDF, designed at DRA, is described. We consider the requirements of parallel computation and discuss the extension of this form for compilation of parallel languages.

12.1 Intermediate languages

Compiler writers have long seen the benefits of intermediate formats that could be used for many source languages and target machines. Language transformations and analyses need only be written for the common format and could be used in many contexts. Additional benefits arise when the intermediate format is not private to the compiler builder but is rather available to equipment manufacturers, software vendors, and software tool suppliers. The Open Software Foundation's ANDF concept is of such an open system intermediate form for programs. I briefly discuss the intermediate language selected and then consider the possibility of extending its range to parallel computation.

12.2 Introduction to ANDF and TDF

In mid-1991 the Open Software Foundation announced that they had selected DRA's TDF as the base technology for their Architecture Neutral Distribution Format (ANDF) (OSF 1993). OSF is working closely with UNIX Systems Laboratories (USL) who have announced their intention to commercialize and distribute TDF. Use of ANDF allows Independent Software Vendors (ISVs) to produce a single version of their product in TDF which will then work on any POSIX system regardless of which processor it is running on. The central idea with ANDF is that the user could buy a shrink-wrapped application in TDF form and install it on any compliant

---

[*]This study is being partially funded by DRA. The views expressed, however, are my own.

Open System platform, thus deriving direct practical benefit from the use of a compliant Open System.

The mode of use is for TDF to be 'produced' by a high-level language 'producer' which can carry out many portability checks. The TDF is then 'installed' to an executable for any system supporting TDF and the open system interfaces required by the program. The installation process uses machine-specific TDF and object libraries. This speeds the provision of application portfolios for a range of processors as well as the provision of compiling systems for new architectures.

TDF is a fully bracketed (tree) language with a tightly compressed concrete format. All bar external textual names are removed so that even if the program is reconstructed it is very difficult to understand the intention of the programmer.

The TDF format is an expression of program semantics which is independent of machine architecture and which can cover the semantics of a number of computer languages. One key problem which the designers of TDF faced was to permit the deferral of machine and environment dependent bindings and decisions to installation. When a C program, for example, is compiled by a conventional compiler, many machine or environment dependent options, such as the size of integers or the order of fields in the FILE structure, are known or provided to the compiler and are incorporated in the binary. The C to TDF producer must produce an abstraction of the program over the possible values of this information which can later be applied to the bindings for a particular processor. This is done by allowing pieces of the TDF tree to be represented by tokens (possibly with parameters). The full TDF tree can be constructed by substituting and expanding the locally held definitions for the tokens during installation.

Code generation from TDF is both fast and effective. The code produced generally compares well with that from native compilers (never more than 5% worse run-time, sometimes better). TDF preserves all relevant information for code generation so that a full range of optimizations can be applied, by TDF transformation, during the code generation process. Since compilation is focussing ever more on correctness-preserving transformations this is a useful framework.

12.3   TDF principles

The success of TDF derives largely from the care and completeness with which it has been designed but certain principles can be inferred which we may use to guide our thinking as we approach the unknown territory of parallel computation.

- Not for human readers. TDF is about semantics, not readability.

- Orthogonality of constructs. Constructs are considered as generators of the program algebra and are chosen to give as simple as possible a set of algebraic laws. For example, leaving aside the difficulties of error interrupt and exception, a small number of constructs deal with store side effects, and a further small number with control transfers.
- Based on computer language abstractions, not machine abstractions.
- No idealizations of computer languages. TDF does not contain constructs relating to mathematical constructs except in so far as these represent the actual semantics of computing machines. Numerical datatypes correspond to (a wide range of) computational representations of numbers.
- Structure serves computation. The scoping and nesting of TDF preserve information that helps to localize optimizations and the program-wide computations that are required for code generation and program transformation.
- Uniform expression language. Allows a great deal of freedom to re-order and optimise.
- Faithful representation of store semantics. All the warts which C programmers could require are represented but no more. This still leaves a good deal of freedom for a range of implementations of store.
- Static cases of general dynamic constructs are singled out where this matters, e.g. loops are distinguished from unstructured jumping constructs and scoped store is distinguished from heap store.
- Avoid over-specification, e.g. sub-expression evaluation and parameter evaluation order are already parallel in the sense that sub-expression evaluations may be freely interleaved. Ordering constraints, specified in any case by equivalent effect, are in general limited to declarations and control transfers; procedures are strict in their parameters.

For our consideration of parallel computation we shall emphasize particularly:

1. The representation of store and its analogues. Variable (and heap) declarations return pointers to 'original spaces'. While pointer addition is supported, the effect of access outside the original space of a pointer is undefined.
2. The representation of constructs as expressions, preserving freedom of order of evaluation.
3. The description of data by datatype, independent of detailed representation (which could even vary dynamically – say from store to register).
4. The use of nested and scoped structure to assist optimization and program-wide computation.

5. The distinct representation of certain statically known constructs different from more general dynamic constructs where this assists optimization and transformation.

12.4 Disclaimer

The remaining discussion is a preliminary review at the start of a substantial piece of work. It is useful to record the state of one's thoughts at the start of an investigation. But the reader should not take the following as more than here described – the recording of a state of mind.

12.5 Representing parallel computation

Parallel computation involves both distribution and concurrency. It may be possible to write programs without reference to either of these dimensions in favourable cases, but I do not believe that they can be ignored in generating code for parallel computers. A number of extant programming languages avoid a general representation of both these dimensions in favour of some restricted or specialized combination. For example, FORTRAN 90 describes computation with a constrained (data-parallel) form of concurrency and HPF (High Performance FORTRAN) adds a constrained form of distribution. Again, occam programs have a PLACED PAR construct but it is necessarily the outermost construct.

We expect to have some notion of place in computation, probably virtualized, and to derive a quotient, which for regular arrays, we call *alignment* and for individual constructs, *collocation* (Lake 1991, Lake 1990). These alignment or collocation features seem to be the most convenient means for programmers to use to indicate their requirements for locality between elements of the program.

12.6 Hardware evolution

Processors are evolving in two diverging directions. Some emphasize high-throughput with multiple instruction issue and deep pipelines giving a large processor state.

Others emphasize fast context switch and share processor resources among independent instruction streams. These latter trends have brought dataflow and multi-threaded computation into a surprising conformity (Papadopoulos and Traub 1991).

We know that MIMD parallel systems are evolving to scalable implementations of cached shared memory, to use of a common memory space supported by non-local memory access, to topology independence of communication and synchronization primitives. A new division has appeared between strongly and weakly consistent architectures. In strongly consistent architectures all processors agree on the order in which memory state changes occur; in weakly consistent architectures they may not agree until after a synchronization point; in BSP (Valiant 1990) architectures the memory state changes requested in one 'superstep' may not be visible to other processors until the next.

For processors with slow context switch, amalgamation of independent threads, elimination of redundant synchronizations, recognition or creation of private local data, pre-fetch of remote data, and cache management are the means for compilers to enhance system throughput.

For processors with fast context switch, recognition or creation of private local data, creation or adjustment of pseudo-concurrency, and cache management are most important.

12.7 Languages for parallel computing

As noted above, the ideal is for a parallel TDF to take its notions from a range of parallel languages, as existing TDF does e.g. for its notion of store and addressing.

12.7.1 *Implicit concurrency*

Many languages permit concurrency in the evaluation of disjoint parts of expressions and in the evaluation of parameters to routine calls. FORTRAN90 even allows parameter evaluation to be delayed past the call itself if the effects are not distinguishable from earlier evaluation. This concurrency must be created and controlled by the compiler.

Functional languages like SISAL and Haskell provide many more opportunities for concurrency since the partial ordering provided by store dependencies is not present. However, these languages do not provide an acceptable performance model for the processing of arrays. It may be that the introduction of bulk processing operators for arrays and other bulk data types in the work of Bird and Meertens (Bird 1988) will allow a more favourable account to be given. Skillicorn (Skillicorn 1990) has shown that these operations can themselves be well suited to compilation for different architectures. For first-class functional languages like Haskell the appeal lies to some extent in the possibility of using normal order evaluation without sacrificing the single evaluation of the parameter, through some version of graph reduction. Infinite structures can be expressed in programs provided that they are only ever partly evaluated. If program evaluation is carried out in parallel synchronization is necessary to communicate the start and finish of evaluation of a delayed expression to other potential users.

12.7.2 *Data parallel concurrency*

Data parallel computation is the simplest expression of concurrency. Its presence in FORTRAN90, substantially influenced by the designers and ambassadors for DAP FORTRAN (now CPP's FORTRAN PLUS), will make this form of concurrency readily available on a wide variety of computers. The HPF (High Performance FORTRAN) (Loveman 1993) dialect, adding the FORALL statement dropped from FORTRAN90 and a constrained set of static and dynamic alignment and distribution statements, lets programmers specify locality (through alignment) and load-balance (through distribution) for MIMD and SIMD architectures. An HPF program has the same single-threaded interpretation as its undistributed version, which makes for easy development. In the compilation of HPF for distributed memory machines, great efforts are made to eliminate unnecessary synchronizations, without sacrificing any of the meaning of the program.

12.7.3 *Explicit concurrency*

There is a long history of concurrent programming, but, with many high-performance computational applications using messaging and process management packages like PVM or the new multi-threaded shared memory UNIXes, the appropriate constructs don't seem to have been provided adequately within a language. With many of these applications, portability among architectures may well be best handled by parameter changes in the algorithm to achieve an appropriate balance of computation and communication at appropriate and relatively coarse granularity.

Concurrency was first studied in the context of operating systems and has direct application to multi-window interactive systems. In these cases the execution is pseudo-concurrent or interleaved. But for parallel computing concurrency has to be combined with distribution. Orthogonal combinations of concurrency with distribution are not so frequent and have not proved satisfactory in performance (possibly because of the hardware or software setting of the points of interest rather than the essential points at issue). The 'diagonal' (Lindsey 1993) combination of distribution and concurrency provided with the first generation of Inmos Transputers was satisfactory only for a restricted range of applications. A range of more or less unsatisfactory 'harnesses' were created to mitigate the situation. A fully orthogonal version of occam has never been ruled out but is not at present available.

Explicit parallelism requires the creation of processes and their interaction. Some of the constructs which mediate interaction in parallel languages are:

synchronous channels between two processes: passed in as parameters or from outer scope or in messages;

asynchronous channels as above: practical store limits mean that unlimited buffering (which is a useful concept) is not available;

task rendezvous combining synchronous messaging and a notion of shared store (although copy-in/copy-out semantics could be imposed);

object methods synchronous/asynchronous – with/without mutual exclusion – private/global naming;

futures closures evaluated in a separate thread: imperative version not as clean as functional futures;

process identifiers and variables allowing processes to interrupt and stop one another;

shared variables e.g. semaphores behaviour depending also on the consistency of store access between processors;

shared objects or object references sharing restricted to objects with specific operations and synchronization protocols;

monitors (with condition variables) mutual exclusion with an optimized interaction with the condition variable and process scheduling.

In all these cases there is a fragment of state which is involved in the operations and must be treated with a particular synchronization protocol.

A question of interest is how these various constructs can be used to represent one another and at what cost.

Many of the more interesting language designs have taken place within the object model. The language μC^{++} (Buhr et al. 1992) provides a rich variety of options of mutual exclusion and object activity within the shared memory framework. POOL (Bronnenberg 1989) is a a strongly typed language with location independent references to distributed objects. FORTRAN-M (Foster and Chandy 1993) provides dynamic and mobile buffered channels between dynamically created FORTRAN processes with disjoint address spaces. It has a clear deterministic subset and provides a small number of operators for non-deterministic programming (PROBE to test whether messages are waiting on a channel, and MERGE to allow messages to be received from any of several channels). With the base language raised to FORTRAN90 this would be an ideal language for multiple SIMD machines. ORCA (Bal 1990) proposes shared objects with mutual exclusion, the shared objects may be remote or even multiply replicated by the run-time system. There is a specific difficulty associated with alternation between guarded methods which must be rolled back if they encounter a deadlock due to a guard on a constituent sub-object. I have proposed the use of location-independent object references with a 'collocating' capability, giving the programmer control over object movement and locality (Lake 1990).

Some of the languages cited above contain objects which have to obey a specific discipline. For example, FORTRAN-M defines ports. Channels are constructed between input and output ports. A port can be moved by sending it over a channel. By this action the port is voided locally. Again, the collocating capability of a reference in MOSS can be assigned or sent to another process, but since it is conserved (the location can only be taken from one other object) it is removed from the source reference when added to the target. Such systems require strong typing or a substitute discipline for their correct operation. If a port or reference was mistakenly overwritten havoc would ensue.

Several of the languages cited contain the notion of a two-level store corresponding to the shared and private portions of the store visible by a process. An interesting notion of store is obtained by allowing references with access capabilities to heap store. With an appropriate set of operations we could ensure that either an exclusive writer or readers had access to any piece of heap store. Capabilities for arrays could be sliced and re-assembled with a certain amount of checking. Again, strong typing is necessary to ensure correct operation of the system. When such capabilities are passed between processes the necessary synchronization for change of access can be assured. Use of such capabilities generalizes the static data access rules of occam which forbid race conditions. The readers/exclusive-writer property is appropriate for deciding whether a copy of data can be made by a give process.

12.7.4 Store consistency

Ada (and Concurrent C^{++} (Gehani and Roome 1989) which follows it in matters of concurrency) provide a great deal of support for concurrency; the weak consistency that Ada provides for variables shared between tasks (consistent only at synchronization points), though awkward from a programmer's point of view can at least be widely realised. In many cases programmers will use C and operating system interfaces to achieve concurrency. In these cases it is not clear yet whether they will be programming to strong or weak consistency. If the former, then substantial compiler assistance will be necessary to achieve reasonable performance on architectures with only weak consistency.

12.7.5 Locality

The discussion in the remainder of this paper will assume that locality is a key determinant of computing performance so that generated code should bring together in time and space instructions which operate on common (and, where this has meaning, local) data. This is just to suggest that the latency of memory response to processor requests will continue to be so much larger than the cycle time that it will be difficult to offset it with concurrency. We use the terms 'place' and 'placed' to refer to the separate loci of computation and the alignment of computational objects with them whether these loci are separate memory spaces or processor-cache combinations.

Shared or common memory machines emphasize the bringing of data to the requesting instruction. They do not seem to have the complementary feature of bringing the next relevant instruction to the data (for example, in choosing where to execute an object method call). It is the combination of these features that makes for high performance. With one or none of them the programmer and compiler must make provision for locality.

A good deal of effort is being put into generation of programs from HPF and more into the automatic derivation of HPF distributions for FORTRAN90 programs. We expect that HPF can be satisfactorily compiled without user intervention, but it is not yet clear to what extent this can be done for FORTRAN90 for distributed machines.

Essentially two constructs in common use allow a programmer to specify locality. Firstly, the use of regular arrays and data-parallel programming allows operations and data to be distributed together, especially if annotations on alignment are provided. Secondly, the use of objects with encapsulated data provides locality for immediate data access. In the first case, communication can be caused by operations between misaligned objects or by array realignment or redistribution. In the second case communication is expressed by interaction between objects which are not collocated. These seem to be the only mechanisms currently available.

12.8 Extending TDF for parallelism

TDF provides a semantic representation of a program whose binding to the target system can be abstracted and then applied in one or more stages. One of the potential steps with parallel machines is the binding to the particular target configuration. This is a problem new to parallel computing and one which has not yet been satisfactorily solved. HPF explicitly allows compilation without knowledge of the number of processors of the target machine. If the distribution of commercial software to parallel machines is to be of importance, then compilation independent of target configuration will be required. High performance may then require an intermediate format rich in semantic information. This will favour the use of intermediate forms like TDF.

Stronger portability occurs when the target architecture can be abstracted. Clearly this is only possible to a certain degree. SIMD programs can be ported across SISD, MIMD, and SIMD architectures with reasonable expectation of efficiency. A data-parallel TDF program could be installed to this range of machines.

Optimization is required for code generation for parallel programs as for sequential. However, analysis and transformation are potentially more expensive for parallel programs where the semantics involves considering all interleavings of program fragments that are not otherwise ordered. It would be a great advantage to have some sort of normalization of a concurrent program, as with the control flow of a sequential program, to a form in which fragments of the program which were tractable to analysis and transformation could be selected. Barrett (1993) has pointed out that programs built with the **par** construct and without data races are amenable to many of the analyses standard for sequential programs. This sort of analysis has also been considered by Harrison (Chow and Harrison 1992).

It is already the case that the only order imposed on the evaluation of a TDF expression, apart from the dataflow ordering, is expressed by the declarations (which order initialization before the scope), sequence and **apply_proc** construct (which orders parameter evaluation before the application).

For the synchronization constructs described above it will be necessary to examine their transformations carefully to determine whether we can find an adequate representation of minimal size in which the whole set can be expressed and for which code generation seems tractable for a range of architectures.

In the case of data-parallel languages the array structure makes possible transformations based on mappings to processors, absence of dependences, and so on. It is my belief that the code generation is most easily carried out if the data-parallel nature of the program can be reflected directly in the expressions being manipulated, giving the maximum freedom of evaluation order to data-parallel manipulations. We shall be considering the introduction of array-values and array-valued expressions. Since TDF expression evaluation is already ordered only by dataflow (except for a few ordered constructs), concurrent parallel expressions fit quite well with this.

We see that it is possible that a small set of new constructs be sufficient for the expression of parallelism. The crucial question is whether efficient and effective code generation is possible from a constrained semantic representation for a range of architectures.

12.9 Existing projects

Two projects have considered TDF as an intermediate for parallel compilation. Our work (Lake and Sloman 1992) has considered the vectorizing restructuring of TDF. Of course the semantic analysis and restructuring capability necessary for vectorization is also required for parallelization or distribution of sequential and data-parallel programs.

We defined an output format for the vectorizer which was more general than the data-parallel output usually selected. We wished to allow the installer latitude in vector code generation. The main construct **vec** denoted a loop in which all dependencies ran forward except within **serial** regions contained within it. The **serial** regions indicated portions of the loop which required serial execution because of the impossibility of eliminating backward running store dependencies. These constructs could be nested. There are advantages in this representation, since the input and output of **serial** regions can be held in vector registers. The serial regions themselves may be regions where information can be passed from one loop iteration to the next in registers. Restructuring to this vector form could be done without prejudicing an installers freedom to choose vector lengths and operation orderings and without allocating vector temporaries.

In the COOTS IED project* a concurrent object language (UC^{++}) (Winder et al. 1991) for distributed systems (with non-mobile processes) is compiled via TDF to a PARSYS Transputer array and a network of DECstations. A number of new constructs were defined for TDF in two areas: firstly, to represent the creation and manipulation of threads and mutual exclusion; secondly, to build the object structure of the whole program, indicating the share of resources and the choice of object classes to be allocated to each processor, and providing for inter-object calls to be constructed either as local or remote calls. All of the new constructs could be defined as TDF tokens, although the modified installers did have to recognize some of them and apply special treatment.

12.10 Distributing programs to disjoint memories

A principal problem in compilation is in distributing programs, whether to separate memory spaces in a message-passing computer or to remote parts of a shared memory. We rephrase this question into asking what we should keep together, then how we can effect the distribution.

The constructs whose locality we have can currently describe are regular arrays of data with various mappings and objects with location. (Other bulk data types may be added.)

The expression of array data location is through mapping, intensional through mapping functions and extensional through mapping arrays. Dynamic remapping is part of HPF – even across routine calls. Alignment or array collocation is emphasized in HPF and provides the locality that we need. Distribution then can act to affect load balance (arranging that a selected subarray still spreads across the machine) and bandwidth required (arranging that nearest communication neighbours are on the same processor often).

The expression of object locality is not clearly agreed. The model that I have found useful is that of collocation by pointer capability. A single pointer to an object may have the collocating property. Movement is transitive through collocation. The transmission of a collocating pointer between objects (say, in a message) implies movement of the associated object in a way that transmission of a non-collocating pointer does not. In this way a dynamic half-way house is defined between the inclusion of one object in another and their separate existence – collocation.

We therefore consider the following recipe for distributed program construction (Lake 1992).

- Take a sequential program with arrays and objects.
- Decide on mappings and mapping strategies for the arrays and objects. These could be either static or dynamic. The aim of the mapping is to bring together in time and space the elements of local

---

*Partners were University College London, DRA Malvern, and Harlequin Ltd.

computations. The term 'rendezvous' applied to data and operations rather than to process best describes what we have to prescribe. A placed expression language would probably do as well as any. Programmer and compiler have to cooperate to achieve the appropriate grain size for these interactions. Naturally in the case of bulk datatypes these rendezvous may be described in bulk with replicating operators outermost so that we see expressions placed over regular arrays of places. These data-parallel operations give rise to elegant algebraic statements of transformations and equivalences. Real programs require both the highly regular mappings of HPF and more general user-specified or explicit mappings of Vienna FORTRAN (Zima et al. 1992).

- Translate the data rendezvous of the mapped expressions into the appropriate mechanisms for the underlying architecture. This could involve the construction of tasks, the creation of messages or remote procedure calls or of routines distributed across computing places.

Note that a surprising difficulty for a virtual shared memory machine is to avoid making all object interactions local. We should distinguish between those that should be local (where the data should be brought together at the rendezvous) and those where the data should stay put and the activation moved to the data's permanent home. The balance is between the cost of process switch and transfer and the cost of data transfer minus the benefit of load sharing.

When a program is distributed we have a choice as to how much replication takes place. Replication can reduce the requirement for communication. In the simplest case of parallel processing the entire program is replicated and different 'data cases', program instances, are run on each processor. A sequential program can be distributed by replicating the entire control flow and DO control variables, distributing the other variables according to taste and, where necessary, replacing the non-local assignments by communication. (A little optimization may be in order as well.) This is the idea behind HPF.

A concurrent program can be distributed to disjoint memories more selectively. An approach which favours distribution is as follows:

- Where variables are shared by processes, encapsulate them in separate processes; access them by channels.

- Distribute the variables according to taste.

- Distribute the processes local to their variables, replicating them where necessary (just as the control flow was replicated for distribution of a sequential program). Note that data accesses via pointers require replicated action (to supply the data) at every processor unless some information is available on the alignment of the value with

other data. (Shared address space alone is not enough as the access represents a synchronization with other programs accessing the datum that must not be lost.)
- Where a non-deterministic construct is to be replicated, choose one of its sites to make the choice (directing communications involved in the choice to temporary variables there) and then recommunicate the choice and any diverted data to ensure that replicated control flow follows a common control path.

For an architecture-independent language, say a language with arrays, objects, alignments, and collocation, compilation to both shared and disjoint memories would be required in the immediate future.

12.11 Distribution and concurrency for TDF

We currently expect to propose two levels of construct in TDF for parallelism and distribution. One will only be derived for disjoint memory systems and forms the lower level from which code generation takes place for them. The process of program distribution to disjoint memories would take place by translation between these two levels. The lower level constructs include a REMOTE constructor for system-wide references and primitives for dealing with messages.

At the slightly higher level TDF is already moderately concurrent. Sub-expressions may be evaluated interleaved. Such interleaving may not be delayed through a control flow construct such as a procedure call. We need to introduce some notion of delayed evaluation or threads to support explicit concurrency or normal order parameter evaluation (by need).

At this level we expect to add constructs in three categories:

1. Constructors for array values, replicated parallel expressions – to complement the sub-expressions of other constructors. The large expressions available will allow many useful re-arrangements and re-orderings. In particular it should be possible to generate vector and array processor codes. A possibility is to introduce **par** creating a parallel tuple and **any** delivering a parallel union, of tuple and union shapes respectively. A non-deterministic expression **first** (resp. **first_true**) to choose one of a set of expressions (resp. one of a set of expressions that returns **true** as a component of its value), is also attractive.

2. Explicit control of concurrency and synchronization. There are already defined thread primitives, for constructing and starting threads based on expressions which might be used. Together with a lightweight **mutex** mutual exclusion primitive and a forcing **await_thread** primitive it is hoped that many operations of concurrent computing, such

as `futures`, `channels`, `semaphores`, and so on, could be constructed and good code produced.
3. Alignment or collocation information with permanent static alignment visible at compile or install time. Static placement could be indicated by labelling expressions with place, but alignment and collocation are not so obviously expressible, especially when dynamic. These attributes must be captured in a form suitable for use in program distribution.

Along with the `par` construct it is helpful to have a `mutex` construct to protect against interleaving. In this way we would express the execution of associative, commutative reductions, by parallel ordering and atomic update of the answer.

We hope that the required functionality will be provided with a small number of additional constructs. A study of transformations involved in code generation for concurrent systems will be necessary to tell whether a small number of constructs will be sufficient to allow efficient code to be generated for a variety of architectures.

12.12 Outlook

Cost, and wide applicability of compiler techniques, favour the use of wide spectrum intermediate forms. A distributable form has the capability to provide portability across parallel machine configurations and some differences of architecture. Such a form enables a new market and working approach for software tools and portable software applications.

Architecture, algorithmics, compilation, and design of languages for parallel systems are still in evolution. We expect that progress in each of these areas will interact to produce more capable and easily used computing systems.

REFERENCES

Bal, H. (1990) *Programming Distributed Systems*, Silicon Press, Summit NJ

Barrett, G. (1993) (private communication)

Bird, R.S. (1988) *Lectures on Constructive Functional Programming*, PRG-69, Oxford University Computing Laboratory, Programming Research Group, Oxford.

Bronnenberg, W. (1989) POOL and DOOM: A survey of ESPRIT 415 subproject A, Philips Research Laboratories, in *PARLE '89 Parallel Architectures and Languages Europe, Vol 1* (eds. Odijk, O., Rem, M. and Syre, J-C.), LNCS 365, Springer-Verlag, Berlin.

Buhr, P.A., Ditchfield, G., Stroobosscher, R.A., Younger, B.M., Zarnke, C.R. (1992) μC^{++}: Concurrency in the Object-Oriented Language C^{++} em *Software – Practice and Experience* **22**(2).

Chow, J. and Harrison W.L. III (1992) *A General Framework for Analyzing Shared-Memory Parallel Programs*, CSRD Report 1239, Univ. of Illinois at Urbana-Champlain.

Foster, I.T. and Chandy, K.M. (1993) *FORTRAN M: A Language for Modular Parallel Programming* (private communication).

Gehani, N. and Roome, W.D. (1989) *The Concurrent C Programming Language*, Silicon Press, Summit NJ.

Lake, T.W. (1990) *MOSS Language*, (internal report) GLOSSA, Reading.

Lake, T.W. (1991) Collocation as Grouping Primitive for a Distributed Object Language, in *First Workshop on Abstract Machine Models for Highly Parallel Computers* (unpublished).

Lake, T.W. (1992) Distributing Computations, in *Software for Parallel Computers*, ed. Perrott, RH, Chapman and Hall.

Lake, T.W. and Sloman, B. (1992) TDF Vectorisation: Towards Open Parallel Systems, in *Proceedings of the Third Workshop on Compilers for Parallel Computers*, ACPC/TR 92-8, Department of Statistics and Computer Science, University of Vienna.

Lindsey, C.H. (1993) A History of Algol 68, *Sigplan Not.* **28**(3).

Loveman, D.B. ed. (1993) High Performance FORTRAN Draft 1.0, obtainable by anonymous file transfer from titan.cs.rice.edu.

OSF Research institute (1993) *ANDF Technology Source Book*, OSF, Cambridge, MA. (I have cited articles by Dr N. Peeling of DRA and Dr S. Mackrakis of OSF.)

Papdopoulos, G.M. and Traub, K.R. (1991) Multithreading: A Revisionist View of Dataflow Architectures b, in *Proc. 18th Ann. Int. Symp. on Computer Architecture*, Toronto, ACM New York.

Skillicorn, D.B., (1990) Architecture-Independent Parallel Computation, *IEEE Computer* **23**(12).

Valiant L.G. (1990) A Bridging Model for Parallel Computation, *CACM* **30**(8).

Winder, R., Wei, M. and Roberts G. (1991) Harnessing Parallelism with UC^{++}, *Proceedings of European C^{++} User Group Winter Conference*, Imperial College. 1991

Zima, H., Brezany, P., Chapman, B., Mehrotra, P., Schwald, A., *Vienna FORTRAN - A Language Specification (version 1.1)*, ACPC/TR 92-4 Dept. of Statistics and Computer Science, Univ. of Vienna. 1992.

13

A FRAMEWORK FOR PORTABLE PARALLEL APPLICATIONS

N. B. MacDonald

Edinburgh Parallel Computing Centre
The University of Edinburgh, UK

Abstract

The lack of a convenient and portable programming interface to parallel computer systems has inhibited the development of parallel applications, and thus restricted the exploitation of parallel computing technology. This paper discusses the role of various efforts to provide a portable interface to parallel computer systems, and presents CHIMP – Common High-level Interface to Message-Passing – in this context. The concepts embodied in the CHIMP model of message passing are described, accompanied with details of the range of parallel systems on which it is implemented and across which it offers source code portability.

An important consequence of this portability is the ability to reuse not just entire application codes, but components of applications. The Parallel Utilities Library (PUL) aims to provide a set of reusable components for implementing parallel applications. Implemented on top of CHIMP, the interfaces to these utilities are themselves portable across a range of platforms, and can be viewed as a set of increasingly abstract models in which to develop applications.

The paper also summarizes experience using CHIMP and PUL to implement parallel applications in collaboration with both industrial and academic groups, and presents plans for future work.

13.1 The need for portability

The lack of a convenient and efficiently implementable portable programming interface for parallel systems has long been an obstacle to the widespread commercial exploitation of parallel computing technology. Given the diverse architectural features found in different parallel machines, it is arguably an open question whether a portable interface which supports an abstraction away from these features could effectively exploit the full power of a range of machines.

Even within classes of machines with similar architectural characteristics, however, portability of applications is rarely achieved due to the prevalence of vendor-specific programming interfaces. The incompatibilities of these systems necessitate varying degrees of porting effort in order to move parallel applications onto new platforms, and this effort inhibits the parallelization of existing packages and the development of new parallel applications. The availability of a common interface would act as a stimulus for application development, which may in turn catalyze the wider exploitation of parallel computing.

The need for portability has been recognized by the industry, and a number of efforts to establish portable parallel programming interfaces are underway. Perhaps the most significant of these is the High Performance Fortran Forum, which has brought together vendors, university researchers and end-users to define High Performance Fortran (HPF)(14), which many vendors have undertaken to support on their systems. HPF is a Fortran 90 superset which includes directives through which the programmer can optimise the distribution of data to exploit a range of computer systems. The data distributions chosen by the programmer may have a significant effect on the performance of the application on a particular architecture, and predicting this effect can require considerable sophistication. However, source code portability allows the programmer to concentrate on choosing good distributions and relieves him or her of the burden of converting much of the source code. Furthermore, portability provides a basis for work on tools to select good distributions, such as research currently being pursued at Rice on Fortran-D and at Vienna on Vienna Fortran.

However, HPF is not a suitable software platform for all parallel applications. The initial HPF compilers are not expected to cope well with imbalanced problems. While developments in compiler technology may well expand the range of applications which can be efficiently tackled using HPF, it seems likely that there will remain a large class of irregular problems which cannot be conveniently expressed. Any emergent data parallel C standard, perhaps based on systems such as Dataparallel C (12), DINO (21) and C* (27), will be similarly disadvantaged.

Efficient implementations of imbalanced and irregular problems can be conveniently expressed in a message passing style. A number of systems have been developed to provide a portable message passing interface. In late 1990, PVM (25) and PARMACS (13) were perhaps the most prominent systems. However, significant differences existed between the models of message passing which these systems embodied and their interfaces were completely incompatible. It was not clear which, if either, would emerge as a *de facto* standard. Both systems were unsupported, and despite the availability of source code in the case of PVM there were some concerns over the portability of the implementation. Furthermore, PVM was only available for networks of UNIX hosts, and not on tightly-coupled multicomputers; and PARMACS only provided bindings for Fortran. These factors, along with a perceived need for more sophisticated functionality, mitigated against the adoption of either package as a basis for commercial work, and early in 1991 researchers at Edinburgh Parallel Computing Centre (EPCC) began work on CHIMP — a Common High-level Interface to Message Passing (28, 4).

The development of CHIMP has three main objectives: *portability, programmability,* and *efficiency*. The principal motivation behind CHIMP is *portability*. We seek source code compatibility across a range of commercially available parallel computers, for both applications and supporting utilities, written in familiar languages such as C and Fortran. Another important objective is *programmability*. The interface must provide sufficient functionality to cover the needs of both applications and utility writers, and at the same time be compact and intuitive enough that it is comprehensible to a wide range of parallel programmers. We believe that it is important that the interface should be constructed in such a way that the communication requirements of parallel library utilities can be hidden from the programmer using such libraries. The final objective is *efficiency*. We wish to construct the interface such that it contains features which permit efficient implementation, in terms of processing and communication resource utilization. This does not imply that we wish to omit functionality which is not, in some sense, efficient since such functionality may nevertheless be required for programmability.

CHIMP seeks to provide only pure messaging. By this we mean that CHIMP is not intended to provide facilities such as parallel reduction. Further, the messaging facility should be restricted to the sending and receiving of single messages. Higher level communication constructs, such as rendezvous, and mixed communication and computation, such as reduction operations, can be provided on top of CHIMP.

CHIMP builds on several years' experience of implementing message passing systems such as Titch (33) and Tiny (8), and developing a wide range of application codes on top of these systems. Implementations of CHIMP are available for T800- and i860-based Meiko Computing Surfaces, the SPARC-based Fujitsu AP-1000, Thinking Machines CM-5, and Unix workstations from Sun, Silicon Graphics, IBM, and Hewlett-Packard. A port to the Intel iPSC/860 is nearing completion, and implementations for a number of other systems, including the Meiko CS-2, are expected in 1993.

Processes running on a node within a MIMD parallel system can use CHIMP to communicate directly with any other process in the application, even if the other process is executing on a node within another parallel system across a network. SIMD parallel systems hosted by one of the supported architectures, such as a Thinking Machines Connection Machine CM-200 (hosted by a Sun workstation) can participate in CHIMP applications using native communication primitives for internal SIMD communication.

CHIMP is available free to academics for non-commercial use, and to industrial and commercial concerns for evaluation. CHIMP is in use in a range of universities in Europe and North America, and for various industrial applications by a number of companies including Shell, Rolls Royce, AEA Technology, Intera Information Technologies and British Telecom.

Section 13.2 presents the concepts underlying CHIMP and the functionality which its interface provides. The current status of the Parallel Utilities Library (PUL), which provides a portable library of reusable parallel application components, is discussed in Section 13.3. An overview of applications projects exploiting CHIMP and PUL is given in Section 13.4 and the paper concludes with a summary of plans for future developments of CHIMP.

13.2 CHIMP concepts

This section discusses a number of alternative design decisions for a pure messaging system and motivates the choices which led to the definition of CHIMP Version 1.0. For details of the language bindings for C and Fortran the reader is referred to the interface specification (4).

13.2.1 *The ISO OSI reference model*

The ISO Open Systems Interconnect Reference Model describes a seven layer architecture for communication systems between general purpose computers (see (26)). Active elements in each layer are referred to as *entities*, and entities in the same layer are described as *peer* entities.

The entities in a particular layer provide a *service* to the layer above, and utilize the services of the layer below. A service is a set of primitives (operations) provided by the layer to the service user. The service defines what the layer is prepared to do for the user but does not specify how this is implemented.

Peer entities exchange information according to a *protocol*. A protocol is a set of rules governing the meaning of packets or messages. It is possible to alter the protocol used, and provided that the services are unaffected, the changes will be transparent to the service user.

Services are available at *service access points*, or SAPs, and are provided by an *interface*. For example, the transport layer provides the transport layer interface (TLI) at transport service access points.

There are some key differences between the hardware of a multicomputer and that of computer networks, arising from the relative physical proximity of multicomputer nodes. In particular, the error rate on interprocessor connections within a multicomputer is negligible and the communication latencies are relatively small. It follows that protocols running over these connections need *not* address error correction and may perform flow control over very small units.

The physical and data-link layers, and to an increasing extent the network layer, are implemented in hardware. In CHIMP we are interested in a communication interface which is broadly comparable to the TLI. Transport is the first true source-to-destination, or *end-to-end*, layer. In general we shall assume that either hardware or software implementations of the network layer are already available.

The kind of service we shall be interested in providing is based on messages – a *reliable datagram* service.

13.2.2 Message transfer properties and models

We can identify a number of desirable properties of a message passing system:

- Message lengths should be notionally unbounded. The user should be allowed to send a message of any length, although in practice message length will be bounded by the memory available to a process. Any restriction on packet lengths which may be imposed by network considerations should not be visible to the user.
- Messages should not be lost.
- The underlying network should be free from internal deadlock states.
- When any messages are sent which cannot be distinguished except by the user data contained, the order in which they are received must be the order in which they were sent.
- In a multiprogrammed machine, where different user programs share the underlying network, messages within any program should not be allowed to indefinitely prevent the delivery of messages within any other program.

Another factor requiring specification for the services of a message passing system is the *model of communication*. Consider the situation where a sending and a receiving agent are separated by some intervening medium, as shown in Fig. 13.1. In order to communicate, the sending agent has to interact with the intervening medium, depositing the message in the medium. The receiving agent also interacts with the medium, removing the message from it. We say that the sending agent communicates with the receiving agent by an *indirect* model of communication. This corresponds to 'asynchronous' messaging.

Can we really talk about a structureless, unbounded space in which any volume of messages can be stored? Our medium is the underlying network, which must have a bounded capacity. In order to transmit messages larger than the available capacity the medium will have to interact with both agents simultaneously, hence very long messages will cause some degree of synchronization between peer agents. Furthermore, the medium itself is constructed from a set of interacting agents and the capacity of the medium may be localized both spatially and temporally. In fact, our 'medium' may have some fairly undesirable properties to which the sender and receiver will be exposed when engaged in the indirect model of communication.

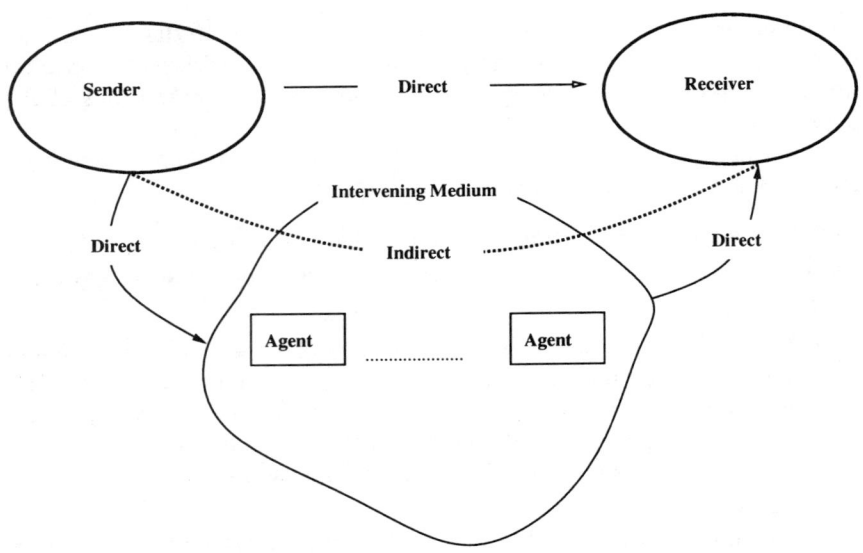

FIG. 13.1. Direct and indirect models of communication

Consider now the interaction of, for example, the sending agent with the medium. Did we invoke a concept of yet another medium intervening in this interaction? We did not, because we implicitly understood the *direct* model of communication between agent and medium. This corresponds to 'synchronous' messaging.

We can implement the direct model between sending and receiving agents in the presence of a medium (underlying network) by use of 'handshaking' protocols. These protocols must provide synchronization between agents and they must also provide security of network resources, in order to hide the properties of the medium.

It may appear that it would be difficult to implement the indirect model if the underlying network enforces peer synchronization. In terms of our discussion of the model, we can clearly allow the extreme of a medium with zero storage capacity, in which case the indirect model can be implemented as the direct model in any circumstances. This is important when we consider hardware networks such as T9000/C104 combination which implement the direct model.

It may also appear that we could have difficulty implementing a mixture of communications using both the indirect and direct models. The problem is that the indirect model may destroy the security of network resources required for the first model. We could require that all processes utilizing the indirect model behave as *eager readers* – processes which will never, under any circumstance, indefinitely refuse to receive any waiting message.

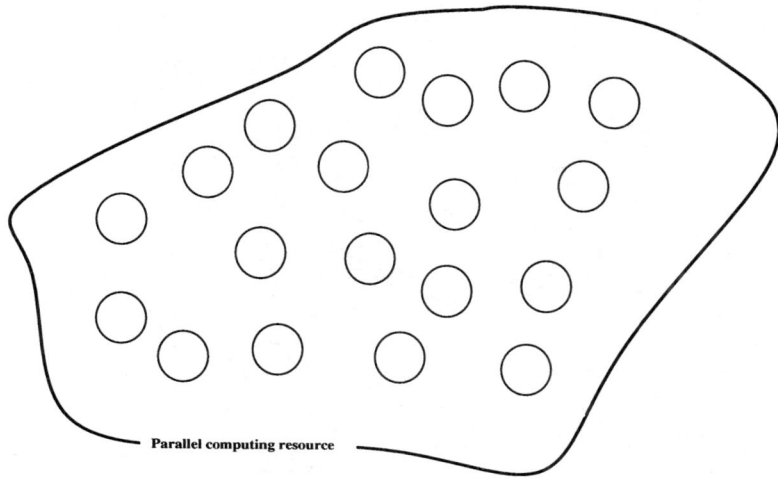

FIG. 13.2. Parallel processes

A communication structure is correct within either model if every process within the structure which uses the indirect model behaves as an eager reader, and it does not deadlock when communications using the indirect model are implemented by the direct model.

Given that the direct model provides security and the indirect model does not, it is reasonable to ask why we might wish to provide the indirect model. Its inclusion could be justified on efficiency grounds. The weaker synchronization properties of the indirect model, coupled with smaller protocol overhead, would allow an experienced parallel programmer to implement structured composite communications, such as rendezvous or exchange, with improved performance. However, we would not expect less experienced parallel programmers to effectively exploit the indirect model, due to its additional complexity, and CHIMP currently provides only the direct model.

13.2.3 Naming schemes

It is clear that a user who wishes to instruct a message passing system to transfer a message from one place to another needs some means of specifying where the message is going to and perhaps where it is coming from. Therefore we need some identification scheme to label source and destination SAPs.

Let us begin by imagining our multicomputer program merely as a bunch of processes (Fig. 13.2). Consider how parallel programmers make use of these processes. From experience, two main types of decomposition are adopted. In *functional decomposition* the problem is split up into differ-

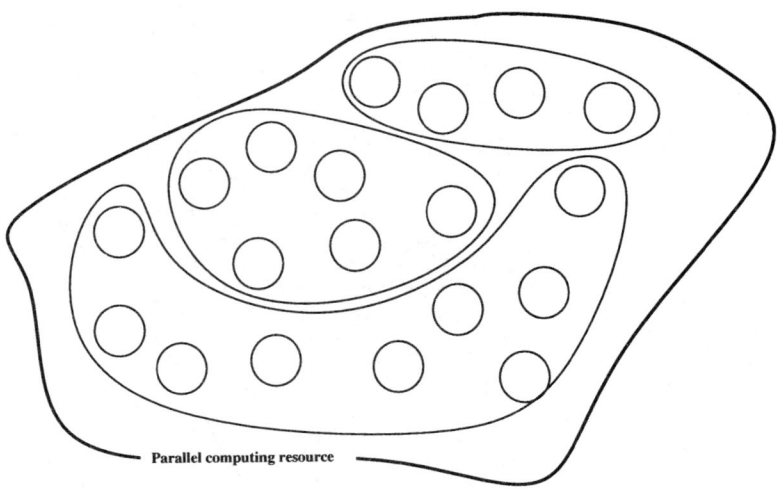

Fig. 13.3. Parallel processes grouped into distinct functional groups

ent functional units running as different processes. In *data decomposition* the same functional process is replicated many times. Hence in application terms the available processes will be grouped into a definite structure (Fig. 13.3).

What implications does this have for a process naming scheme? Should the interface to a message passing system provide the user with a symbolic or numeric method of labelling processes? Symbolic names allow functionally different processes to be distinguished in a very natural way. However, naming functionally identical processes symbolically is awkward. On the other hand, numeric labels support arithmetic on labels, and their use as array subscripts, but do not conveniently differentiate between functionally different processes. In practice, it is rare for real problems to be decomposed in a purely functional or data parallel fashion. Generally, there is a mixture of these two decompositions, with a number of functional groups each consisting of more than one process sharing the same function and sharing the load. This leads to the conclusion that the naming scheme can usefully consist of a symbolic component identifying the functional group and a numeric component identifying the group member.

There is one further step in complexity which we believe to be important. Thus far, processes have been grouped into disjoint sets. Let us instead look at the different algorithms which they may be performing. A particular algorithm may be replicated across a number of processes but some members of this group may have other algorithms to perform in addition, possibly together with processes outwith this initial group. This leads

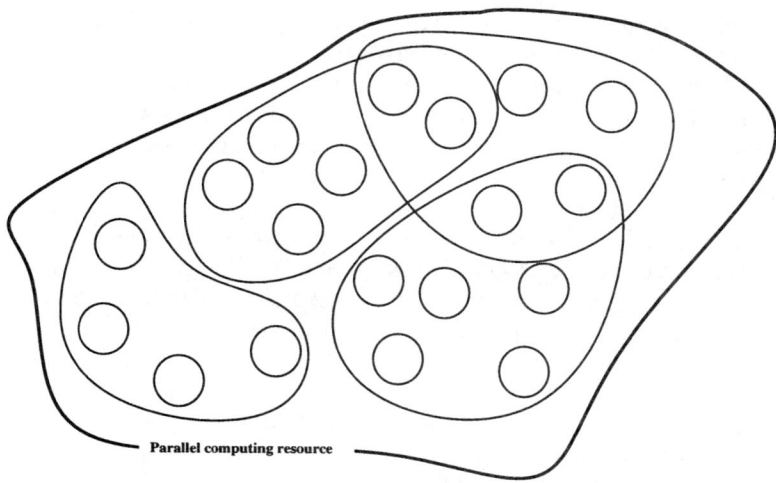

FIG. 13.4. Parallel processes placed into overlapping algorithmic groups

to a more general picture (Fig. 13.4) where the same process belongs to different algorithmic groups and each process has a number of (symbolic, numeric) aliases in which the symbolic portion identifies a group, and the numeric component is an identifier within that group.

The final step in this discussion is that for these multiple process names to be useful, messages should in principle be addressable from any of these names to any other name. To tie this in with OSI terminology we associate these names with SAPs, and processes address messages from one SAP to another. SAP pairs can be thought of as channels in a certain sense.

This scenario rests well with the requirements of parallel library writers. They wish to have functional groups within their library which will communicate through the same message passing interface as the users, but to make this structure transparent to the user.

13.2.4 *Message primitives*

Send and receive The concept of one process sending a single message to, or receiving a single message from, another process is central to message passing. We have suggested that instead of using process names these messages would be directed through SAPs, but the simple notion of sending one message from source to destination is unchanged. Considerably more functionality can be provided while remaining within the remit of pure messaging.

There is a subtlety when we talk about sending a *single* message. What we mean is that one physical block of data is transmitted from the sender. However, this *same* block can be delivered to more than one receiver without being in conflict with the single message philosophy. The concept of multicast to which we are alluding here is used in many applications. Multicast could be implemented by a loop of simple sends to each of the intended recipients, but a more efficient solution is possible. The message always has to be copied the same number of times, but we can minimize the bandwidth requirements and buffer storage in the network by using a *multicast tree*. To do this most efficiently, tables are set up *a priori* at all the members of the multicast group. We must identify these groups beforehand and have the necessary tables in place. These groups usually correspond to a set of processes running a copy of the same algorithm. The SAP algorithmic groups discussed above imply an intrinsic group structure, and these tables can already be set up.

Similarly, although there is some asymmetry, there is more we can do on the receiving side and remain within the confines of sending and receiving single messages. We can allow the user to receive a message from any one of a number of sources, as opposed to always specifying the source. This is the concept of *selection* or *alternation*, and is also common in many applications. In a simple implementation of selection, the process would specify the list of source from of which it is prepared to receive a single message. It would then check through this list one by one, determining which of the sources are waiting to send. If it finds another process which is ready, it will receive from that source. If none are ready it waits until one becomes ready and processes that. The time complexity of this method is $O(n)$, where n is the number of sources in the selection group.

We can, however, do much better than this if we previously know the selection group and utilize a queue (which can be implemented as a linked list) of ready sources within the group. The selecting process tests if there is a ready source in the group queue. If so it processes that and removes it from the queue. If there is not, it waits until there is, and processes that. The time complexity of this method is $O(1)$. Again, we often find that these groups correspond to the intrinsic algorithmic SAP groups which have already been described, allowing our optimizations to be performed for selection.

Similar queueing structures have been implemented in hardware (23). This is a crucial area for the development of message passing systems and will continue to be important to monitor closely.

Blocking and non-blocking operations Up to this point the reader may have been assuming that the type of procedure calls we have been describing are ones which wait until the transfer is complete before allowing users to continue with their own calculations. We describe such procedures as *blocking*.

The commonly-used technique of overlapping communication and computation to reduce idle time is known as latency hiding, and improves performance. This functionality can be provided by allowing the user to initiate a send, or receive, which proceeds in the background, and then wait for it to finish. If the environment provides facilities for the creation of concurrent *threads* then the provision of blocking procedures is sufficient for this purpose.

However, this functionality might be more conveniently provided in a message passing interface through the provision of *non-blocking operations*, especially since in some environments thread creation may not be available or may be very expensive. CHIMP provides support for both non-blocking and blocking operations.

13.2.5 *Configuration issues*

The configuration of hardware resources and placement of processes is, strictly speaking, outside the scope of a message passing interface. On the other hand the configuration of the message passing interface, that is. the specification of the SAPs required by each process, is certainly within this scope. This distinction has often been overlooked, largely because of a tendency to program very close to the hardware level.

There are a number of possible approaches to configuration.

- In a *static* configuration, the message interface is described along with the process configuration information in a configuration file. From the user's point of view this has the advantage that SAP configuration validation can be performed without requiring the necessary parallel resources to be available, and the disadvantage that the file must be modified to make even small changes in the SAP configuration.
- A *quasi-static* configuration requires the user to write a configuration program which specifies the SAP configuration along with the process configuration. This provides increased flexibility to the user, coupled with the disadvantage of the difficulty of writing the configuration program.
- A third alternative is *pseudo-dynamic* configuration, in which the SAP interface is configured from within the processes of the parallel program at run time using a procedure library. There is a point when the interface is undefined in every process, followed by a phase of SAP configuration in every process, followed by a point when the interface has been defined in every process. The interface may not be further defined or modified after this sequence of events. This offers few advantages to the user, except that the SAP configuration is in some sense more tightly bound to the instantiated processes, and the disadvantage that no configuration validation can occur until run time.
- A *fully dynamic* interface allows the processes of the user program to configure SAPs at any point in their execution, using a run time

library. This provides the ultimate freedom of expression to the programmer, and the greatest possible opportunity for programming errors.

From CHIMP's point of view there are significant problems with a static strategy. We would either be obliged to modify existing vendor-specific tools (which would require access to vendors' source code), or to define yet another configuration language and implement yet another configuration utility which translates our SAP configuration description into the input files of existing configurers. In both cases this introduces a substantial workload in matters outwith the scope of CHIMP. Moreover, the adoption of a static SAP configuration would preclude the exploitation of dynamic process creation, should it become available.

To CHIMP, a quasi-static strategy is unattractive for the same fundamental reasons as the static strategy. The obstacles have simply been moved from the configuration language into the configuration library.

Unless dynamic process creation is available, a fully dynamic configuration strategy offers few clear advantages over the pseudo-dynamic approach. There are a number of additional, and as yet unresolved, issues in the dynamic strategy which require further investigation.

Pseudo-dynamic SAP configuration, however, is an attractive development strategy for CHIMP, since it does not preclude the use of dynamic hardware and placement configuration facilities, and provides a convenient path toward a fully dynamic interface configuration.

Since dynamic process creation is not widely supported on tightly-coupled multicomputer platforms, a static process creation model is assumed. CHIMP currently requires the programmer to provide a textual configuration file which specifies the processes which are to be executed in the application. This file can be parametrized on a number of arguments which will be extracted from the command line.

13.3 PUL

The source code portability provided by CHIMP is being exploited by the Parallel Utilities Library (PUL) under development at EPCC. The main aim of the PUL programme is the development and maintenance of a library of utilities which promote code reuse in parallel programs, supporting application development and easing the burden of code migration from serial to parallel systems. PUL aims to achieve these objectives by providing support for frequently used parallel programming techniques, or paradigms, within a modular suite of libraries (7). The parallel infrastructure of an application can then be drawn from, and implemented within, the PUL modules, removing the burden of writing, debugging, and optimizing such codes from the application programmer. Implementing these modules using

CHIMP ensures vendor-independent portability for both PUL and applications which use it.

PUL modules are being developed in three separate strands: *non-specific* modules which implement generic functionality independent of the underlying mapping; *paradigm specific* modules which provide support for stereotyped decomposition and mapping strategies; and *application domain specific* modules which provide facilities which are only useful in particular types of application. In principle, any application program can make use of any number and mixture of PUL modules.

PUL modules provide a functional interface for programs written in both C and Fortran. A *procedural* interface is provided which encapsulates data flow, and provides maximum flexibility for applications. Where appropriate, *skeletal* interfaces are provided which also encapsulate control flow; each skeleton offering high level support to a more restricted class of algorithms. The skeletal and procedural modes of operation are inter-operable across processes within an application, thus allowing the programmer to import serial code and run it in parallel quickly, before isolating and optimizing parts of the application later.

Five modules are currently under development: PUL-EM (29) provides extended message passing functionality such as scan operations; PUL-GF (5) is a global file access utility allowing parallel processes to sensibly share files; PUL-TF (31) is a classical task farm in which a source process generates subproblems to be executed by one or more workers, the results being combined by a single sink process; PUL-RD (6) supports regular domain decomposition of rectilinear arrays of arbitrary dimensions; PUL-SM (30) will provide support for unstructured mesh problems, such as finite element methods.

To date, PUL-EM, PUL-GF, PUL-TF and PUL-RD have been released for internal use within EPCC. These modules have been implemented on Sun 4 workstations, Silicon Graphics Iris workstations, and T800- and i860-based Meiko Computing Surfaces. The intention is that PUL will be available on all systems running CHIMP.

13.4 Applications experience with CHIMP and PUL

Since 1991, EPCC's collaborations with industry have been greatly augmented under the auspices of the DTI/SERC Parallel Applications Programme. Many projects which have been or are being undertaken exploit the portability offered by CHIMP and PUL

The ANSE supersonic fluid dynamics code developed by Rolls Royce includes a solver for Navier Stokes equations on irregular meshes incorporating an associated turbulence model. Researchers at EPCC and in the School of Computer Studies at the University of Leeds are developing a parallel implementation of ANSE using CHIMP and PUL, initially targeted on Silicon Graphics workstations.

AEA Technology's ASTEC code models three-dimensional subsonic fluid dynamics using a finite-element method on unstructured meshes. EPCC has worked with AEA to develop a CHIMP implementation of ASTEC for use on an i860-based Meiko Computing Surface and Sun workstations.

In a project with Intera Information Technologies, Caplin Cybernetics Corporation and IBM, a parallel implementation of Intera's ECLIPSE oil reservoir simulation code is being developed using CHIMP. The target architecture for the parallel implementation is a network of IBM RS6000 workstations connected by serial optical channels, but CHIMP's portability will allow the application code to exploit a wider range of parallel systems.

EPCC and British Telecom are using CHIMP to develop a MIMD implementation of GAS (Generic Application Services), a software package developed by British Telecom to support network management applications. The GAS software is an implementation of the Open System Interconnection/Network Management Forum (OSI/NMF) Application Services Specification. The OSI/NMF standards and protocols provide common connectivity and inter-operability between different networks or associated equipment for the purposes of network management. This project seeks to demonstrate the potential benefits of parallel processing technology to achieve the higher computing performance and capacity essential for the management of the more complex and higher bandwidth networks of the future.

Parallel algorithms for Geographical Information Systems are being implemented using CHIMP and PUL in a project involving Meiko Scientific, Oracle UK, Digital Equipment Corporation, ESRI UK, Laser-Scan, Smallworld Systems Ltd, GIMMS (GIS) Ltd, and British Gas Scotland.

The Euphrates project seeks to exploit previous collaborative research with Shell which has investigated three-dimensional time-varying seismic processing on distributed memory MIMD architectures. The aim of the project is to make the results of this earlier work available in a production environment. CHIMP, PUL-TF, and PUL-RD are being used to provide portability across Silicon Graphics workstation networks and more tightly coupled multicomputer platforms.

CHIMP and PUL have also been used to develop demonstrator implementations of the UK Home Office codes for fingerprint encoding and matching.

A wide range of academic research projects are also using CHIMP and PUL on a variety of systems.

A parallel framework for genetic algorithms has been implemented on CHIMP (17) and used to develop genetic algorithms for optimizing the choice of shares in financial portfolios (24) and for routing pipelines to minimize economic and environmental cost (1).

CHIMP has been used in the implementation of Neurosys, a simulator for layered feed-forward neural networks (15), and with PUL, in a parallel library for irregular mesh problems (18, 20), and non-linear optimization codes (19).

CHIMP underpins a number of parallel programming language implementations. Concurrent ML has been developed on Sun workstations and T800-based Meiko Computing Surfaces (2), while CHIMP versions of Dataparallel C exist on a number of platforms (22).

A number of applications use CHIMP to allow communication between a high performance parallel computer and a remote visualization resource. Calculation intensive code running on a Connection Machine CM-200 communicates over a network with visualization code running on a Silicon Graphics workstation in applications implementing a shallow water model of the Earth's atmosphere (3) and a simulation of the Scandinavian ice sheet during the last ice age (32, 10). Another project seeks to develop a CHIMP application which will allow researchers in the University of Edinburgh's Institute for Cell and Molecular Biology to exploit EPCC's Connection Machine for transparent remote enhancement of microscope images (16).

Work is also proceeding on a large number of other CHIMP- and PUL-based applications, including distributed circuit simulation, inter-visibility applications, reaction diffusion systems, indoor radio propagation models, sequence comparison and radar simulation.

13.5 Further work

The formation of the Message Passing Interface (MPI) Forum, an international group set up along the lines of the successful HPF Forum with a view to define a standard message passing interface, is potentially of great importance. Working from an initial proposal (9), the group aims to promulgate an interface definition by September 1993. Contributing to the deliberations of the MPI forum will continue to be an important aspect of the CHIMP project.

The CHIMP interface is currently being extended, with the intention of providing a fully dynamic interface with some level of support for heterogeneous systems and greater transparency of resources in process configuration. Prototype implementations of this extended message passing model, which will be completely compatible with existing codes, are currently under development.

An effort is also underway to build on preliminary work on visualizing the execution of CHIMP programs (11) with the aim of providing an X-based tool for monitoring the behaviour and performance of CHIMP and PUL applications.

13.6 Acknowledgements

Many individuals at EPCC have been involved in the work which is reported in this paper. In particular, the paper owes much to the earlier work of Lyndon Clarke, Hamish Mills, and Arthur Trew.

REFERENCES

1. R. Allenson. Genetic Algorithms with Gender for Multivariate Optimisation. Report EPCC-SS92-01, Edinburgh Parallel Computing Centre, The University of Edinburgh, Edinburgh, EH9 3JZ, UK, September 1992.
2. R. A. F. Bhoedjang. Porting Concurrent Poly/ML to the Computing Surface. Report EPCC-SS92-02, Edinburgh Parallel Computing Centre, The University of Edinburgh, Edinburgh, EH9 3JZ, UK, September 1992.
3. R. Brockie. Implementation and Visualisation of the Shallow Water Meteorological Model. Report EPCC-SS92-05, Edinburgh Parallel Computing Centre, The University of Edinburgh, Edinburgh, EH9 3JZ, UK, September 1992.
4. R. A. A. Bruce and J. G. Mills. CHIMP Version 1.0 Interface. Technical Report EPCC-KTP-CHIMP-IFACE 1.5, Edinburgh Parallel Computing Centre, The University of Edinburgh, Edinburgh, EH9 3JZ, UK, 1992.
5. S. R. Chapple. PUL-GF Prototype User Guide. Technical Report EPCC-KTP-PUL-GF-PROT-UG, Edinburgh Parallel Computing Centre, The University of Edinburgh, Edinburgh, EH9 3JZ, UK, 1992.
6. S. R. Chapple. PUL-RD Prototype User Guide. Technical Report EPCC-KTP-PUL-RD-PROT-UG, Edinburgh Parallel Computing Centre, The University of Edinburgh, Edinburgh, EH9 3JZ, UK, 1992.
7. L. J. Clarke. PUL Concepts I. Technical Report EPCC-KTP-PUL-CONC-I, Edinburgh Parallel Computing Centre, The University of Edinburgh, Edinburgh, EH9 3JZ, UK, 1991.
8. L. J. Clarke and G. V. Wilson. Tiny: An Efficient Routing Harness for the Inmos Transputer. *Concurrency: Practice and Experience*, June 1991.
9. Jack J. Dongarra, Rolf Hempel, Anthony J. G. Hey, and David W. Walker. A Proposal for a User-Level Message Passing Interface in a Distributed Memory Environment. Technical Report ORNL/TM-12231, Oak Ridge National Laboratory, Engineering Physics and Mathematics Division, Mathematical Sciences Section, Oak Ridge, Tennessee 37831, October 1992.
10. I. Flockhart. Visualisation of the Scandinavian Ice Sheet. Report EPCC-SS92-09, Edinburgh Parallel Computing Centre, The Univer-

sity of Edinburgh, Edinburgh, EH9 3JZ, UK, September 1992.
11. D. G. Hanley. Monitoring and Visualising Multicomputer Application Performance. Report EPCC-SS92-10, Edinburgh Parallel Computing Centre, The University of Edinburgh, Edinburgh, EH9 3JZ, UK, September 1992.
12. Philip J. Hatcher and Michael J. Quinn. *Data-Parallel Programming on MIMD Computers*. MIT Press, 1991.
13. R. Hempel. *The ANL/GMD Macros (PARMACS) in FORTRAN for Portable Parallel Programming using the Message Passing Programming Model*. Gesellschaft für Mathematik und Datenverarbeitung mbH, Postfach 1240, 5205 St. Augustin 1, West Germany, November 1991. Version 5.1.
14. High Performance Fortran Forum. Draft High Performance Fortran Language Specification. Technical report, Center for Research on Parallel Computation, Rice University, Houston, TX, USA, November 1992.
15. R. B. Hofmann. Neurosys User Guide: A Portable Library for Parallel Feed-Forward Neural Networks. User Guide EPCC-UG92-03, Edinburgh Parallel Computing Centre, The University of Edinburgh, Edinburgh, EH9 3JZ, UK, 1992.
16. A. C. Hume. Biological Image Deconvolution. Report EPCC-SS92-12, Edinburgh Parallel Computing Centre, The University of Edinburgh, Edinburgh, EH9 3JZ, UK, September 1992.
17. G. P. Jones. Parallel genetic algorithms for the travelling sales rep problem. Master's thesis, Department of Computer Science, The University of Edinburgh, Edinburgh, EH9 3JZ, UK, September 1992.
18. S. Keras. A Navier-Stokes Equations Solver for Incompressible Viscous Fluid. Report EPCC-SS92-14, Edinburgh Parallel Computing Centre, The University of Edinburgh, Edinburgh, EH9 3JZ, UK, September 1992.
19. M. I. Marr. Parallelisation of Nonlinear Optimisation Algorithms. Report EPCC-SS92-18, Edinburgh Parallel Computing Centre, The University of Edinburgh, Edinburgh, EH9 3JZ, UK, September 1992.
20. K. D. Murphy. Distributed Irregular Meshes in Computational Fluid Dynamics. Report EPCC-SS92-21, Edinburgh Parallel Computing Centre, The University of Edinburgh, Edinburgh, EH9 3JZ, UK, September 1992.
21. Matthew Rosing, Robert B. Schnable, and Robert P. Weaver. The dino parallel programming language. *Journal of Parallel and Distributed Computing*, 13(1):30–42, September 1991.
22. B. K. Seevers. Dataparallel C on the Meiko Computing Surface. Report EPCC-SS92-23, Edinburgh Parallel Computing Centre, The University of Edinburgh, Edinburgh, EH9 3JZ, UK, September 1992.

23. Charles L. Seitz. In *VLSI and Parallel Computation*, chapter 1. Morgan Kaufman Publishers, Inc., 1990.
24. J. Shapcott. Index Tracking: Genetic Algorithms for Investment Portfolio Selection. Report EPCC-SS92-24, Edinburgh Parallel Computing Centre, The University of Edinburgh, Edinburgh, EH9 3JZ, UK, September 1992.
25. V. S. Sunderam. PVM: A Framework for Parallel Distributed Computing. *Concurrency: Practice and Experience*, 2(4), December 1990.
26. Andrew S. Tanenbaum. *Computer Networks*. Prentice-Hall International, second edition, 1989.
27. Thinking Machines Corporation, Cambridge, MA, USA. *Programming in C**, June 1991. Incorporates 'C* Programming Guide' and 'C* Users Guide'.
28. A. S. Trew, J. G. Mills, and L. J. Clarke. CHIMP Concepts. Technical Report EPCC-KTP-CHIMP-CONC 1.2, Edinburgh Parallel Computing Centre, The University of Edinburgh, Edinburgh, EH9 3JZ, UK, June 1991.
29. S. Trewin. PUL-EM Prototype User Guide. Technical Report EPCC-KTP-PUL-EM-PROT-UG, Edinburgh Parallel Computing Centre, The University of Edinburgh, Edinburgh, EH9 3JZ, UK, 1992.
30. S. Trewin. PUL-SM Prototype Functional Specification. Technical Report EPCC-KTP-PUL-SM-PROT-FS, Edinburgh Parallel Computing Centre, The University of Edinburgh, Edinburgh, EH9 3JZ, UK, 1992.
31. S. Trewin. PUL-TF Prototype User Guide. Technical Report EPCC-KTP-PUL-TF-PROT-UG, Edinburgh Parallel Computing Centre, The University of Edinburgh, Edinburgh, EH9 3JZ, UK, 1992.
32. M. C. Warren. Modelling the Scandinavian Ice Sheet on the Connection Machine. Report EPCC-SS92-25, Edinburgh Parallel Computing Centre, The University of Edinburgh, Edinburgh, EH9 3JZ, UK, September 1992.
33. S. Wilson and M. Norman. TITCH: Topology Independent Transputer Communications Harness. Technical Note EPCC-TN91-08, Edinburgh Parallel Computing Centre, The University of Edinburgh, Edinburgh, EH9 3JZ, UK, 1991.

14

USING A FUNCTIONAL NOTATION TO SPECIFY ABSTRACT SIMULATION MODELS

H.L. Muller[*], P.H. Hartel and L.O. Hertzberger

University of Amsterdam,
Kruislaan 403,
1098 SJ Amsterdam,
The Netherlands

Abstract

Simulation is the process where a real world entity is modelled and implemented on a computer. It turns out that all simulations in all application domains use only a few different simulation algorithms. In this article four of these algorithms are derived informally. Starting with a definition of a simulation specified in a functional language, a demand driven simulation algorithm, continuous time, discrete time, and discrete event algorithms are subsequently derived. The advantage of using a functional language to specify these algorithms is that it is relatively simple to reason about several aspects, as correctness, efficiency, expressiveness, and potential parallelism.

The algorithms are not functionally equivalent; they have increasing expressiveness and increasing efficiency at the cost of decreasing potential parallelism. This is illustrated with the help of a trivial example from the field of logic circuit design, although the algorithms can be used for other (large) applications without modification. The four simulation algorithms have been implemented and executed to verify the efficiency considerations in practice.

14.1 Introduction

Many simulation tools have been developed over the past decades. Some tools are specifically written for one application (the simulation of a certain type of microprocessor) and other tools are useful for a specific application domain, such as fluid dynamics, digital signal processing architectures, or queueing problems. Despite the rich variety of simulators, all are based on the same simulation principles, which will be discussed

---

[*]Now with: Computer science department, University of Bristol, 10 Priory Road, Bristol BS8 1TU, UK. E-mail: muller@acrc.bristol.ac.uk

in detail. Most simulators have one problem: the vast amount of CPU time needed. To provide insight in the fundamental problems involved in developing efficient simulation methods, a formal, functional description of each of the basic simulation algorithms is given. These descriptions are analyzed with respect to the expressiveness, efficiency, and potential parallelism. Because a formal description is used to specify the algorithms, it is possible to reason about these properties (Kelly 1989, Sijtsma 1989).

Reasoning about a description and its properties is facilitated by viewing the description as an abstract machine. Execution on an abstract simulation machine corresponds to executing a simulation. The abstract machine view of simulation algorithms allows a clear separation to be made between the specification of the simulation systems and the specification of the simulated object. The first plays the role of the abstract machine, the latter plays the role of an abstract machine program, as is explained in Section 14.2.

Section 14.3 presents the essence of simulation. The resulting algorithm will never be used for real simulations because of grave inefficiencies. From this inefficient algorithm, a simulation algorithm can be derived with a continuous time scale, as presented in Section 14.4. Continuous time simulators are mostly used to model physical processes, where variables change continuously in time. The algorithm can also be turned into a discrete time simulator (Section 14.5), which is more efficient when processes with discrete steps are to be simulated, as for example logic circuits, or problems with queues. In Section 14.6 the event driven simulator is presented, which is the most efficient simulation algorithm for discrete time problems. The paper is concluded with a discussion of the most important properties of the four algorithms.

14.1.1 Running example and notation

Throughout the paper, the *flip-flop* is used as a running example. The flip-flop is one of the elementary circuits that can be used for data storage. The schematics of the flip-flop are depicted in Fig. 14.1, together with the truth table. Although a flip-flop is not a complex circuit, it possesses all hard problems for simulation: it contains a loop, it contains state, and it will not stabilize unless used in the right way. The flip-flop has two inputs labelled $\overline{\text{Set}}$ and $\overline{\text{Reset}}$, and two outputs Q and $\overline{\text{Qb}}$. For a stable situation both $\overline{\text{Set}}$ and $\overline{\text{Reset}}$ are kept high, giving a high signal on one of the outputs and a low signal on the other: $\overline{\text{Set}} = \overline{\text{Reset}} = 1 \Rightarrow Q = \neg \overline{\text{Qb}}$. Driving $\overline{\text{Set}}$ low for a while causes Q to become high regardless of its previous state. Driving $\overline{\text{Reset}}$ low for a while causes Q to become low (and $\overline{\text{Qb}}$ high) regardless the previous state. Driving $\overline{\text{Set}}$ or $\overline{\text{Reset}}$ low for a short period of time brings the flip-flop in an oscillating state for an indefinite period of time: it causes a short pulse to start racing through the two nand gates.

INTRODUCTION

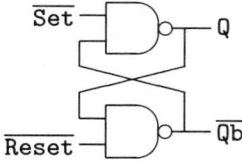

Set	Reset	Q	Qb
1	1	Q	Qb
0	1	1	0
1	0	0	1
0	0	1	1

FIG. 14.1. The flip-flop circuit and truth table

In Miranda (TM of Research software Ltd.) the high and low signals are represented by H and L respectively, while the X stands for an undefined signal that can be either high or low. All program fragments use the following definition for the type representing the state of a wire called threestate and the definition of the nand characteristics (refer to (Bird and Wadler 1988) for an introduction to functional programming, and to (Turner 1985, Turner 1990) for the definition of Miranda):

```
> timestamp == num        || The time is stored as an integer
> threestate ::= X | L | H   || undefined, low, high
>
> nandfun :: threestate -> threestate -> threestate
>
> nandfun H H = L          || The only way to get 'Low' out of a nand
> nandfun L b = H          || 'Low' on left port forces a high output
> nandfun a L = H          || 'Low' on right port forces a high output
> nandfun a b = X          || all other inputs result in undefined
```

The function nandfun takes two values of the type threestate and produces a value of the same type. When applied to two high input values (nandfun H H), the result is a low output (L), a low value on the first or the second parameter results in a high output, while all other inputs (for example nandfun H X) result in undefined, X (note that a and b are free variables).

The flip-flop is simulated, under the assumption that the nand gates introduce a fixed delay of 3 time units. The $\overline{\text{Set}}$ and $\overline{\text{Reset}}$ wires are driven by the two clock signals that are depicted in Fig. 14.2, together with the expected output signals on Q and $\overline{\text{Qb}}$. The signals on the Q and $\overline{\text{Qb}}$

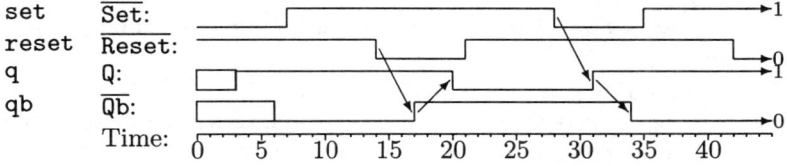

FIG. 14.2. The inputs and outputs of the examples

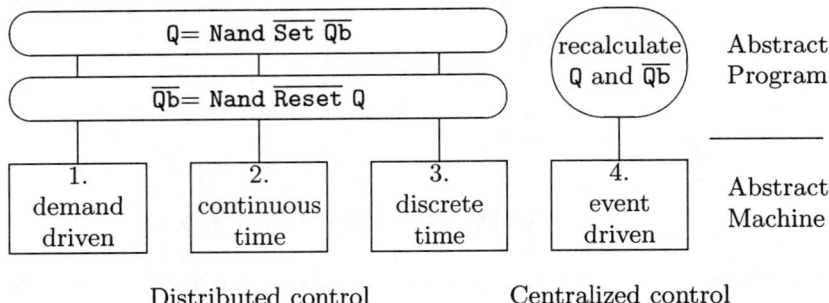

FIG. 14.3. The simulations as two different abstract programs on four different abstract machines

wires are initially undefined, which is shown by the shaded areas. The first change is a transition of the $\overline{\text{Set}}$ wire from low (0) to high (1) at time $t = 7$. This has no further effect on the state of the circuit. The second change at time $t = 14$ does have an effect on the state of the circuit, which is shown by the two arrows, that represent the causality of the sequence of events. The transition on the $\overline{\text{Reset}}$ wire first causes a transition on the $\overline{\text{Qb}}$ wire with a delay of 3 time units. This in turn causes a transition on the Q wire with another delay of 3 units. The system is simulated through another transition of the $\overline{\text{Set}}$ wire and another transition on the $\overline{\text{Reset}}$ wire.

14.2 An abstract machine program for the flip-flop

A system of interacting components, such as the flip-flop, can be described by using recursion equations. Assuming that the signals $\overline{\text{Set}}$ and $\overline{\text{Reset}}$ are defined externally, two equations are necessary to fully define the flip-flop:

$$\begin{aligned} \text{Q} &= \text{Nand } \overline{\text{Set}} \quad \overline{\text{Qb}} \\ \overline{\text{Qb}} &= \text{Nand } \overline{\text{Reset}} \text{ Q} \\ \overline{\text{Set}} &= \ldots \\ \overline{\text{Reset}} &= \ldots \end{aligned}$$

The true interpretation of the signal wires is the ensemble of all possible values on the signal wires at every possible instant in time. Simulation provides an approximation to the values at approximated time instances. Because all values of interest are approximated, simulation on a computer is feasible.

The recursion equations that describe a system, which is to be simulated, can be viewed as an abstract program. Such a program may be fed to a certain number of abstract machines that differ in the way in which simulation is performed. All abstract machines should give correct results. The results may be different only in the level of detail provided

```
> wire == timestamp -> threestate
>
> || ================ ABSTRACT MACHINE PROGRAM =================
> q, qb :: wire              || Definition of the q and qb
> q         = nand1 set  qb  || Define upper nand
> qb        = nand1 reset q  || Define lower nand
>
> || ============ INPUT OF ABSTRACT MACHINE PROGRAM =============
> set, reset :: wire         || Definition of set and reset
> set   t = L,  if t mod 28 < 7 || This defines a clock
>         = H,  otherwise       || 7 ticks low, 2 high
> reset t = set (t+14)       || reset is shifted set.
>
> || ==================== ABSTRACT MACHINE =====================
> delay :: timestamp -> timestamp
> delay t     = t-3
>
> nand1 :: wire -> wire -> timestamp -> threestate
> nand1 x y t = nandfun (x tbefore) (y tbefore), if t > 0
>             = X,                               otherwise
>               where
>                 tbefore = delay t
```

FIG. 14.4. The source code in Miranda for a demand driven simulator

or in the time it takes to perform a simulation. Fig. 14.3 shows the four different abstract simulation machines. Each provides an implementation of the **Nand** abstract machine instruction. This implementation performs the simulation of the circuit based on the data structures used by the particular abstract machine. Although each abstract machine uses a different data representation and mode of execution, no changes should be required to the abstract program. We shall work through each of the four basic simulation algorithms and see how faithful we can be to this principle.

14.3 Demand-driven simulation

The Miranda program of Fig. 14.4 shows a demand driven simulator. The approximations to the real values of Q, $\overline{\text{Qb}}$, $\overline{\text{Set}}$ and $\overline{\text{Reset}}$ are represented by the functions q, qb, set and reset respectively. The type signatures show that these functions map approximated time (of type timestamp) onto approximated values (of type threestate). As required, the abstract program (i.e. the recursion equations) are literally present in the Miranda program. The nand1 function and the associated data structures represent the abstract machine.

The state on the wires at a moment t in time is defined in terms of the output of the nand gate at time t, while the nand-function at time

t depends on the state at the input wires at time (*delay t*). When the simulator is asked for the state of Q at time 40, all states of Q and $\overline{\text{Qb}}$ are recursively calculated until the moment the simulator comes to a well-defined state (when one of the input signals is low, the output of the nand is fixed regardless its previous state), or until time zero, when the signal is X (by definition). In this example, the calculation of q 40 requires the values of set and qb of three steps earlier (the delay of the nand gate), the value of set 37 (which is H) and the value of qb 37. To calculate qb 37, the value of reset 34 (H) and the value of q 34 is required. This depends on the value of set 31, which is L; consequently q 34 equals H, qb 37 equals L, and q 40 thus equals H, which is the answer.

An interesting aspect of the demand-driven simulator is that by expressing it in a lazy functional language, the mode of evaluation required by the simulation is the natural mode of evaluation provided by lazy evaluation. The language implementation guarantees that the computations as described above proceed in exactly the way described. Most importantly, no states are calculated that are not strictly required, and all expressions can be evaluated in parallel. In spite of this apparent efficient behaviour, demand driven simulation is not used in practice because of three serious drawbacks. Firstly it outputs only the state at a given moment in time, the question: 'What is the first time that Q will become high after time T?' cannot be answered directly; only an exhaustive search can provide the answer. To print the history of state changes on both q and qb one would apply each to all time stamps of interest, as is written in Miranda with the map function:

```
> map q  [0..27] = [X,X,X,H,H,H,H,H,H,H,H,H,H,H,H,H,H,H,H,H,L,L, ...
> map qb [0..27] = [X,X,X,X,X,X,L,L,L,L,L,L,L,L,L,L,L,H,H,H,H,H, ...
```

The second drawback is that the recursive calculation places a huge demand on the memory, since a whole stack of calculations is built before they are evaluated. Thirdly, in a more complex circuit, where Q is used in more than one place, Q will be recalculated each time leading to an exponential time consumption. All three drawbacks can be relieved by reversing the order of computations, thus by starting with the state at time zero, and by processing forward in time. By memoizing (Hughes 1985) the old states in (lazy) lists, states in the past can be referred to. There are two ways to maintain these lists: with implicit timing information, giving a *continuous* time simulation, or with explicit time stamps, resulting in a *discrete* time simulation. These methods are presented in the next sections.

14.4 Continuous-time simulation

A simulation of the flip-flop with an implicit continuous time increment is an improvement over the demand driven simulator (Vree 1989). The wires are now represented by infinite lists (streams) of states. The ith element

```
>  || ================ ABSTRACT MACHINE PROGRAM =================
>  q, qb :: [threestate]
>  q    = nand2 set   qb
>  qb   = nand2 reset q
>
>  || ============ INPUT OF ABSTRACT MACHINE PROGRAM =============
>  set, reset :: [threestate]
>  set  = [L,L,L,L,L,L,L,H,H,H,H,H,H,H,H,H,H,H,H,H,H,H,H,H,H,H]
>  reset= [H,H,H,H,H,H,H,H,H,H,H,H,H,L,L,L,L,L,L,L,H,H,H,H,H,H]
>
>  || ==================== ABSTRACT MACHINE =====================
>  delay :: [threestate] -> [threestate]
>  delay xs              = X:X:X:xs
>
>  nand2, nand2' :: [threestate] -> [threestate] -> [threestate]
>  nand2' x       []     = []                    || terminate
>  nand2' []      x      = []                    || terminate
>  nand2' (x:xs) (y:ys)  = nandfun x y : nand2' xs ys || apply nand
>  nand2 xs       ys     = delay (nand2' xs ys)
```

FIG. 14.5. The flip-flop in a continuous time simulator.

of such a list represents the value of the wire at time $i \times \delta_t$, where δ_t is the time increment. The components are modelled by functions working on these lists, as synchronous processes (Kahn 1974). A nand-process in the flip-flop thus takes two *lists* of states as input parameters and produces a list of states as output. The source code for a continuous simulation of the flip-flop is shown in Fig. 14.5. Comparing this with Fig. 14.4, shows that the abstract program has not been altered, although the abstract machine and the data structures involved are now completely different.

The function nand2 consumes states from two input streams, and produces the output stream with help of the function nandfun as defined earlier. The function nand2 starts with three undefined states on the output list (X:X:X) to model a delay of three time steps. As before, the four wires are named q, qb, set, and reset, and are connected by means of two nand gates. By providing two input lists for the $\overline{\text{Set}}$ and $\overline{\text{Reset}}$, the program computes the approximated output values on the streams Q and $\overline{\text{Qb}}$:

```
>  q  = [X,X,X,H,H,H,H,H,H,H,H,H,H,H,H,H,H,H,L,L,L,L,L,L,L,L,L,L]
>  qb = [X,X,X,X,X,X,L,L,L,L,L,L,L,L,L,L,L,H,H,H,H,H,H,H,H,H,H,H]
```

Although the lists q and qb are mutually dependent, the algorithm does not deadlock because the start elements of the lists are defined (X:X:X). The productivity theory of (Sijtsma 1989) provides a framework to reason about liveness of functional programs. In terms of this theory, the function nand2 is +3-productive, indicating that a (cyclic) network of nand2 functions is productive (which means that the network will keep producing elements as long as input is provided on the input lists). Since the functions operating on the streams are independent, the simulation can be parallelized easily.

An essential property of continuous time simulators is that the time step is constant: all processes are synchronous, the states are produced synchronously, implying that all components of the simulation use the same time step. This type of simulator is used in simulations where the state would change continuously in time, but where the state is necessarily modelled with small discrete steps because the time cannot be incremented continuously. This technique is amongst others applied in (Vree 1989) for the simulation of water heights. Other examples include the simulation of electrical currents in a circuit or the positions of planets in a solar system. All such systems can be simulated with the same technique. Most of these problems do not have discrete state values, like the high, low, and undefined values used in the flip-flop simulation, but a real value, estimated by a floating point number.

Modelling processes with a non-constant time step with a continuous simulation model results in a waste of computing power, since it is not necessary to recalculate the state continuously: stable parts need not be simulated. This feature is provided by a *discrete-time* simulator.

14.5 Discrete-time simulation

To define a discrete-time simulation, again streams are used to model the state of a particular wire in the circuit. In contrast with the continuous simulator, the streams now have explicit time stamps, to allow the functions to operate asynchronously on these streams. The processes may consume an element from one input list, without consuming an element from the other input lists. There is thus no direct relation between the position in the stream and the time stamp.

In Fig. 14.6 the flip-flop is specified with a discrete time model. The stream modelling a wire consists of data structures of the form `Until T V`. The meaning of this data structure (called a tuple for short) is that the wire will be in state V until time T; at time T the state changes instantaneously into the value described by the next tuple. By definition, the time stamps of the tuples in a stream are increasing: the time proceeds forward. The nand function operating on these streams generates an output tuple for the state up to the lowest time-stamp in its input streams. Since all other streams have a higher time stamp, the states of all these streams are defined by their first tuple.

By adding a constant to the time stamp of the output tuple, the delay of the nand-gate is modelled. In contrast with the continuous time simulator, the lists consist of the changes only: the length of the lists does not depend on the granularity of the time.

```
> ||  ================ ABSTRACT MACHINE PROGRAM ================
> q, qb :: [state]
> q    = nand3 set   qb                    || The upper nand
> qb   = nand3 reset q                     || The lower nand
>
> ||  ============ INPUT OF ABSTRACT MACHINE PROGRAM =============
> set, reset :: [state]
> set   = [Until  7 L, Until 28 H ]        || A clock period
> reset = [Until 14 H, Until 21 L, Until 28 H ]
>
> ||  ==================== ABSTRACT MACHINE ======================
> state  ::= Until timestamp threestate   || The state of a wire
>
> select2 :: [state] -> [state] ->
>            (timestamp,threestate,threestate,[state],[state])
> select2 xs ys = (yt,xv,yv,xs ,ys'), if yt < xt
>               = (xt,xv,xs',ys ), if xt < yt
>               = (xt,xv,yv,xs',ys'), otherwise
>                 where
>                 (Until xt xv):xs' = xs
>                 (Until yt yv):ys' = ys
>
> delay :: timestamp -> timestamp;
> delay x      = x + 3
>
> nand3, nand3' :: [state] -> [state] -> [state]
> nand3' [] ys  = []                       || Terminate
> nand3' xs []  = []                       || Terminate
> nand3' xs ys  = Until (delay t) (nandfun xv yv) : nand3' xs' ys'
>                 where
>                 (t,xv,yv,xs',ys') = select2 xs ys
>
> nand3 xs ys   = Until (delay 0) X : nand3' xs ys
```

FIG. 14.6. The source code for a discrete time simulator

The representation of the flip-flop in Fig. 14.6 satisfies the requirement that the abstract program should remain unchanged while the abstract machine uses another data structure and corresponding **Nand** function.

The correctness of the discrete-time simulator can be established by showing that the simulator does not deadlock, generates correct outputs, and makes progress in time. The simulator will not deadlock because of the following invariant: the consumption of a tuple from an input stream will always result in a new tuple on the output stream. By using the theory of (Sijtsma 1989) again, the function **nand3** is +1-productive so the simulator will not deadlock. The simulator generates correct output values as long as all streams are ordered on their time stamp. It can be proved from the definition of nand3' that when the time stamps on the input lists are increasing, the output lists are ordered as well, hence all streams are ordered. Because a non zero delay is added to the time stamp, the simulator makes progress in virtual time.

The invariant 'each consumed tuple produces an output tuple' is also the weakness of the approach. Due to the loop in the definition of the flip-flop, tuples with increasing time stamp but identical state will start racing around the Q and $\overline{\text{Qb}}$, while the state of the circuit does not change. This is best observed on the calculated value of q:

```
> q = [Until   3 X, Until   6 H, Until   9 H, Until 10 H, Until 12 H,
>       Until 15 H, Until 16 H, Until 18 H, Until 20 H, Until 21 L,
>       Until 22 L, Until 24 L, Until 26 L, Until 27 L, Until 28 L,
>       Until 30 L, Until 31 L]
```

Although q is high from tick 3 until tick 20, there are seven tuples stating that q is still high at ticks 6, 9, 10, 12, 15, 16, and 18. The tuple at tick 12 is caused by the tuple at tick 6: Until 6 H in q causes a tuple Until 9 L on qb, which in turn causes the tuple Until 12 H on q. In the same way, the tuple at tick 9 causes the tuple at 15, and the tuple at tick 10 causes the tuple at 16.

The seemingly obvious solution of deleting tuples with an identical state is incorrect: it violates the invariant, and leads to an immediate deadlock of the program. Some of the redundant tuples can be avoided by *relaxing* the invariant: in the case of a nand, a low signal on one input fixes the output to a high value, regardless of the other input (it may even be undefined), on which the tuples may thus be ignored. When the nand is reprogrammed to ignore tuples on one channel as long as the other channel is low, far fewer tuples are generated. Still, the stable flip-flop has tuples running around, since both $\overline{\text{Set}}$ and $\overline{\text{Reset}}$ are high in the stable situation, requiring a redundant tuple to float through Q and $\overline{\text{Qb}}$.

The redundant tuples floating around are essentially caused by the distributed nature of the approach: the two nand circuits and wires operate completely autonomously which makes an empty tuple essential for the progress in the simulation (in (Chandy and Misra 1979) a similar effect is observed, which is solved by sending explicit NULL messages to keep a distributed simulator running). The problem can be solved by centralizing the solution, giving rise to a simulator known as an event-driven simulator.

14.6 Event-driven simulator

With respect to the previous simulator, two major changes are implemented: an explicit state of all wires is maintained, and there is a global list of things to happen in the future, the so-called *events*. The events leads to new states, and possible new events. The Miranda source of the

```
> wire    ::= Set | Reset | Q | Qb
> state   == [(wire,threestate)]              || List of association pairs
> event   ::= At timestamp wire threestate
> || ================= ABSTRACT MACHINE PROGRAM =================
> recalculate  :: wire -> (timestamp -> wire -> state -> event)
> recalculate  Q    = nand4 Set Qb
> recalculate  Qb   = nand4 Reset Q
>
> dependencies :: wire -> [wire]
> dependencies Set   = [Q]   || Wire Set used in def of wire Q
> dependencies Reset = [Qb]  || etc. Can be derived automatically
> dependencies Q     = [Qb]  || from definition of recalculate,
> dependencies Qb    = [Q]   || explicit for the sake of simplicity
> || ============ INPUT OF ABSTRACT MACHINE PROGRAM =============
> set           = [ At 0 Set L,   At  7 Set H,  At 28 Set L ]
> reset         = [ At 0 Reset H, At 14 Reset L, At 21 Reset H ]
> initialevents = merge set reset
> initialstate  = [(wire,X) | wire <-[ Set, Reset, Q, Qb ] ]
> || ==================== ABSTRACT MACHINE ======================
> select :: state -> wire -> threestate
> select st want    = hd [old | (have,old) <- st; want = have]
> update :: state -> wire -> threestate -> state
> update st want new = (want,new) : [(w,v) | (w,v) <- st; want ~= w]
>
> sim :: [event] -> state -> [event]
> sim []      st = []                              || Termination
> sim (e:es) st = sim es st, if select st wire = what || IGNORE
>              = e : sim es' st', otherwise       || Process event
>              where
>              (At time wire what) = e
>              es' = merge es (sort more)         || New events
>              st' = update st wire what          || New state
>              more = [ mkevent out | out <- dependencies wire ]
>              mkevent wire = (recalculate wire) time wire st'
> delay :: timestamp -> timestamp
> delay t = t+3
>
> nand4 :: wire -> wire -> timestamp -> wire -> state -> event
> nand4 x y t w st = At (delay t) w (nandfun (select st x) (select st y))
>
> main = sim initialevents initialstate
```

FIG. 14.7. The source code for an event driven simulator

event driven simulator is listed in Fig. 14.7. The circuit description has been embedded in the definition of recalculate. This has altered the circuit description although the essential aspects have been retained.

An event is represented by a three tuple `At time which what`. The tuple tells at what time, which wire will get what value. The list of events is sorted on increasing time stamp, so the event with the lowest time is handled first. The time of the lowest event is also the current time of the simulation, and since the time should increase, newly generated events should have a time stamp larger than or equal to the current time, also known as the causality condition: an event can have consequences for the future, but no consequences for the past. The function `sim` traverses the list of events recursively, each time producing a new state based on the old state and the consumed event.

An event may change the state of the circuit, and when the state of one of the wires is changed, all components connected to that state are required to recalculate their output value, generating new events for their output wires. The new events are merged into the old event list before calling the simulator to consume the rest of the events. Note that in this example program the circuit is specified twice. In the definition of `recalculate` and in the definition of `dependencies`. This last function tells which wires have to be recalculated on a change of a state on a specific wire. This last function is in fact the inverse of the former, and can be derived automatically, but both are defined for the sake of simplicity.

An event that does not cause any change in the state may be ignored, as is done on the line marked *IGNORE*. This optimization is not allowed in the discrete-time simulator of the previous section, because it would cause a deadlock of that algorithm. The event-driven simulator does not deadlock because all events are managed centrally: there are still future events to continue the calculation. Without this optimization, the event-driven simulator would have the same poor performance as the previous discrete-time simulator.

The output of the event driven simulator contains the state changes on all the wires as a sorted list:

```
> main = [At 0 Set L, At 0 Reset H, At 3 Q H, At 6 Qb L, At 7 Set H,
>         At 14 Reset L, At 17 Qb H, At 20 Q L, At 21 Reset H,
>         At 28 Set L, At 31 Q H, At 34 Qb L]
```

14.7 Discussion: summarizing the simulation algorithms

The four algorithms presented in the previous sections describe the basic principles of simulation. The second algorithm is the one used in all continuous time simulations, because of the fixed time step. The discrete-time and event-driven algorithms are used for problems with varying time steps. However it is possible to simulate a continuous-time problem with a discrete-time simulator, or vice versa, at the cost of decreasing performance. Discrete simulations are sometimes parallelized by using the discrete-time algorithm; it has a distributed nature, but other algorithms can be parallelized as well (Misra 1986, Overeinder *et al.* 1991). There are some

DISCUSSION: SUMMARIZING THE SIMULATION ALGORITHMS   255

important differences in terms of efficiency, expressiveness, and potential parallelism. A comparison on base of these aspects is presented below.

14.7.1 The efficiency

The demand-driven algorithm is inefficient (both in space and in time). In the worst case, the algorithm needs execution time that is exponential in the in the number of components and the length of the virtual time (construct a flip-flop with *three*-input nand gates and connect the second and the third input of the upper nand to $\overline{\text{Qb}}$, and the first and the second input of the lower nand to Q: the values of q and qb will be calculated twice, recursively, leading to an exponential time behaviour).

The time requirements of the continuous-time algorithm are linear in the length of the simulation run, and inversely proportional to the time step δ_t. A discrete simulation can be performed using a continuous simulator (δ_t should be set to the greatest common denominator of all delays in the circuit), but it is inefficient, since there are many unnecessary recomputations.

Although the discrete-time algorithm deals better with problems with a non constant time step, inactive parts of the model need still to be recomputed to prevent the simulator from deadlocking. The worst-case time behaviour is for this reason identical to that of the continuous-time algorithm.

The event-driven algorithm deals with *all* inefficiencies, the execution time is linear in the number of executed events, the number of changes of the state in the circuit. For this reason the event driven algorithm is widely used for discrete-time simulations.

14.7.2 The expressiveness

All algorithms can be used to simulate all types of problems, although simulations of a discrete-time problem with a continuous-time simulator or vice versa are rather inefficient. There is another difference between the algorithms that is related to the correctness of the simulation algorithm. In the example of the flip-flop a constant delay is introduced by the nand gate: the nand gate delays the output signal with 3 clock ticks. However, in realistic circuits it is common that the delay is not fixed, but that the delay depends on the state of the circuit. The example of Fig. 14.8 shows a more realistic scheme of how a simple buffer behaves. In this example, the buffer needs 5 time steps to drive a signal high, and only 1 time step to bring the signal back to low again. A pulse entering the buffer will result in a shorter pulse on the output of the buffer.

The first algorithm computes in a demand driven fashion. Consequently, the delay has to be known before the state is calculated. This implies that the delay can only depend on the time in that part of the program, and not on the state. The buffer of Fig. 14.8 can thus not be

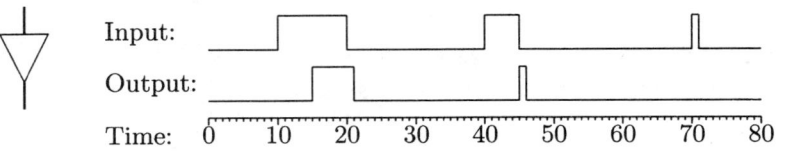

FIG. 14.8. A simple buffer circuit (Output := Input), that needs more time to get a signal high, than to bring it to low again. Consequently, pulses are shortened, while a short pulse disappears.

modelled using this algorithm.

The continuous-time algorithm introduces the delay as a fixed number of X's at the start of the output list. It is possible to make the delay dependent on the state of the wire, by checking if the wire contains a swap from L to H (or vice versa), and to generate a variable number of output elements as response (extra output elements for more delay, fewer output elements for short delay). Still, it is possible to prove that the simulator does not deadlock in that case (see (Muller 1993) for more details).

In the discrete-time algorithm, the function delay adds the constant 3 to the time. This function may be changed to an arbitrary function as long as delay satisfies two properties that follow directly from the correctness proof of the algorithm: progress in virtual time should be guaranteed (delay $t > t$) and the streams should be ordered ($t_x > t_y \Rightarrow$ delay $t_x >$ delay t_y). The first condition forbids the simulator to run backwards (in fact the causality condition), and the second property guarantees that the lists remains ordered: an unordered list of output tuples has no meaning. A delay that directly depends on both state and the time is incorrect: suppose that the delay function returns $(t+5)$ on a high input, and $(t+1)$ on a low input, in that case the second property is not satisfied (since delay H 10 $\not>$ delay L 12).

In the event-driven algorithm, the only constraint is that an event generated at time t should have a time stamp greater than or equal to t (the causality condition). Consequently, two consecutive events I_1, I_2 on an input, with $\text{Time}(I_1) < \text{Time}(I_2)$, can cause two output events O_1, O_2 with $\text{Time}(O_1) > \text{Time}(O_2)$, implying that the second event has overtaken the first one. This is sometimes the intended result, but most of the time, the result is disastrous. Although the event list provides the most flexible solution in terms of the delay, it is also the most dangerous implementation: it offers the designer the possibility of shooting in his own foot. It does not only support assignments, it even contains assignments *somewhere in the future*. Functional programmers advocate that it is hard to reason about assignments; it is illustrated here that it is even harder to reason about future assignments. An event list should be used carefully so that common causality rules are not violated.

14.7.3 *The potential parallelism*

The first simulation algorithm has an exponential complexity, the parallelization is thus trivial (ordinary divide-and-conquer suffices). The continuous- and discrete-time simulators are trivially parallelized, as well. Both lead to process networks (Kahn 1974) that can be mapped onto a multiprocessor machine. The communications and computations are decentralized, so as long as the communication graph of the process network can be mapped on the physical machine, both the continuous and discrete time simulators are perfectly scalable. Depending on the number of start elements on the various lists, the communication and computations can even be pipelined. The event-driven simulator is not parallel at all. The central event list needs to be stored somewhere, and even events with an identical virtual time cannot be executed in parallel in this implementation (because the order in which events are applied might result in different values). For this reason discrete simulations are always parallelized using the discrete-time algorithm. Heuristics are then applied to reduce the number of redundant synchronizations (see for example (Vries 1990)), but in worst case the algorithm is as inefficient as a continuous-time simulation.

14.8 Conclusion

The abstract machine models of four simulation algorithms have been specified. These are for a demand-driven, a continuous-time, a discrete-time and an event-driven simulation. Three of the models have a distributed control that allows for a parallel implementation of the simulation problem, while the fourth model has a centralized nature. To extend the insights on the abstract machine models, they have been specified formally as functional programs. These formal definitions allow us to reason about the performance (parallelism, efficiency) and correctness (deadlock freedom, simulation results).

Demand-driven simulation is inefficient, and is consequently directly parallelizable. Continuous-time simulations are trivially parallelized (in fact many applications that are currently executing on today's parallel machines are continuous simulation problems). Parallelizing discrete time simulators is much harder; the version that is easily distributed is inefficient (in the worst case as inefficient as the use of a continuous-time scale simulator), the one that is efficient cannot be distributed because of the global control. Parallelization of discrete-event problems in general is hard, but for certain classes of simulation problems (which can be formulated using the discrete time algorithm without loosing too much efficiency) an efficient parallel simulation is feasible.

14.9 Acknowledgements

We thank Wim Vree for his comments on a draft version of the paper.

REFERENCES

R. S. Bird and P. L. Wadler, (1988). *Introduction to functional programming*. Prentice Hall, New York.

K. M. Chandy and J. Misra, (1979). Distributed simulation: A case study in design and verification of distributed programs. *IEEE transactions on software engineering*, **SE-5**(5),440–452.

R. J. M. Hughes, (1985). Lazy memo-functions. In J.-P. Jouannaud, editor, *2nd Functional programming languages and computer architecture, LNCS 201*, pages 129–146, Nancy, France, Springer-Verlag, Berlin.

G. Kahn, (1974). The semantics of a simple language for parallel programming. In J. L. Rosenfeld, editor, *Information processing*, pp 471–475, Stockholm, Sweden, Aug. North Holland.

P. H. J. Kelly, (1989). *Functional programming for loosely-coupled multiprocessors*. Pitman, London.

J. Misra, (1986). Distributed discrete-event simulation. *Computing Surveys*, **18**(1),39–65.

H. L. Muller, (1993). *Simulating computer architectures*. PhD thesis, Dept. of Comp. Systems, Univ. of Amsterdam.

B. Overeinder, L. O. Hertzberger, and P. M. A. Sloot, (1991). Parallel discrete-event simulation. In W.-J. Withagen, editor, *3rd Computer systems*, pp 19–30, Eindhoven, The Netherlands, Univ. of Eindhoven.

B. A. Sijtsma, (1989). On the productivity of recursive list definitions. *ACM transactions on programming languages and systems*, **11**(4),633–649.

D. A. Turner, (1985). Miranda: A non-strict functional language with polymorphic types. In J.-P. Jouannaud, editor, *2nd Functional programming languages and computer architecture, LNCS 201*, pp 1–16, Nancy, France, Springer-Verlag, Berlin.

D. A. Turner, (1990). *Miranda system manual*. Research Software Ltd, 23 St Augustines Road, Canterbury, Kent CT1 1XP, England.

W. G. Vree, (1989). *Design considerations for a parallel reduction machine*. PhD thesis, Dept. of Comp. Sys, Univ. of Amsterdam.

R. C. de Vries, (1990). Reducing null messages in Misra's distributed discrete event simulation method. *IEEE transactions on software engineering*, **SE-16**(1),82–91.

15

STATISTICAL MODELLING AS A TOOL FOR STUDYING THE PERFORMANCE OF PARALLEL SYSTEMS

R. Candlin and J. Phillips*

Department of Computer Science
University of Edinburgh
King's Buildings
Mayfield Road
Edinburgh EH9 3JZ, UK

Abstract

We describe how parallel programs with the same type of synchronization pattern may be characterized in terms of a small number of parameters whose values can be easily observed. Unlike most models of parallel programs, our models suppress individual details of the computation and look for broad similarities between programs. We have shown that this is a feasible approach which leads to consistent quantitative results and useful practical applications.

We have conducted simulation experiments to discover how performance in a given hardware environment depends on the values of these parameters, and on the group to which the programs belong. Our approach has been to simulate the execution of synthetic programs on an accurate model of a machine. We can then control the parameter values to ensure that behaviour over a large part of 'program space' can be systematically examined. The use of traditional factorial design techniques in conjunction with the analysis of variance has enabled us to discover quickly and quantitatively the relative importance of the various parameters in determining performance. Furthermore, using the same techniques, we have been able to study the interaction between these parameters and the performance of a representative process migration strategy.

We provide a brief account of our experimental methodology, and illustrate it by describing an experiment to determine the factors affecting processor utilization for a CSP-type program on a transputer-based machine. We then show how the same type of experiments can be used to analyse the behaviour of a representative process migration strategy in the same environment.

*Supported by a studentship from the Science and Engineering Research Council

15.1 Introduction

Parallel machines dispose of a great deal of processing power but are difficult to use effectively. A programmer changing from a sequential to a parallel machine undergoes a culture shock: low-level features that are hidden on a sequential machine (and whose use is in fact discouraged because of the lack of portability of the resulting program) suddenly become a predominant feature in program development. Thus the programmer becomes concerned with such problems as placement and load-balancing that should properly be the responsibility of the systems software. At present, it is still largely the case that programmers have to hand-tune programs to obtain satisfactory performance, and this process is both time-consuming and off-putting to the average user.

It is instructive to consider the analogy with virtual memory in conventional machines. In the early days of sequential programming, programmers had to manage the use of memory themselves by overlaying code and data. Virtual memory freed them from that task and allowed them to concentrate on the problem to be solved in a machine-independent way, thus reducing development time and improving portability. Virtual memory 'works' because of the statistical properties of memory accesses, which are similar for a very large number of programs, thus giving good performance overall, although there may still be programs that would benefit from hand-tuning. We have wondered whether there might be similar behavioural patterns in parallel programs which could be utilized in developing systems software for parallel machines.

In order to build an accurate performance model of a given program, we not only have to take into account the structural and dynamic properties of the program, but we also have to consider its interaction with the machine on which it is running, and the run-time software. The whole system is extremely complicated, with dynamic properties that are in general poorly understood. Although it is an artificial construction, and in principle we know how the component parts work, the interactions between one part and another cannot usually be predicted. The system is therefore rather like an unknown system that we might encounter in the natural world, and it seems a good idea to study it in the same kind of way: by making observations of its behaviour under varying conditions. We might also hope that, as in many natural systems, inner complexity may be hidden, if we can average over a large number of occurrences.

These are the ideas that lie behind the work described in this paper. The question we ask is: 'Can we look at the macroscopic properties of a parallel program, without concerning ourselves with the detailed instruction sequence, and make a useful quantitative prediction about its performance?'. If we can find an empirical relation between program properties and performance which is valid for a large number of programs, we shall have a useful practical tool. Also, by finding out which factors are unimportant, we may be able to make helpful simplifying assumptions about behaviour, and be in a position to construct an analytical model.

In order to answer the question posed above, we have first to find a useful set of parameters with which to characterize a program. This necessarily involves experimentation, and we do not expect to arrive immediately at the best choice. We have therefore developed a methodology which is applicable to the quantitative study of parallel program performance in general. The main features of this approach are listed below.

- A parallel program is characterized in terms of a small set (say, 5 - 10) of parameters that refer to the program as a whole, for example: average grain size, average degree of the program graph.
- Classes of programs which have a similar synchronization behaviour are approximated by a set of program templates which represent activities such as 'compute', 'send', or 'receive'. These activities are all modelled by time delays.
- Synthetic programs with defined properties can be randomly generated, and their execution simulated on an accurate machine model. There has not yet been any attempt to simplify the description of the machine in the same way as the program model has been simplified. This is to ensure that we do not introduce spurious variability because of inadequacies in our machine model.
- Factorial experiments, followed by an analysis of variance, allow quick identification of the factors which are most important from the point of view of determining performance, and provide data from which a performance prediction formula can be derived for *all* programs within the given class and parameter range.

Unlike most models of parallel programs which describe in some detail the dynamic properties of *specific* programs, we have looked for models which emphasize the similarities between programs. Thus many of our parameters refer to time-averaged properties, since time-averaging essentially implements a low-pass filter and reduces effects due to rapid fluctuations. We have found that we can obtain consistent and meaningful results, at least for some classes of programs, even with very simple models. Having once obtained a program model, we can make use of it in further studies on run-time behaviour. As an example, we describe a set of experiments which are concerned with studying the relation between program characteristics and the effectiveness of process migration. We are able to interpret program behaviour in terms of the values of certain program parameters, and predict the type of program whose performance will be improved under a particular migration strategy.

In this paper we describe the framework within which our experiments are conducted and show how to find out what factors affect processor utilization for a particular class of parallel program. We illustrate the practical application of these results by showing that the same methodology can be used to study the behaviour of these programs under a representative process migration strategy.

15.2 Synthetic programs

15.2.1 *Program models*

Although most performance models have been based on variants of the task graph, we have preferred a process graph representation of a parallel program. Here the nodes represent processes and the edges indicate one-to-one communications channels between processes. The reasons for this choice are outlined below.

- The overall design of many parallel programs is structured as a set of sequential modules intermittently sending messages to one another in a fixed communication pattern: the CSP model (Hoare 1985).
- Many programs have a strong iterative structure, with long-lived processes that loop over a compute phase followed by a communications phase. The task graph does not seem a natural representation for programs of this type, as each iteration would constitute a separate task.
- The node and edge weights are readily obtainable, from the source code or by preliminary profiling.
- A weighted process graph represents a summation over time of the task graph and shows where the demand for processor and link resources will, on the average, be greatest.
- The process graph is a much cruder description than a task graph of a parallel program, but precisely because it takes out detail, it may be more suitable for generalizing about program behaviour.

We thus assume a weighted process graph as the basis of our program model. To take a particular case, a graph for an occam (Inmos Ltd 1988) program will have a static structure of top-level parallel processes connected by unidirectional channels. Node weights will represent time-averaged properties of the individual processes, such as number of instructions executed between channel I/O operations. Edge weights will represent properties of a channel such as average message length.

The next step is to compress the information in this graph, and summarize the weight distributions by a small number of statistics. This gives us a manageable number of macroscopic parameters for the graph. For example, we might summarize the information in the graph by eight parameters. Two of these take into account the size and connectivity of the graph: the number of nodes and the average node degree. Of the other six, three refer to the distribution of processing load, and three to the distribution of data transfers (i.e. they refer to the distributions of node and edge weights respectively). In each case they represent the overall mean value of a weight, averaged over the graph, the variation in this weight over the graph, and the variation at a given node or edge over time.

In addition to program properties that can be associated directly with the graph we have to consider the patterns of synchronization between processes. Although in principle we could model synchronization delays at a given node by a probability function, we do not know what form of function would be appropriate and we feel that it is better in the first instance to model certain common classes of synchronization patterns explicitly. We do this by associating a program template with each process. At any point in time, the template describes how the process can select the next activity according to the particular type of program being modelled. The set of possible activities has been taken to be: compute, send on a channel, or receive on a channel. These activities appropriately describe a compute-bound single-user program.

15.2.2 *The use of synthetic programs*

The reasons for using synthetic programs to study performance are quite compelling: they allow us to choose parameter values first, and then to construct a program with those parameters afterwards. This means that we can range over parameter space systematically and quickly, rather than being limited to a small set of values that may be unevenly distributed. A further advantage is that we are not restricted by the way in which programmers happen to have written their programs, which may have been structured more to conform to some particular machine architecture than to the natural structure of the application. The justification for using synthetic programs rests on our hypothesis that programs in a given class will exhibit common performance characteristics. In this case, conclusions can be drawn from experiments on *any* program from the class. The validity of this assumption will be tested when we see how much variation there is between programs from the same class and with the same parameter values.

An instance of a synthetic program is generated from a set of model parameters. There are two important points to note. First of all, obviously, only one out of the set of programs that are consistent with the model will be actually generated on any one occasion. Secondly, we need to know much more about the probability distributions of various program properties than in fact we do. This means that the forms of probability distribution that we assume in order to generate synthetic programs are really extra, hidden parameters to our model. We have assumed the following: that unweighted graphs are uniformly distributed, that means are normally distributed, and that variances are distributed as chi-square.

15.3 Experimental approach

The tool which underpins this research is described in (Candlin and Skilling 1991). It consists of several components: a modeller MIMD (Skilling 1991) written in Simula (Simula Standard 1987) based on DEMOS (Birtwistle 1986), an experiment support tool eg (Skilling 1991) and procedures for certain statistical analyses. For more sophisticated statistical procedures we have used the package GENSTAT (Genstat 5 Committee 1988). MIMD can be extended easily to support new hardware and software models as required, and permits modelling at different levels of detail. The overall control of an experiment is provided by eg, which also comprises a synthetic program generator.

The level of detail at which we model a program, a machine and the interaction between the two is a matter of choice. We have chosen to simplify the program, but to model the operation of the machine and its run-time support software in some detail. The experiments described in this paper are based on the hardware and the run-time environment of the Meiko Computing Surface (Meiko UK Ltd 1989) in the Edinburgh Parallel Computing Centre *. Thus the compute-time delays are obtained from timing experiments on the Meiko Computing Surface. The delay in sending a certain number of bytes across a link using a particular software communications harness was also measured experimentally (Candlin and Luo 1989). Other aspects of machine operation have been modelled explicitly: in particular, the processor scheduling mechanism, and the operation of the communications harness. Time slicing occurs as it would in a real system by maintaining a queue of active processes for each processor. Inter-processor message hops are explicitly modelled by passing the data through communications processes running on each processor and messages are queued until a link is free. We therefore take into account delays due to competition for resources and can estimate their importance. The experimental validation of this machine model is described in et al. (Candlin et al 1989).

15.4 Experimental design

The standard procedures of experimental design as described in for example et al.(Hunter 1978) and (Jain 1991) have not been used frequently in computer performance studies, although there are examples of their successful application (Namce et al. 1987, Lyon et al. 1993, Candlin et al. 1991). However, there is a large body of experience which performance specialists could utilize to advantage. Basically, the idea is that a relatively small number of experiments, each with a different combination of parameter values, can give the same amount of information about a system as a

---

*The Edinburgh Parallel Computing Centre is supported by major grants from the Computer Board, the Department of Trade and Industry, the Engineering and Physical Sciences Research Council, and industry

much larger number of experiments in which only one factor is allowed to vary at a time. Furthermore, factorial experiments give information about interactions between factors which is not obtainable from other types of experiment.

Factorial designs allow different models to be explored quickly, and the relative importance of the factors to be determined by an analysis of variance. In a two-level factorial design, each factor can appear at either a high or a low value, and each run takes a particular combination of high and low values of each parameter. A full factorial experiment allows the contributions to the variance of a response variable to be estimated for every factor, and for all interaction terms. A simple linear regression equation can easily be fitted from the results, giving an empirical, quantitative relation between a performance metric and factor values. This equation can then be used to predict the performance of an arbitrary program whose parameter values are known.

The approach can be extended to non-linear models by using factorials at intermediate levels of some or all parameters.

15.5 Factors affecting the performance of loosely-synchronized parallel programs

We now present an example to show how statistical methods can be used to derive performance prediction models for a particular class of programs. These are programs consisting of a number of long-lived, top-level, sequential processes which are loosely synchronized by message-passing. We model each process by a template which consists of an iterative loop. Within the loop, each process delays for a randomly-chosen 'compute' time, and then sends a message of random length to each of its neighbours. Different processes can have different mean compute times, and different channels can have different mean message lengths. In each case these are assumed to be time-averages of the behaviour over the whole execution of the program.

The performance metric which is studied here is U, the percentage utilization measured over all the processors. This metric has been suggested as a measure for evaluating the effectiveness of load-balancing strategies by other authors (Hac and Johnson 1986, Zhou 1988). In Section 15.7, we shall show how we can relate processor utilization under a typical process migration strategy to program parameter values.

15.5.1 *Experiment description*

Previous experiments (Candlin and Phillips 1993, Candlin et al. 1991) have indicated that a small set of model parameters can sufficiently describe the performance of programs consisting of similar processes. We accordingly start from the assumption that the same set of parameters is also appropriate for a program with the same synchronization structure, but where the processes and channels have different weights from each other. We accordingly assume the following parameter set:

$$\{N, c, \mu_{cg}, \sigma_{cg}, \mu_{mg}, \sigma_{mg}, \sigma_c, \sigma_m\}$$

- N and c define the size and connectivity of the process graph.
- Time-averaged compute times are distributed normally as $N(\mu_{cg}, \sigma_{cg})$ over the nodes of the program graph.
- Time-averaged message lengths are distributed normally as $N(\mu_{mg}, \sigma_{mg})$ over the edges of the graph.
- σ_c^2 and σ_m^2 describe the variance, from one iteration to the next, in compute times and message lengths respectively.

By using non-zero values of σ_{cg} and σ_{mg}, non-uniform patterns of activity can be generated, since individual nodes and edges will be allocated different mean compute times and mean message lengths. The degree of non-uniformity present depends upon the sizes of σ_{cg} and σ_{mg}, relative to their respective means. For example, if $\mu_{cg} = 5000$ and $\sigma_{cg} = 3000$, then mean compute times would vary greatly across the nodes of the process graph. If, however, $\mu_{cg} = 200000$ and $\sigma_{cg} = 3000$ then the degree of variation would be far less significant, since the standard deviation is small compared to the mean. It is better, therefore, to take as parameters $\sigma_{cg\%}$ and $\sigma_{mg\%}$, which express σ_{cg} and σ_{mg} as proportions of μ_{cg} and μ_{mg} respectively. For example, if $\sigma_{cg\%} = 10$, σ_{cg} would be equal to 10% of the value of μ_{cg}.

A two-level full factorial experiment was carried out varying six of the eight program parameters described above. The values were set so as to include a large subset of possible real programs. The levels used were as follows:

Graph shape
    N: 32 or 142
    c: 4 or 12

Computation times in clock ticks
    μ_{cg}: 3000 or 300000
    $\sigma_{cg\%}$: 10 or 80

Message lengths in bytes
    μ_{mg}: 10 or 5000
    $\sigma_{mg\%}$: 10 or 80

The remaining two parameters, σ_c and σ_m, were set to 10 and 1 respectively, to ensure that the programs generated displayed strongly time-invariant behaviour.

A number of other parameters had to be fixed in order to define the experiment fully. The hardware simulated was a 4×4 mesh of transputers (Inmos Ltd 1989). A round robin placement strategy was used and each simulation run was executed for $20,000,000$ clock cycles (1 sec) of the

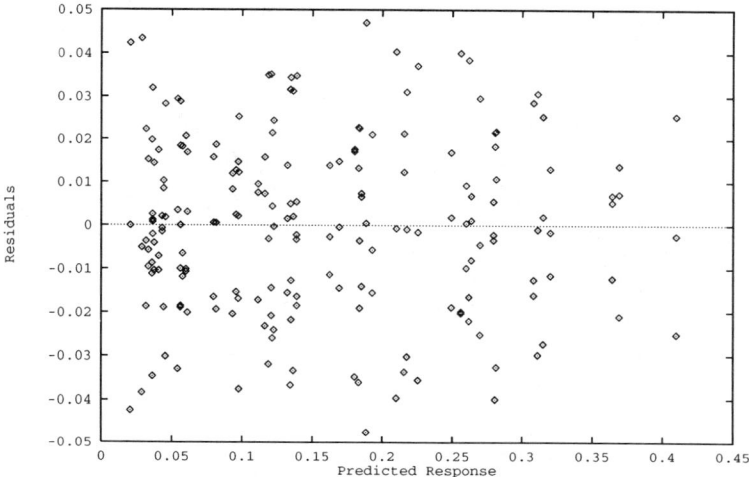

FIG. 15.1. Residual scatter plot for U'

simulated machine. Three replications were run, in which each replicate differed both in the random graph and random number seed used.

15.5.2 *Results*

Three assumptions must be satisfied in order for an analysis of variance to be valid. Firstly, the residuals must be IID* normal variates with zero mean, secondly the residuals must have constant variance and lastly the underlying model must be structurally adequate. If tests show that these assumptions are not valid, a transformation of the response variable may be appropriate. In this case, we applied a transformation due to Guerrero and Johnson which is particularly suited to response variables which are expressed as percentages (Atkinson 1987, Guerrero and Johnson 1982).

The transformed variable U' was calculated as follows:

$$U' = \left(\frac{U}{100-U}\right)^{0.08} - 1 \tag{15.5.1}$$

Fig. 15.1 shows a residual scatter plot and Fig. 15.2 shows a residual normal quantile-quantile plot for U'. These plots indicate that the normality and constancy of variance assumptions hold, since there is no discernible trend in Fig. 1, and Fig. 2 shows an approximately straight line.

The analysis of variance table for U' is given in Table 15.1. The differences in mean values between parameter settings are large compared to

---

*independent and identically distributed

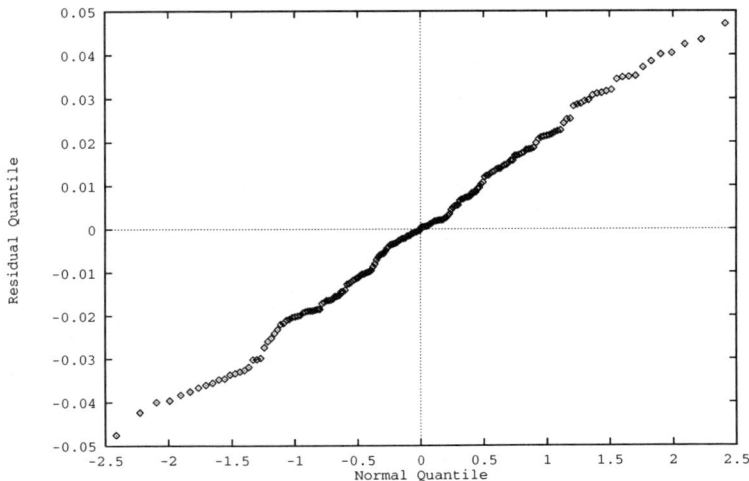

FIG. 15.2. Normal quantile-quantile plot for U'

Table 15.1 Analysis of variance table using response U'

Source	D o F	Sum squares	Mean squares	F ratio	F prob.
Between replicates	2	0.00124	0.000621	1.013	0.366
Between treatments	63	1.96	0.0311	50.273	0.001
Residual	126	0.0773	0.000613		
Total	191	2.038			

the differences in mean values between replicates. This is reflected in the F values, which indicate that the null hypothesis, that different replicates produce the same mean value of U' over all trials, can be accepted at the 0.1 per cent significance level. Similarly, the null hypothesis, that different parameter settings produce the same mean value of U', can be rejected at the 0.1 per cent significance level.

The effects due to all individual parameters and their interactions were estimated. Many of these were small and made little contribution to the variability. All effects which contributed at least 1 per cent are shown in in Table 15.2. In this experiment, the dominant terms are N, the number of nodes, and $\sigma_{cg\%}$, the standard deviation of process compute times. These, together with their first order interaction, account for 84.1 per cent of the variation in U'. The μ_{cg} and μ_{mg} effects are of less importance, accounting for 1.54 per cent and 1.39 per cent of variation respectively. The remaining two parameters, c and $\sigma_{mg\%}$, have very little influence indeed.

Table 15.2 Estimates of effects contributing at least 1% to the variability of the response U'

Effect	Estimate	t-value	% var
(gm)	0.1557	87.0936	
N	0.0505	28.27	24.05
μ_{cg}	0.0128	7.15	1.54
$c\mu_{cg}$	-0.0114	-6.39	1.23
σ_{cg}	-0.0783	-43.83	57.80
$N\sigma_{cg}$	-0.0155	-8.65	2.25
$\mu_{cg}\sigma_{cg}$	-0.0140	-7.81	1.84
μ_{mg}	-0.0122	-6.80	1.39
	Standard error = 0.00179		

15.5.3 Discussion

The small differences between replicates, and the fact that the chosen model explains all except 3.85 per cent of the variation over the experiment, gives us confidence that our results are generally applicable to programs with this type of synchronization behaviour.

We can draw other conclusions from the effects table, and relate them to our knowledge of the type of program and the machine environment. Large synchronization delays in a parallel program will inevitably lower the utilization of the machine that the program is being executed on, since a processor can do nothing while the processes it has been allocated are blocked. One would expect, therefore, processor utilization to drop as the value of $\sigma_{cg\%}$ increased, as this corresponds to a mismatch in process compute-times. This indeed appears to be the case since the sign of the $\sigma_{cg\%}$ effect is negative. The relative unimportance of μ_{cg} tells us that it is the degree of variation in mean compute times, rather than their magnitude, which has the greater effect on processor utilization.

The number of nodes, N, is also of some importance. This is because the larger the number of processes on a processor, the less chance there is of it finding itself idle. Consequently, mean processor utilization tends to increase as the number of processes increases: an assumption which is confirmed by the positive sign of the N effect.

The factors related to the lengths of messages, μ_{mg} and $\sigma_{mg\%}$, are of no great predictive value. This is an indication that it is the act of sending a message, rather than the length of the message, which has the greatest impact on performance. This is a consequence of the synchronous nature of the communications mechanism. The degree of the process graph, c, also has very little affect on performance, due to the parallel nature of message transfers and the characteristics of the transputer.

The results we have obtained are therefore in accordance with what we would expect. The interesting aspect of this experiment is that we can now *quantify* the effect of each program parameter.

15.6 A model for performance prediction

It is possible to use a designed experiment and accompanying analysis of variance to fit a polynomial response surface to a set of observations, using the method of orthogonal polynomials (Montgomery 1991). This technique allows one to decompose the effects of individual factors and their interactions into linear, quadratic, cubic, and possibly higher-order components. The contribution of each component to the overall regression equation can be tested since individual sums of squares can be produced. The method of orthogonal polynomials relies on the levels of the factors used in the experiment being chosen so that they are equally spaced. The generation of the response surface follows directly from an analysis of variance and the results are equivalent to those produced using conventional least squares techniques.

Since the two dominant parameters influencing the performance of the class of parallel program simulated were N, the number of nodes, and $\sigma_{cg\%}$, the standard deviation of compute times (expressed as a percentage of the mean compute time), it would seem reasonable to suppose that a performance prediction model using just these two parameters might provide useful estimates of performance, regardless of the levels of other factors.

To develop a model expressing U as a function of N and $\sigma_{cg\%}$, a full factorial experiment was executed with N set at six equally spaced levels. Again, a Guerrero-Johnson transform was applied to the response variable to obtain a transformed response U'.

Fig. 15.3 is a plot of the surface showing the mean values of U' over the three replicates, at each possible combination of N and $\sigma_{cg\%}$. The plot is rather difficult to interpret, although, as expected, it has a general downward slope as N decreases and $\sigma_{cg\%}$ increases.

A response surface was fitted to the experimental results from tables of contrasts and orthogonal ploynomials generated by GENSTAT. Full details of this procedure are given in (Montgomery 1991). A polynomial regression equation was obtained which expresses U' in terms of the predominant factors as:

$$10^5 U' = 14782 + 92.9N + 0.001(N-87)^4 - 3.38(N-87)^2 - \\ 208.47\sigma_{cg\%} + 4.74(\sigma_{cg\%} - 45)^2 - 0.028(N-87)^2(\sigma_{cg\%} - 45)$$
(15.6.2)

The resulting predicted response surface for U' is illustrated in Fig. 15.4.

A MODEL FOR PERFORMANCE PREDICTION

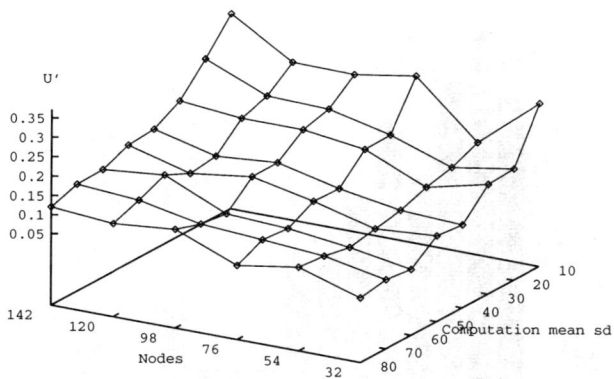

FIG. 15.3. Surface generated by experiment

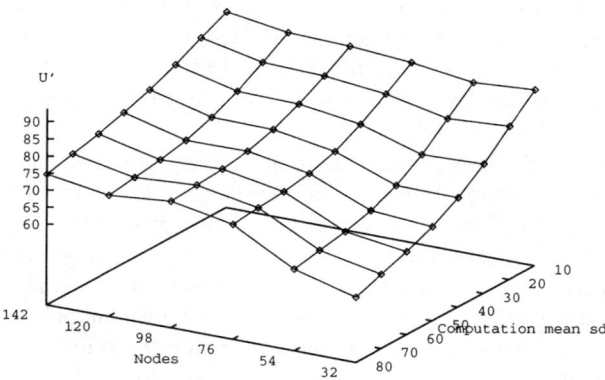

FIG. 15.4. Surface fitted to observed data

15.6.1 *Discussion*

To test how well the regression equation holds for arbitrary programs, 200 simulation trials were carried out, where, for each individual trial, the values of the six parameters were selected at random within the range of

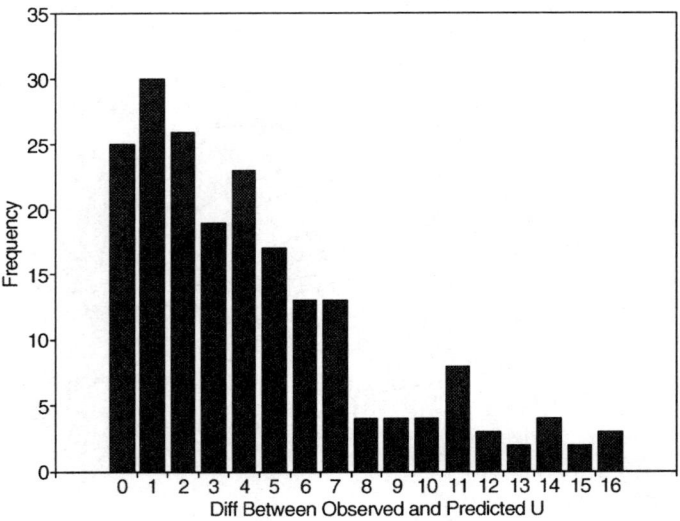

FIG. 15.5. Differences in observed and predicted responses for 200 Random Trials

values explored by the preliminary two-level factorial experiment.

Fig. 15.5 is a histogram showing the distribution of the modulus of the differences between the predicted values of U and those obtained by direct simulation. The median value is 4 per cent and the upper quartile value is 6.8 per cent.

Current work shows that a more accurate model can be obtained if we include two additional model parameters that are related to delays due to synchronization and to the sharing of processor resources. Unlike the original set of model parameters, these two new parameters reflect the effect of the interaction of the program and the machine environment, and cannot be set to specified levels in advance of the experiment, although they can be measured during the experiment. In order to estimate their influence quantitatively, these parameters have to be treated as 'nuisance variables', and their effects obtained from an analysis of covariance (Candlin and Phillips 1993). They provide a good explanation of differences in behaviour for programs which have the same set of values for the original parameters. However, it does not seem possible to deduce their values from the program structure in advance.

15.7 Analysis of a process migration strategy

The particular class of migration strategy assumed here is dynamic, in the sense that migration decisions are made according to the current state of the machine, and distributed, in that there is no global point of control. The program is assumed to consist of a set of top-level parallel processes which are the candidates for migration. Migrations take place between processors which are immediate neighbours. The algorithm is similar to those described by a number of other authors (Qian and Yang 1991, Willebeek-LeMair and Reeves 1991, Saletore 1990). Neighbouring processors periodically exchange load information so that they can evaluate whether their load is low **L**, medium **M**, or high **H**, relative to that in their neighbourhood. The percentage utilization of the processor has been chosen as a measure of load.

There are certain choices to be made about the mode of operation of a strategy, for example: the selection of threshold values, the method of choosing the destination processor, the selection of the migration candidate, and the frequency with which migrations are considered.

The experiment described here is designed to see how the usefulness of a particular migration strategy varies with program parameter values. In fact two sets of experiments were conducted: one in which the program was known to be badly mapped originally, and one in which the mapping resulted in a slightly to moderately unbalanced program. As would be expected, the improvement due to the application of the strategy was very marked in the badly balanced case, and it is more interesting to see if the strategy can also bring about improvement for a moderately unbalanced program. Accordingly, it is the results for this second experiment which are presented here.

15.7.1 *Analysis of an example strategy*

15.7.1.1 *Implementation of the Migration Heuristic* The way in which the strategy was implemented for this experiment is described briefly below.
Threshold definition. The thresholds categorizing **H**, **M** and **L** processors are defined, for each local neighbourhood, as being 5 per cent either side of the mean processor utilization within the neighbourhood.
Period of application. The migration strategy is invoked after a fixed time period of 2,000,000 clock cycles of the simulated machine (i.e. every 100 millesoconds).
Process selection policy. The policy attempts to satisfy two objectives. Primarily, it attempts to select as many processes for migration as there are immediate neighbours in the **L** state. If this goal is not achievable, it will try to select the maximum number of processes possible. The secondary aim is to ensure that the processes chosen are those which are using the greatest proportion of the available compute time. Both of these objectives are subject to the restriction that the processor should not move directly from the **H** state to the **L** state as a result of migrating the processes in question.
Process location policy. Given a set of N immediate neighbours in

the **L** state and a set of no more than N candidate processes, then the process location policy will randomly map the processes to the processors in a one-to-one fashion.

Inundation policy. It is assumed that there are always adequate resources to service an incoming process and a **L** processor will never reject any migrating processes sent to it. We do not attempt to prevent a processor moving directly from the **L** state to the **H** state as a result of accepting migrants.

15.7.1.2 *Experimental details* We assume the usual set of program parameters:

$$\{N, c, \mu_{cg}, \sigma_{cg}, \mu_{mg}, \sigma_{mg}\, \sigma_c, \sigma_m\}$$

but in the next set of experiments we only vary the two most important parameters with repect to processor utilization: N, the number of nodes, and $\sigma_{cg\%}$, the standard deviation of compute times expressed as a percentage of the mean compute time. To test the performance of the migration strategy, a two-level full factorial experiment was carried out with three factors: N set at 32 or 144, $\sigma_{cg\%}$ set at 10 or 80 and the migration strategy turned on or off.

The remaining parameters were set at reasonable intermediate values. The mean compute time across the nodes of the process graph, μ_{cg}, was set to 40,000, corresponding to a small to medium grained program (where, for example, several floating-point operations are carried out on each of 100 data items in each iteration of the compute phase of a process). The degree of the program graph, c, was set to 8, allowing 4 incoming and 4 outgoing channels per process on average. The mean message length, μ_{mg}, and standard deviation of message lengths, $\sigma_{mg\%}$, were set to 1000 and 10 respectively. Finally, σ_c and σ_m were set to 10 and 1 respectively.

A number of other parameters had to be fixed in order to fully define the experiment. Firstly, a migrating process was assumed to occupy 2,000 bytes of space. The hardware was set to a 4×4 mesh of transputers. A restricted random initial placement was used to ensure that, as far as possible, each processor received the same number of processes. Each simulation run was executed for 100,000,000 clock cycles (5 seconds) of simulated machine time. To increase confidence in the results obtained, six replications were executed at each parameter rather than the three that were taken in the previous experiment. Here we are interested in picking up smaller effects, and as we have a smaller number of parameters, the time required for an experiment is not excessive. Each replicate differed both in the random graph and random number seed used.

15.7.1.3 *Results* As before, a Guerrero and Johnson transform was applied to U giving:

Table 15.3 Estimates of effects for exploratory experiment using U'

Effect	Estimate	t-value	% Var
(gm)	0.407		
s	0.079	9.79	9.53
N	0.136	16.84	28.19
sN	0.037	4.6	2.10
$\sigma_{cg\%}$	-0.174	-21.64	46.57
$s\sigma_{cg\%}$	0.063	7.8	6.05
$N\sigma_{cg\%}$	0.01	1.2	0.14
$sN\sigma_{cg\%}$	0.049	6.02	3.61
		Total	96.19
	Standard Error = 0.0081		

$$U' = \left(\frac{U}{100-U}\right)^{0.18} - 1 \qquad (15.7.3)$$

Table 15.3 presents a table of effects for the response U', where the presence of the label s in the first column means that the migration strategy is active. We can see that all three main effects have a significant impact on performance. Furthermore, the sign of the s effect is positive, indicating that having the migration strategy turned on increases the estimated value of U' and so improves performance*. The various interactions of s with the other two parameters are also important, accounting for more than 10 per cent of the variation in U'. Their signs are also positive.

The above analysis seems to suggest that, generally speaking, the migration strategy improves performance. It would be useful to know the extent of this improvement and the characteristics of the programs for which the potential gains are greatest. This can be achieved by examining the data in greater detail using paired t tests.

A paired t test (Montgomery 1991) is a standard statistical technique used to compare two alternatives. The aim of the test is to decide whether the mean values observed under each of two treatments differ significantly. The observations must be paired in the sense that there should be a one-to-one correspondence between observation i under the first treatment and observation i under the second treatment.

As well as considering the data as a whole, similar tests can be carried out for various subsets of the data. For example, by fixing N and/or $\sigma_{cg\%}$ at certain values, one can partition the data in a number of different ways. This approach allows one to gain an understanding of the conditions under which the migration strategy is most effective.

*Since U' is a monotonically increasing function of U.

Table 15.4 Results of paired t tests for exploratory experiment using U'

Data	$\overline{U}$	$\overline{U}_s$	n	Reject H_0: $\overline{U}' \geq \overline{U}'_s$?
overall	76.22	86.35	24	Yes 1%
$\sigma_{cg\%}=10$	91.37	92.48	12	No
$\sigma_{cg\%}=80$	61.07	80.22	12	Yes 1%
$N=32$	70.95	78.53	12	Yes 1%
$N=144$	81.49	94.17	12	Yes 1%
$\sigma_{cg\%}=10$ $N=32$	87.79	89.83	6	No
$\sigma_{cg\%}=80$ $N=32$	54.11	67.23	6	Yes 1%
$\sigma_{cg\%}=10$ $N=144$	94.96	95.14	6	No
$\sigma_{cg\%}=80$ $N=144$	68.02	93.20	6	Yes 1%

Table 15.4 contains the results of a number of t tests for various partitions of the data. The first column specifies which particular subset of the observed data values is being tested. For example, the t test referred to in the first line considers all observations, the second considers only those where $\sigma_{cg\%}$ was set to 10, and so on. The second column gives the observed mean values of U across the six replications when the migration strategy is inactive. The third column gives the corresponding figures when the strategy is active. The fourth column gives the number of pairs of observations on which each test is based. The final column gives the results of the t tests which are carried out in terms of the transformed metric U'. The null hypothesis, H_0, is always that

$$\overline{U}' \geq \overline{U}'_s$$

where $\overline{U}'$ is the mean value of U' across all six replications where the migration strategy is inactive, and $\overline{U}'_s$ is the mean value of U' for the replications where the migration strategy is active. In order to be able to conclude that any improvements in performance are due to the migration strategy, and not due merely to sampling fluctuations, one must be able to reject the null hypothesis, H_0. For various subsets of the data, the final column indicates whether H_0 can be rejected, and if so, at what significance level.

Considering the data as a whole, we can accept the alternative hypothesis, that the migration strategy improves processor utilization at the 1 per cent significance level. A more detailed examination reveals that the performance benefits are not significant for reasonably uniform programs: that is, those trials when $\sigma_{cg\%}=10$. For all other partitions of the data, however, we can be confident that the migration strategy improves processor utilization.

To show that utilization is a good measure of performance, we also measured the mean number of message transfers carried out on each channel during a run. This metric, which is directly related to program behaviour, was closely correlated with processor utilization.

15.8 Conclusions

We have demonstrated the feasibility of describing the performance of a CSP-type parallel program in terms of a small number of parameters that refer to *macroscopic* properties of the program, without taking account of the details of its instruction sequence. We have obtained a statistical model describing the performance behaviour of this type of program in a particular environment and shown that a quantitative prediction of processor utilization can be obtained that is within 7 per cent of the observed value for 75 per cent of programs.

We have also shown that we can interpret the response of a program to a particular migration strategy in terms of the values of the program model parameters. A current experiment is concerned with finding the best choice of strategy parameters for this migration strategy.

Other performance metrics can be studied in the same way: for example, we have carried out experiments on rates of computation and message transfer.

Although our results refer to a particular class of program, and to a specific environment, we see the methodology as being generally applicable, and as being a powerful method for getting quick estimates of performance. It provides a basis for making decisions about program structure, and about the likely success of improving performance by process migration. Since machine environments are stable over long periods, it would be worth considering making studies for a given environment, and incorporating the results into operating systems that support process migration transparently.

Empirical results like these can help in the derivation of analytical models. In the first place, they show which are the important variables to incorporate into any analytical model. Secondly, they provide a measure against which predictions from an analytic model can be tested.

We have become convinced of the practical value of using formal experimental design and analysis techniques from the very beginning. There seem to be a number of benefits. The experimenter is forced to decide what questions the experiment is supposed to answer *beforehand*, which aids clarity of thought. The time required to obtain the answers is greatly reduced, compared to the time required for conducting a large number of runs in which only one factor at a time is varied. Quantitative information can also be obtained about the interactions between parameters. The techniques permit a rapid screening of models to find important parameters, which can then be followed by a more detailed investigation once preliminary results have been obtained.

REFERENCES

Atkinson, A. C. (1987). *Plots, Transformations and Regression.* Oxford Statistical Science Series 1, Oxford University Press, Oxford.
Birtwistle, G. M. (1986). *Discrete Event Modelling on Simula*. Macmillan, London.
Box, G. E. P. and Cox, D. R. (1964). An analysis of transformations. *Journal of the Royal Statistical Society B* **26** 211–243.
Candlin, R. and Luo, Q. (1989). Communications patterns in Occam programs. In *Parallel Computing 89*, 569–574. Elsevier, Amsterdam.
Candlin, R. and Skilling, N. (1991). A modelling system for the investigation of parallel program performance. In *Computer Performance Evaluation*, 397–409. Elsevier, Amsterdam.
Candlin, R. and Phillips, J. G. (1993). in *Proceedings of the International Conference on Decentralized and Distributed Systems, ICCDS '93*.
Candlin, R. Luo, Q. and Skilling, N. (1989). The investigation of communications patterns in Occam programs. In *Developing Transputer Applications*, 99–108 . IOS Press, Amsterdam.
Candlin, R. Fisk, P. and Skilling, N. (1991). A statistical approach to finding performance models of parallel programs. *Proceedings 7th UK Computer and Telecommunications Workshop*, pp 15–27. Springer-Verlag Workshop Series, Berlin.
Genstat 5 Committee. (1988). *GENSTAT 5 Reference Manual.* Clarendon Press, Oxford.
Guerrero, V. and Johnson, R. (1982). The use of the Box–Cox transformation with binary response models. *Biometrika* **69**, 309–314.
Hac, A. and Johnson T. (1986). A study of dynamic load-balancing in a distributed system. *Proc. ACM SIGCOM Symposium on Communications, Architectures and Protocols* pp 348–356.
Hoare, C. A. R. (1985). *Communicating Sequential Processes*. Prentice-Hall International Series in Computer Science, London.
Hunter, W., Box, G. and Hunter, J. (1978). *Statistics for Experimenters*. Series in Probability and Mathematical Statistics, John Wiley and Sons, New York.
Inmos Ltd (1988) *Occam 2 Reference Manual.* Prentice Hall International, London.
Inmos Ltd (1989) *The Transputer Databook*, Inmos Ltd, Bristol.
Jain, R. (1991) *The Art of Computer Systems Performance Analysis*, J.Wiley, New York.

Lyon, G., Snelick R. and Kacker, R. (1993).Time-perturbation tuning of MIMD programs. In *Computer Performance Evaluation '92: Modelling Techniques and Tools*. Edinburgh University Press, Edinburgh.

Meiko UK Ltd (1989) *Computing Surface Reference Manual*. Meiko UK Ltd, Bristol.

Montgomery, D. C. (1991). *Design and Analysis of Experiments* (3rd edn), J. Wiley, New York.

Nance, R.E., Moose, R.L. and Foutz, R. V. (1987) A statistical technique for comparing heuristics: an example from capacity assignment strategies in computer network design. *Comm. ACM* **30**, 430–442.

Qian, X. and Yang, Q. (1991) Load-balancing on generalized hypercube and mesh microprocessors with LAL. *IEEE International Conference on Distributed Computing Systems*, pp 402–409.

Saletore, V. (1990) A distributed and adaptive dynamic load-balancing scheme for parallel processing of medium grain tasks. *Proc. Fifth Distributed Memory Conference*, pp 994–999.

SIMULA Standard SS 63 61 14, (1987). SIS, Stockholm.

Skilling, N. (1991a). **eg**, University of Edinburgh, Department of Chemical Engineering Internal Report.

Skilling, N. (1991b). MIMD, University of Edinburgh, Department of Chemical Engineering, Internal Report.

Willebeek-LeMair, M. and Reeves, A. (1991). A localized dynamic load-balancing strategy for highly parallel systems. *3rd IEEE Symposium on the Frontiers of Massively Parallel Computation*, pp 380–383.

Zhou, S. (1988). A trace-driven simulation study of dynamic load balancing. *IEEE Transactions on Software Engineering* **14**, 1327–1341.

16

RESCHEDULABLE COMMUNICATIONS

Geoff Barrett

Inmos Ltd.
PACT
10 Priory Road, Bristol BS8 1TU, UK

Abstract

Modern compilers and processors owe their performance to the fact that sequential programming models allow many degrees of freedom in the scheduling of memory operations. This is because, although the programming model is very concrete about the order in which assignments take place, the semantics is very abstract in this respect. In this paper, we explore the impact of rescheduling communications in message-passing programs and suggest some properties of a semantics which successfully abstracts away from synchronization.

16.1 Introduction

One of the ways in which modern microprocessors have obtained the performance improvements witnessed over the past few years is by introducing parallelism into the execution of instructions. In particular, this is achieved by pipelining. Pipelining allows instructions which are in different stages of execution to be executed at once. For instance, the pipeline may consist of a number of stages which perform address decoding, operand fetches, operations, result stores, and so on.

For a sequential program, the only barrier which prevents all of the instructions being executed simultaneously is the fact that many of the instructions depend on the results of previous instructions. It is therefore necessary to stall some of the instructions while earlier instructions complete. With a well-designed instruction set, it should be possible to ensure that the only reason for stalling an instruction is to ensure that reads and writes on each register happen in the right order. This would mean that the instruction pipeline could be arbitrarily deep (although its useful depth would most likely be constrained by the number of registers). Figure 16.1 shows a processor pipeline.

The analogy of this situation in a parallel program is where the instructions of individual parallel threads depend on the results produced by

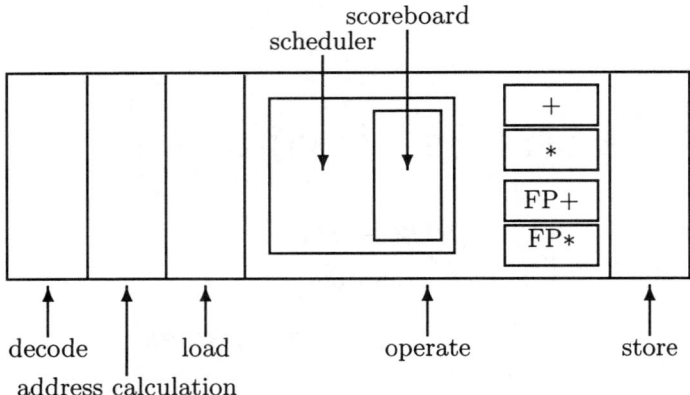

FIG. 16.1. A processor pipeline

instructions executed in other threads. Now, a thread must stall when all of its instructions are stalled.

In order to gain a better understanding of the problem, we shall first investigate some of the properties of sequential programs which allow instruction rescheduling to take place, and then we will see how far these principles can be applied to a communicating program.

An important feature of any modern instruction set design will be the benefit which can be obtained by applying pipelining techniques in an implementation. In the design of a multi-processor architecture, it will therefore be important that the implementation of communications between processors can benefit from these techniques also. In order to take full advantage of such an instruction set, it is important that the semantics of higher-level programming models is invariant under the rescheduling of communication and synchronization instructions, just as they are invariant under the rescheduling of memory operations.

16.2 The semantics of sequential programs

Excluding the effect of operations on volatile memory, the semantics of a sequential program can be given by its effect on the state of a computer. For many high-level programming languages, this can be described purely in terms of its effect on the value of program variables. Each program can, therefore, be modelled as a function (or relation in the case of non-deterministic programs) between the values of variables before program execution and the value of variables after program execution.

This semantics is totally abstract with respect to the order of assignments within a sequential program. For instance, if x and y are distinct variables, then assignments to them of values which do not depend on x or y can happen in any order. Algebraically, this is represented as follows:

$$x := a \,;\, y := b \quad \equiv \quad y := b \,;\, x := a$$

This rule is a consequence of more general rules which deal with multiple assignments. These rules are similar to an *operational semantics* for programs, in that they may be interpreted as a way of executing the program.

The algebraic rules divide into two types. The first type of rule shows how to perform computational steps in the program. The second type of rule shows how to alter the form of the program until a computational step is applicable. The first computational rule is*:

$$xx := ee \,;\, xx := ff \quad \equiv \quad xx := ff\,[xx := ee] \qquad (16.2.1)$$

where $ff\,[xx := ee]$ denotes the expression obtained by substituting the corresponding expression in ee for each occurrence of a variable in ff†. For instance, $x + 2\,[x := y]$ is equal to $y + 2$. This rule shows how to execute an assignment. If we consider the first assignment as a description of the state of the program variables, then this rule specifies that the second assignment is executed by substituting the values of the current state into the expressions for the values in the new state, evaluating the expressions and assigning them to the program variables.

The other rules for assignment show how to manipulate the program so that the first rule can be applied:

$$P\,;(Q\,;R) \quad \equiv \quad (P\,;Q)\,;R \qquad (16.2.2)$$

$$xx := ee \quad \equiv \quad xx, x := ee, x \qquad (16.2.3)$$

$$(x \text{ not in the list } xx)$$

$$x_1, ..., x_n := e_1, ..., e_n \quad \equiv \quad x_{\sigma(1)}, ..., x_{\sigma(n)} := e_{\sigma(1)}, ..., e_{\sigma(n)} \qquad (16.2.4)$$

where σ is a permutation of $1, ..., n$. Applying these rules from left to right, the first rule allows a program to be transformed into a form where the first two assignments are next to each other. The second two rules allow the assignments to be extended and rearranged so that they assign to the same variable list. The second rule allows a multiple assignment to be extended by assigning a variable its own value. For instance:

$$x, y := 0, 1 \quad \equiv \quad x, y, z := 0, 1, z$$

Using these rules, it is possible to prove the commutativity rule above:

---

*The possibility of expressions which cause exceptions invalidates this rule and provides part of the reason why traps are difficult to implement on pipelined processors.

†In all of the algebraic rules, names in lowercase *italic* stand for arbitrary lists of expressions; names in teletype are program identifiers; and names in uppercase *italic* are arbitrary fragments of program text.

$$x := a\ ;\ y := b\ \equiv\ x, y := a, y\ ;\ x, y := x, b$$
$$\{\text{by rules 16.2.3 and 16.2.4}\}$$
$$\equiv\ x, y := x, b\,[x, y := a, y] \qquad \{\text{by rule 16.2.1}\}$$
$$\equiv\ x, y := a, b \qquad \{\text{evaluation of substitution}\}$$
$$\equiv\ y := b\ ;\ x := a \qquad \{\text{by symmetry}\}$$

We shall investigate the semantics of conditional constructions using an if, then, else type construct. The program $P_1 \nmid b \nmid P_2$ tests the condition b and behaves like P_1 if b is true and like P_2 otherwise. It is, therefore, defined by the following rules:

$$xx := ee\ ;\ (P_1 \nmid b \nmid P_2)\ \equiv\ (A\ ;\ P_1) \nmid b[A] \nmid (A\ ;\ P_2) \qquad (16.2.5)$$
$$\text{where } A = xx := ee$$
$$P_1 \nmid \text{TRUE} \nmid P_2\ \equiv\ P_1 \qquad (16.2.6)$$
$$P_1 \nmid \text{FALSE} \nmid P_2\ \equiv\ P_2 \qquad (16.2.7)$$

The first of these rules specifies that the condition is evaluated in the current environment. The other rules show which branch to take according to the value of the condition.

The next rule shows how to manipulate the program so that the previous rules can be applied:

$$(P_1 \nmid b \nmid P_2)\ ;\ Q\ \equiv\ (P_1\ ;\ Q) \nmid b \nmid (P_2\ ;\ Q) \qquad (16.2.8)$$

The semantics of loops can be given with a single construction $\chi : P$ where χ is a function which maps jump labels to programs and P is a program as before except that now it may contain statements of the form $\text{goto}\ l$. The rules for this construction are as follows:

$$\chi(l) \text{ defined} \implies \chi : (\text{goto}\ l\ ;\ P)\ \equiv\ \chi : \chi(l) \qquad (16.2.9)$$
$$P \equiv Q \implies \chi : P\ \equiv\ \chi : Q \qquad (16.2.10)$$
$$\chi : (xx := ee\ ;\ P)\ \equiv\ xx := ee\ ;\ \chi : P \qquad (16.2.11)$$
$$\chi_1 : (\chi_2 : P)\ \equiv\ (\chi_1 \oplus \chi_2) : P \qquad (16.2.12)$$

The function $\chi_1 \oplus \chi_2$ maps a label l to $\chi_2(l)$ unless this is not defined, in which case it is mapped to $\chi_1(l)$. These rules allow an execution strategy whereby the program is reduced to the form $\chi : xx := ee\ ;\ (\text{goto}\ l\ ;\ P)$ using the previous rules and rule 16.2.12 and then:

$$\chi : xx := ee\ ;\ (\text{goto}\ l\ ;\ P)\ \equiv\ xx := ee\ ;\ \chi : (\text{goto}\ l\ ;\ P) \qquad \{\text{rule 16.2.11}\}$$
$$\equiv\ xx := ee\ ;\ \chi : \chi(l) \qquad \{\text{rule 16.2.9}\}$$
$$\equiv\ \chi : (xx := ee\ ;\ \chi(l)) \qquad \{\text{rule 16.2.11}\}$$

The most important feature of the sequential semantics is that the sequential execution of assignments can be transformed into a single assignment by representing the computation in the structure of the expressions. This form of the program abstracts away from the order in which the computation is performed. Note, however, that if some of the variables correspond to volatile memory then the semantics is no longer valid — the order of reads and writes on the variable are important. What makes the sequential, non-volatile case easy is that the behaviour of the memory is modelled as part of the semantics. It is assumed that the only important property of the memory is its state after the computation has been performed. This is because the only *observable* effect of the program is to alter the state of the memory. The semantics of parallel processes is complicated by the fact that synchronization allows intermediate memory states to be observed.

16.3 Parallel semantics

It is useful to be clear at this stage that we are interested in the semantics of *closed* parallel systems. This means that we maintain the advantage of only being able to observe the state of memory before and after the execution of a program. The sorts of programs which interest us are those which achieve the same effect as a sequential program but do it by employing parallel execution. We are not, therefore, interested in the semantics of communication *per se*, as is, for instance, the semantics of CSP or CCS. We shall refer to this class of parallel programs as *effectively sequential programs*.

The approach which we will take is that of developing a parallel program from the sequential program by applying transformations. The first step in this transformation is to partition the program variables and transform the program into a form in which each assigned expression only depends on the value of variables in a single partition. Assignments which assign an expression to a variable of the same partition indicate a purely local computation, whereas assignments which assign to a variable of a different partition indicate a communication. At the conceptual level, this program may now be executed by all of the processing elements concurrently, with each processor only performing the computations associated with its own variable partition.

This approach works well for computations which only involve arrays whose indices can be statically partitioned. For more complicated index calculations, such as pointer-like manipulations, the parallelization procedure is more complicated. This area is not yet well understood and we will not propose a solution in this paper. The aim here is to understand how a simple message passing model can provide an abstract machine model which is suitable for the compilation of the statically partitionable case and then to show how well this model can be implemented in a pipelined architecture.

However, before we move on to discuss the message passing model, let us first state some properties of parallel programs which are derived from sequential programs by data partitioning:

- Programs are deadlock-free. In other words, the only way in which a program can fail to reach the end of its computation is if it contains a non-terminating loop. No program will stop because its individual components are waiting for communications from each other.
- Programs are deterministic. In other words, there is only one possible state which can result from any one initial state.

16.3.1 The occam message passing model

The occam model of message passing is based upon the synchronized communication of values over a channel which connects two independent threads. Channel communication implements a distributed assignment in a very straightforward way. This is illustrated by the algebraic law:

$$c?x \parallel c!e \quad \equiv \quad x := e$$

where c stands for a channel which no concurrent process can use for input or output. This rule states that an input and an output on the same channel which happen in parallel are equivalent to the assignment of the output value to the variable specified in the input. For a complete set of laws of occam, see (Roscoe and Hoare 1986).

An occam program which is made up purely of parallel processes and channel communications is deterministic. As with sequential programs, only one state can result from any one initial state. However, not all of these programs are deadlock-free. This can be seen from the following program:

$$(c_1?x_1 \, ; \, c_2!v_1) \parallel (c_2?x_2 \, ; \, c_1!v_2)$$

This program will deadlock immediately because the two processes do not agree on the order in which communications will happen over the two channels c_1 and c_2. We may therefore deduce that this program is not correctly derived from a sequential program by data partitioning.

As well as parallel processes and channel communications, occam has a construct which selects between several communication options. This construct is used to provide arbitration between inputs. The use of this construct can introduce non-determinism. For instance, consider the following program:

$$[c_1?x \rightarrow c_2?x \,\square\, c_2?x \rightarrow c_1?x] \parallel c_1!1 \parallel c_2!2$$

This program assigns either 1 or 2 to x depending on which of the outputs is accepted first. Therefore, this program is non-deterministic and we may deduce that it is not correctly derived from a sequential program.

Both of the examples above can be used to show that occam programs are sensitive to the rescheduling of their communications. In the first case, rescheduling the communications can avoid the deadlock. In the second case, rescheduling the communications leads to different results.

The following two rules show how to deal with communication in a single thread:

$$xx := ee\,;c!e \quad \equiv \quad c!e\,[xx := ee]\,; xx := ee \qquad (16.3.13)$$
$$xx, x := ee, e\,; c?x \quad \equiv \quad c?v_c\,; xx, x := ee, v_c \qquad (16.3.14)$$
$$v_c \text{ unique}$$

The first rule shows that output can be implemented by evaluating the expression in the current state, outputting the value on the channel and then continuing in the same state. The second rule shows that input only changes the current state of the input variable.

The standard way in which to give the semantics of parallel threads is to use the alternative construct as follows:

$$c!e\,; P \quad \equiv \quad [c!e \to P] \qquad (16.3.15)$$
$$c?e\,; P \quad \equiv \quad [c?e \to P] \qquad (16.3.16)$$
$$P \parallel Q \quad \equiv \quad [g_{p_l} \to (P_{p_l} \parallel Q) \,\square \qquad (16.3.17)$$
$$\tau \to (x_m := e_m\,; (P_{p'_m}\,; Q_{q'_m}))\,\square$$
$$h_{q_n} \to (P \parallel Q_{q_n})]$$
$$\text{where } P = [g_i \to P_i]$$
$$Q = [h_i \to Q_i]$$
$$[g_i \to P_i]\,; Q \quad \equiv \quad [g_i \to (P_i\,; Q)] \qquad (16.3.18)$$

where p_l and p'_m partition i; q'_m and q_n partition j; g_{p_l} is a communication outside the alphabet of Q; h_{q_n} is a communication outside the alphabet of P; and, either $g_{p'_m} = c_m?x_m$ and $h_{p'_m} = c_m!e_m$ or vice versa. These rules allow all the communications options to be brought to the front of the program so that, at the outermost level where there are no longer any channel communications, the scheduler can decide which of the communications should happen first.

In many ways, this last set of rules is too general for programs which are derived from sequential programs by data partitioning because the scheduling decisions appear in the semantics of the program as non-deterministic choices. In fact, it does not matter what decisions are made in the scheduler, the outcome will always be the same. The problem is that we have had to introduce a new construct in the algebra which is capable of expressing more behaviors than effectively sequential programs can display. In the following sections we shall go some way towards remedying this situation by introducing some new rules which are always true for effectively sequential programs but which fail for other occam programs. But first we shall prove an important property of deadlock-free programs without arbitration.

16.3.2 Buffer tolerance

A program is said to be *buffer tolerant* when the introduction of arbitrary amounts of buffering on its channels does not affect its resulting behaviour. This is clearly an important property for pipelined communications because pipelining is very closely related to the introduction of buffers.

First of all we note that a state which results from executing any program with unbuffered channels is a possible result state of the program with buffered channels. This is because the unbuffered behaviour is similar to the buffered behaviour when the act of putting a value into the buffer and taking it out again always happens atomically, that is, with no intervening communications.

Therefore, we only need to consider what extra result states the buffered program may have. Clearly, a program may only have extra result states if some part of the program is executed which would not otherwise have been executed*. This happens either when a previously untaken branch is taken or when execution proceeds further than it would previously have done. Since the program is arbitration-free, it is not possible for the buffering of communications to cause a new branch to be taken. Therefore, the only way in which more behaviors can be introduced is by allowing processes which are stopped at a communication to proceed whereas previously they would never have done so. This means that the only extra behaviors of the buffered program come from threads which were previously deadlocked. Since we are only considering deadlock-free programs, this cannot happen.

An important property of occam programs without arbitration is that the introduction of buffering cannot introduce new deadlocks. As we showed above, the enabling of extra communications cannot cause extra branches in the code to be reached, it can only cause threads to progress further than they would have done. Therefore, the more buffering which is introduced, the fewer deadlocks are possible. This is not, however, true of programs with arbitration because the buffering of an output may cause a previously untaken branch of an alternative to be taken.

---

*Occam programs do not allow more than one thread to write to a variable, so the sequence of values held in a particular variable does not depend on the relative timings of threads.

16.3.3 Buffered communications

The fact that deadlock-free programs without arbitration are buffer tolerant allows us to introduce buffering on channels without altering the semantics of the program. A program with the value e in the buffer of channel c is written $P / c!e$. The rules which govern this construct are as follows:

$$(c_1?x \,;\, P) / c_2!e \;\equiv\; \begin{cases} x := e \,;\, P, & c_1 = c_2 \\ c_1?x \,;\, (P / c_2!e), & c_1 \neq c_2 \end{cases} \quad (16.3.19)$$

$$(c_1!e_1 \,;\, P) / c_2!e_2 \;\equiv\; c_1!e_1 \,;\, (P / c_2!e_2) \quad (16.3.20)$$

$$(c!e \,;\, P) \parallel Q \;\equiv\; P \parallel (Q / c!e) \quad (16.3.21)$$

c is an input channel of Q

The first part of rule 16.3.19 specifies that an input from a channel whose buffer is full can complete immediately by assigning the contents of the buffer to the input variable and emptying the buffer. The rest of rule 16.3.19 and rule 16.3.20 show that inputs and outputs do not affect the values in buffers for other channels. Rule 16.3.21 shows that an output can be implemented by putting the output value into the channel buffer of the thread which will perform the input. A full set of rules for processes which are infinitely buffered can be found in (Josephs et al. 1989). A similar calculus is offered in (Josephs and Udding 1990).

These rules are enough to allow an input to be delayed behind another communication so long as there is something in the channel buffer. For instance, if the second communication is an output:

$$\begin{aligned}
(c_1?x \,;\, c_2!e_2) / c_1!e_1 \;&\equiv\; x := e_1 \,;\, c_2!e_2 & \{\text{by rule 16.3.19}\} \\
&\equiv\; c_2!e_2 \,[x := e_1] \,;\, x := e_1 & \{\text{by rule 16.3.13}\} \\
&\equiv\; c_2!e_2 \,[x := e_1] \,;\, (c_1?x / c_1!e_1) & \{\text{by rule 16.3.19}\} \\
&\equiv\; (c_2!e_2 \,[x := e_1] \,;\, c_1?x) / c_1!e_1 & \{\text{by rule 16.3.20}\}
\end{aligned}$$

Thus, so long as e_2 does not depend on x, then the output can be executed first. However, this rule is not entirely satisfactory because this is precisely the case when we are not interested in rescheduling the communications. In this case, the input can complete immediately anyway because the buffer is already full. It is when there is no value in the buffer that we want to get on with the output rather than wait. What we need to be able to do is to delay the input so long as some buffer space has been allocated.

In order to be able to give a proper semantics for effectively sequential programs, we shall depart from the traditional style of semantics and introduce a new buffering construct. The notation $g \twoheadrightarrow P$, where g is a communication of the form $c!e$ or $c?x$, is intended to denote a process which has buffer space for the communication g. It satisfies the following rules:

$$g\,;P \quad \equiv\rangle \quad g \ggg P \qquad (16.3.22)$$
$$(g \ggg P) \parallel Q \quad \equiv \quad g \ggg (P \parallel Q) \qquad (16.3.23)$$
if the channel of g is not connected to Q

The first rule here is the assertion of buffer tolerance. The rule means that the expression on the right can replace the expression on the left, but not vice versa. In other words, we can always *introduce* buffering on a channel–buffering can be removed but there is a complicated side condition. The second rule allows a buffered communication in a branch of a parallel composition to be scheduled immediately.

The commutativity rules for the buffering construct effectively define the degree of slackness which is available in the implementation of the program. Notice that rule 16.3.23 gives the following commutativity rule:

$$
\begin{aligned}
g_1 \ggg g_2 \ggg (P \parallel Q) \quad &\equiv \quad g_1 \ggg (P \parallel (g_2 \ggg Q)) \\
&\quad \{\text{by rule 16.3.23 assuming } g_2 \text{ not connected to } P\} \\
&\equiv \quad (g_1 \ggg P) \parallel (g_2 \ggg Q) \\
&\quad \{\text{by rule 16.3.23 assuming } g_1 \text{ not connected to } Q\} \\
&\equiv \quad g_2 \ggg g_1 \ggg (P \parallel Q) \qquad \{\text{by symmetry}\}
\end{aligned}
$$

Because of the disjointness rules, it is necessary that the channels and variables used in g_1 and g_2 are disjoint. In the standard semantics, communications which originate from different threads may commute whereas communications originating from the same thread may not. This is because the only buffering which is present in the program arises from the concurrent execution of threads. However, in buffer-tolerant programs, we may introduce arbitrary amounts of buffering on each channel without affecting the result of the computation. Therefore, the hypothesis of the commutativity rule may be weakened to allow any pair of communications to commute so long as they do not interfere with each other; in other words, so long as neither communication inputs to a variable which is used by the other:

$$g_1 \ggg g_2 \ggg P \quad \equiv \quad g_2 \ggg g_1 \ggg P \qquad (16.3.24)$$
if the communications do not interfere

Finally, we may give the semantics of the parallel construct in terms of the buffering operator:

$$(c?x \ggg P) \parallel (c!e \ggg Q) \quad \equiv \quad x := e\,;(P \parallel Q) \qquad (16.3.25)$$
$$(g \ggg P)\,;Q \quad \equiv \quad g \ggg (P\,;Q) \qquad (16.3.26)$$

These rules allow us to give a semantics to concurrent threads without having to resort to the non-determinism of the alternative construct. The

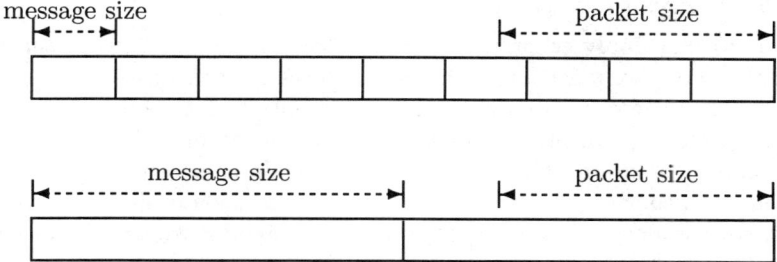

FIG. 16.2. Aggregated and packetized communications can be automatically matched to machine granularity

commutativity rule allows the communications to be rearranged so that either an external communication is at the head of one of the parallel components, or else both components have a communication on the same channel at their heads.

16.3.4 *Aggregating communications*

An aggregate communication is composed of several smaller communications. In many computer systems, an aggregate communication can be achieved more efficiently than sequential execution of the individual communications. This is because there is a cost associated with setting up the communication. An aggregate communication amortizes a single set up cost across all the individual communications.

Aggregate communications have the following definition:

$$c!ee_1 \frown ee_2 \equiv c!ee_1 \,;\, c!ee_2 \qquad (16.3.27)$$
$$c?xx_1 \frown xx_2 \equiv c?xx_1 \,;\, c?xx_2 \qquad (16.3.28)$$

These rules simply equate the aggregate communication to the sequence of individual communications. However, applying these rules from right to left allows us to implement a sequence of communications by a single, aggregated communication.

In practice, these rules cannot be applied very often unless the program is buffer-tolerant because the semantics insists that communications over separate channels are strictly interleaved. This means that two communications on the same channel which are separated by a communication on a different channel cannot be aggregated. However, with buffer-tolerant programs, channel communications can be rescheduled in order to bring communications over the same channel together.

Figure 16.2 shows how these rules allow messages to be automatically matched to machine granularity.

16.3.5 *Programs with deadlock*

So far, we have only considered deadlock-free programs. This is because the rules which we have derived do not preserve the semantics of programs which may deadlock. In the standard semantics of occam, deadlock is viewed positively. In other words, deadlock is sometimes a "good thing" because the program will not continue on to do something bad. When dealing with control processes, this is a very reasonable position. However, as we stated above, we are only interested in programs which achieve state changes through parallel execution, not programs which perform communication or control tasks. Therefore, deadlock can be viewed as a "bad thing" and it is reasonable to avoid deadlocks by rescheduling communications. In this way, a program can always be improved by adding more buffers.

16.4 Conclusion

In this paper, we have shown how buffer-tolerance can be used to justify the rescheduling of communications in the implementation of a parallel program. This property allows us to provide a semantics of communicating processes which is abstract with respect to synchronization. The semantics which we are aiming for is subtly different from the semantics of infinitely buffered communications — in that scenario, it is not always possible to reschedule an output before an input. However, we have shown how it is possible to reschedule communications except where doing so would create a data flow hazard.

In developing the semantics, we have restricted attention to a subset of occam which is deadlock-free and **ALT**-free. In order to address the deadlock restriction, we have to investigate the following questions:

- Is there a decision procedure for occam programs which recognizes a subset of deadlock-free occam programs which includes effectively sequential programs?
- If not, is there a semantics for programs which may deadlock in which the addition of buffer space only improves programs?

If the answer to the first question is positive, then we can use these rules to provide compile time checks. This would restrict the range of admissible occam programs but remove a common source of programming errors.

In order to address the arbitration restriction, the following questions must be investigated:

- What is the sequential construction whose parallel implementation requires arbitration?
- Is there a form of arbitration which guarantees buffer-tolerance and can be used to express all buffer-tolerant programs which contain **ALT**?

These two questions are closely linked because any sequential construction whose parallel implementation requires arbitration is likely to guarantee buffer-tolerance because the sequential model does. However, it may turn out that this use of arbitration does not imply the full expressive power of buffer-tolerant `ALT` programs.

Finally, we have avoided the issue of parallelizing programs which manipulate pointers. In many cases this can be achieved by using task queues and a global data base which is updated in an exclusive write fashion. However, not all programs may be suitable for this approach. The question this raises is:

- What further synchronization primitives should be added to the message passing model in order to provide a distributed implementation of pointer assignment?

REFERENCES

Josephs, M. B. and Udding, J. T. (1990) Delay-insensitive circuits: An algebraic approach to their design. In J. C. M. Bæten and J. W. Klop, editors, *CONCUR '90*, pp 342–366. Springer Verlag, Berlin.

Josephs, M. B., Hoare, C. A. H. and Jifeng, H. (1989) A theory of asynchronous processes. Technical Report PRG-TR-6-89, Programming Research Group, Oxford University.

Roscoe, A. H. and Hoare, C. A. R. (1986) The laws of occam programming. *Theoretical Computer Science*, **60**,177—229, 1986.

17

SOME PRACTICAL CONSIDERATIONS FOR OBJECT-ORIENTED PROGRAMMING ON DISTRIBUTED MEMORY PARALLEL COMPUTERS

Mike Livesey and Colin Allison

Division of Computer Science
University of St. Andrews
St Andrews, Scotland, UK

Abstract

The object-oriented programming paradigm portrays objects as encapsulated entities which communicate by message passing and execute concurrently with respect to each other. Object-orientation itself however does not offer any guidance as to how concurrency control issues should be addressed. It is important that a model of concurrency is supported and adopted in an object-oriented environment so that those programming within it do not have to resort to *ad hoc* techniques which are not reusable, may not be safe, and are unlikely to be compatible with other environments. The *Warp* protocol described in this paper is an attempt to provide a uniform approach to concurrency control in distributed object-oriented environments. It has many highly desirable properties such as optimism, liveness, fairness, fault-tolerance, and scalability but makes assumptions about the resources available at a site that may not be valid in MIMD systems, for example, fast access to local non-volatile storage. This paper describes the original Warp protocol as implemented on a network of workstations with local disks and examines the possibilities of adapting it to work on parallel systems where nodes have limited resources.

17.1 Introduction

The object-oriented paradigm allows for concurrent execution of any or all objects in a system. Instantiated objects are typically seen as disjoint entities that only communicate by message passing. By virtue of this feature the paradigm appears to provide a natural and intuitive programming model for parallel and distributed systems. However, object-orientation (o-o) does not in itself adopt any particular model of parallel computation and does not require any communication mechanism other than general message passing. This contrasts strongly with parallel programming languages such as Occam or Parallel C where explicit models of parallelism have been adopted and particular mechanisms provided in support of these models. Because the general concurrency and parallelism implicit in o-o is not reflected in explicit support mechanisms, extensive o-o applications may be programmed without awareness of the often subtle synchronization problems present. As soon as objects interact, even in a small group, the problems of unprincipled parallel computation can occur. These include:

- deadlock
- termination
- non-fairness
- starvation

Beyond these basic pitfalls are some broader issues that a model of distribution and concurrency suitable for use in an o-o environment should address:

- scalability: a parallel program that consists of a few distinct active objects or *processes* may employ a synchronization mechanism that cannot be scaled to tens or hundreds of processes
- global *coherence*, by which we understand some model of distributed computation that involves constraints on the view of the global system state visible to a user or running program
- conceptual uniformity and distribution transparency
- efficient realization

The Warp protocol is an attempt to provide a uniform approach to concurrency control in distributed o-o environments and as such it can be adopted by any parallel abstract machine model requiring support for multiple concurrent object interactions. It has many highly desirable properties such as scalability, fairness, liveness, and safety but was designed originally to work in an o-o environment over a network of workstations with virtual memory and local disks. The question we address in this paper is whether it can be modified to run on a distributed memory MIMD architecture which is unlikely to have either virtual memory or local disk storage.

The remainder of this paper is organized as follows:

- We briefly review the object-oriented paradigm with respect to system resources, concurrency and parallelism
- We describe the Warp protocol in detail
- We outline modifications to the protocol for an MIMD architecture
- Status
- Summary

17.2 Object-orientation, system resources, and concurrency

17.2.1 *What are objects?*

Objects are characterized by various features:
- they belong to and are specified within a hierarchical type system
- every object is a uniquely identifiable instance of a *class*, which is essentially a type definition
- subtyping and inheritance allow for objects to be derived from and share features with other objects
- each object has an *identity* that persists over time independently of changes of the state of the object
- objects execute in response to *messages*, the operations which are executed are called *methods*
- an objects state is encapsulated and can only be accessed via an interface
- interfaces are the only visible parts of objects

A programming system is said to be *object-based* if it supports all these features other than inheritance and a hierarchical type system (Chin and Chinson 1991). For more general information on the o-o paradigm and terminology refer to (Wegner 1987). The key features of interest with respect to parallelism are that objects encapsulate private state data and can be treated as single units which interact solely by message passing.

17.2.2 *Inter-object communication*

By virtue of its generality the idea of message passing in an o-o environment is not prejudiced with regard to the choices that any object or group of objects may make in selecting a mode of interaction. The pattern of communication is any-to-any but particular objects may adopt a more restricted approach such as point-to-point. An o-o messaging system is essentially asynchronous but objects may choose a synchronous mode of communication or an RPC interface. Support for basic messaging is essential, and support for concurrency is highly desirable in all but the most trivial of programs involving more than one object.

17.2.3 *Concurrent invocation of objects*

This is of course the main theme of the paper. Specific models of concurrency can be made explicit as part of an objects type definition, or class. The Raven system (Acton and Nufeld 1992) for example assumes each class has a concurrency property which can have the value *controlled* or *uncontrolled*. For a controlled object the Raven runtime system automatically provides a lock-based access mechanism which supports the multiple-readers, single-writer model. As with many lock-based schemes Raven has a potential for deadlock and does not provide a distributed deadlock detection mechanism due to the anticipated cost being too high.

An alternative approach taken by the Apertos (Hirotsu and Tokoro 1992) o-o system is to base the safety of concurrent invocations on *atomic transactions* which are *serializable* (Oszu and Valduriez 1991). Apertos does this by having each object invoke a transaction manager when involved in a transaction. The transaction managers maintain dependency graphs and communicate out-of-band to help prevent deadlock and ensure termination. Subsequent requests to join other transactions are brokered by the manager. Unlike Raven the management overhead is not considered too high but it is not clear that this approach avoids the problems of centralization and scalability that were being addressed — it is rather a case of allowing any object to play the role of central manager. Nevertheless serializable transactions do provide a safe approach to concurrency management and an intuitive programming construct. We discuss transactions in more detail in section 17.3.2. and go on to show how a transaction service can be realized using the Warp protocol.

17.2.4 *Granularity*

A recurring problem in o-o design is the choice of the granularity for objects in the system. Should a single integer or character be treated as a single object? The management overhead for such a fine grain is likely to be prohibitively expensive and few if any programmers would find any advantage in explicitly programming at that level. A SmallTalk-based implementation of a distributed o-o system which treated each character as a distinct object proved too slow to be usable (Dollimore *et al.* 1991). With respect to distributed MIMD machines should it be possible to map a single object onto multiple processing nodes? A similar question arises with respect to representation of system resources and configuration options: if we wish to represent system resources as objects, should a grid of 256 processing nodes be treated as a single object or as 256 distinct objects? In the former case internal components of the object are assumed to share the same address space and this may mean creating a distributed shared memory platform before any problem-oriented programming can begin. In the latter case we treat each processing element and its memory as a site where one or more objects may be invoked. This approach is more suitable for most existing distributed memory MIMD systems and we accordingly limit our current investigation by assuming that objects cannot span nodes.

17.2.5 *Address space, virtual memory, and non-volatile storage*

O-o systems require each object to have a unique persistent identifier. In practice this results in a need for a relatively large sparse address space. Persistent object ids have to be mapped on to virtual memory addresses at some point and a major cause of concern for system builders is the extra level of indirection incurred on every object reference and the maintenance of inter object references. Schemes such as using virtual addresses as object ids (Chase and Levy 1992) and *pointer swizzling* (Wilson and Kakkad 1992) have been devised to improve performance on computers with memory management units (MMUs). MIMD machines do not typically have MMUs on individual nodes and do not provide either large address spaces or virtual memory. The availability of a large address space is relevant to concurrency if an object cloning technique is used to spawn multiple instances of an object as is the case with Warp. The type of MIMD machine we have in mind is the common case where a number of processing elements are interconnected on one or more boards and the system is hosted by a general purpose computer such as a Unix box or PC which provides I/O services. Machines such as the Meiko In-Sun Computing Surface and Fujitsu AP1000 fall into this category. Each processing element has some local memory but does not have local disk and does not have virtual memory support. Access to non-volatile storage is channelled through the host computer and schemes such as virtual memory or memory mapped files are impractical. In summary, the cost of access to non-volatile storage from an MIMD node is potentially very high.

17.3 The Warp protocol

The protocol is based upon a universal *backtrack* mechanism, and provides a system of stable processes. Coherence is modelled by the notion of a *transaction*—an atomic disseminating computation. The transaction model of coherence is not new (c.f. (Schmuck and Wylie 1991)); the novelty lies in the protocol, which provides a flexible kind of transaction and is also able to support a variety of different coherence models. The term "backtracking" carries the connotation of search — in our context a search for a coherent computation path. The term also implies a fundamentally optimistic approach. A computation is allowed to proceed eagerly, relying on backtracking to remedy the effects of over-eagerness.

A primary component of backtracking in a distributed system is *rollback*—finding and (effectively) returning to a consistent global *snapshot* (Chandy and Lamport 1985, Mattern 1990). Any form of rollback must obviously be supported by state checkpointing. In a distributed system the checkpoints would ideally be made locally, yet a snapshot implies a global constraint on such local checkpoints. The information in a global snapshot resides in transit messages as well as local state. If transit messages are lost, the system loses guaranteed message delivery. So rollback based on local checkpointing alone is prey to the *domino* effect, where a rollback avalanches back to the initial state because no other combination of local checkpoints constitutes a complete snapshot. The only defence against the domino effect is to log inputs as they are consumed, so that they can be made to reappear as transit messages when rollback occurs.

However, there still remains the problem of how to return to a particular snapshot — the rollback protocol. This combines with the *retry* protocol —

how to run forward again—to give the complete backtrack protocol. In our system this protocol is maximally optimistic. Not only is the checkpointing performed locally and asynchronously, but also the rollback. Each separate process in the system rolls back to its own local checkpoint and retries independently. Every "wave" of backtracking is triggered from a single process, which defines a local state that the "goal" snapshot must contain. The rollback protocol, which derives from the Time Warp distributed simulation system (Jefferson 1985), forces the rest of the system back to a snapshot by sending *antimessages* corresponding to all the messages it has originally sent during computation (we say that it "unsends" its messages). An antimessage rolls its target process back to the state in which it consumed the original "positive" message—the target process then unsends all its messages since that point, thereby propagating the rollback. When the rollback wave has finished, the resulting local states form a snapshot: no process has rolled back further than necessary, and any event causally dependent* on the initial backtrack point is connected to it by a chain of communications and will therefore also have been undone by the wave.

However, propagation is only one of three issues concerning backtracking. The other two are:

- triggering backtrack
- immunity from backtrack (*commitment*)

Eagerness is limited by the outside world. Messages to print a page or launch a missile cannot be unsent. A process that represents a driver for some real device must therefore always operate with maximal pessimism or *conservatism*—it can do nothing until it is guaranteed to be immune from further backtrack. Such a guarantee is called commitment.

In the light of our experience with an earlier version of the backtrack protocol (Livesey and Allison 1992), this paper describes a revised architecture aimed at a lower level of implementation, though the main emphasis is on the model. Because the backtrack protocol is Time Warp without explicit virtual timestamps we call the architecture "Warp".

17.3.1 *The Warp architecture*

The primary components of the Warp architecture are shown in Fig. 17.1. A major design decision has been to base the Warp architecture on the notion of stable process rather than stable data structures as in (Brown and Rosenberg 1991, Dasgupta and Chen 1991). Processes do not share address space or clocks. They communicate by asynchronous message passing, and all data resides in the virtual address spaces of processes. The process is therefore the unit of:

- stability and recovery

---

*in the sense of Lamport(1978)

- backtrack
- logical storage
- persistence and global naming — the unit to which persistent object IDs (PIDs) attach.

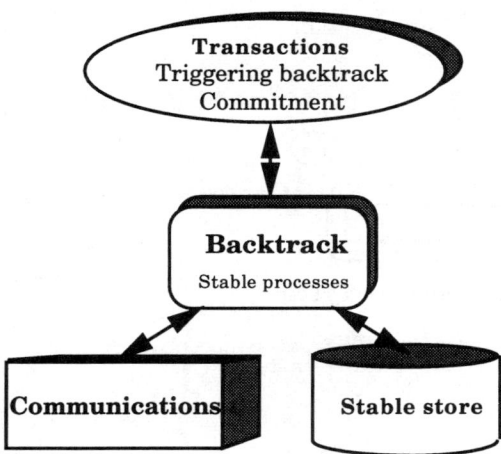

FIG. 17.1. The Warp architecture

Each process in the system consists logically of two components: the *client*, representing the application computation, and the *server*, implementing the generic features of the Warp architecture. Each operates in its own region of the process virtual address space, the client running at user level and the server as a microkernel.

The communication behaviour of a process is controlled by three message repositories (see Fig. 17.2(a)): the *input pool* contains waiting messages; the *input log* contains consumed messages; and the *output log* contains those messages already generated by the process, whether sent or not. Notice that the input and output logs are necessarily sequences because a process is a single locus of control. The logs together with state checkpoints constitute a complete record of the process history. The queues are managed by the server, and client communication is via system calls to the server. The effect of a client input is to move some message from the pool to the input log. The means by which this message is determined from the input instruction issued by the client is called the *selection* protocol.

17.3.1.1 *Backtracking* We have already described how backtrack is propagated by means of antimessages. Fig. 17.2 shows a schematic representation of backtrack at a typical process.

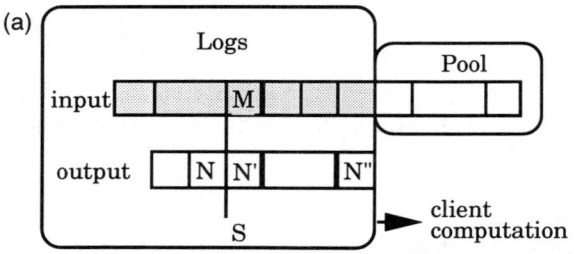

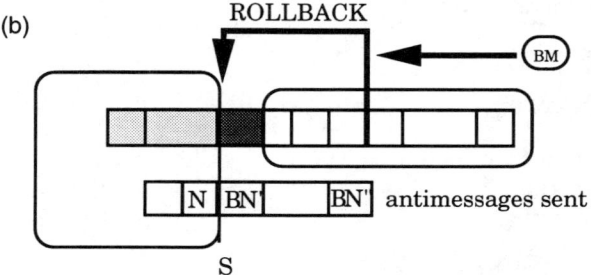

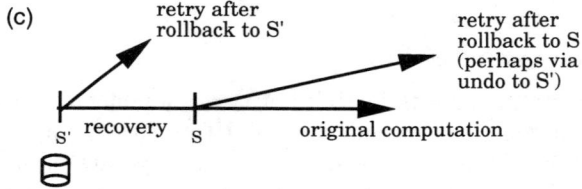

FIG. 17.2. The backtrack protocol

Figure 17.2(a) is the situation before backtrack occurs. Figure 2(b) shows the rollback caused by the receipt of the antimessage $\neg M$ corresponding to the positive message M. Fig. 2(c) shows how backtrack results in a tree-structured computation over time, because generally a different retry computation ensues after each rollback.

Notice that in Fig. 17.2(a) the message M is already in the recipient's past. Were M still in the input pool, it would simply be cancelled out by $\neg M$ without causing rollback.

For subsequent retry to take place, the client state S at the backtrack point must be restored. It may happen that S' itself was not checkpointed, in which case the nearest previous checkpoint (e.g. S' in Fig. 17.2(c)) must be restored and the client run forward using the appropriate past inputs to reconstruct S — we say that the process *undoes* to S', and that S is *recovered* from S'. Notice that this scenario does not require the output messages between S' and S to be unsent, and also that for recovery to be an exact replica of the original computation, the selection protocol must be deterministic.

17.3.2 Transactions

Transactions are a global mutual exclusion primitive, and as such constitute a fundamental coherence mechanism. They are identifiable units of computation which disseminate through the system from process to process. Transactions are orthogonal to processes, in the sense that one transaction may touch many processes and many transactions may be hosted at a single process at any given time.

The purpose of mutual exclusion is to guarantee that all the computation that disseminates round the system associated with a particular transaction is atomic, that is, not interleaved causally with the computation from any other transaction. This is equivalent to the database concept of serializability. Transactions thus operate *competitively*, coming into conflict with each other when they try to communicate with the same process. Whenever conflict occurs, one of the competitor transactions must lose. Two of the primary functions of the transaction protocol are to prevent deadlocks and to ensure *liveness*—that a transaction does not keep backtracking indefinitely but will eventually commit (provided its computation is actually finite).

Transactions are identified by unique totally-ordered transaction IDs (TIDs). The ordering of TIDs is used to ensure liveness, by treating the TID of a transaction as its age, with oldest having highest priority. Moreover, every message in the system carries a unique TID stamp. A distributed mechanism for generating TIDs from local clocks in an unbounded system is described in Livesey (1991).

17.3.3 The Computational Model

A process can host many *visiting* transactions, all executing simultaneously. However, when more than one transaction is present, the thread of each visitor T executes on its own private *clone* of the process (the T-*clone*), which lives only for the duration of its *parent* transaction T. To distinguish it from its clones, the original copy of the process is called the *master*. When a process originates a transaction, it is the *root* of that transaction.

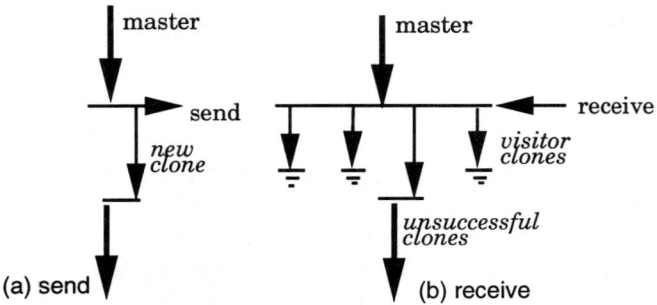

FIG. 17.3. Clone creation

Only a *T-clone* can execute on behalf of T, that is, send or consume messages stamped with the TID T. Also, the commitment protocol described below forbids the *T-clone* of any process other than the root of T spontaneously to send messages on behalf of T. It follows that a master cannot itself perform communications; it must create a clone, or clones, to do so. If the master is attempting input, it must create a clone for every transaction represented in its input pool; if it is attempting output, it must create a clone for a new TID. The two scenarios are shown in Fig. 17.3.

Only when one of its visitors commits will the master resume execution, from the point where the visitor clone terminates. A clone cannot terminate until it executes a special *end of transaction* (ETX) system call. It is worth noting that ETX is the only manifestation of the transaction mechanism at the program level.

17.3.4 *Commitment*

When a transaction commits, each of its clones replaces the master of the corresponding host process. Commitment is obviously impossible without some coordination amongst all the component computational threads of a transaction. To this end, each process keeps track of every transaction that visits it, until the transaction commits. At any given time, the visitors to a process are competing for its attention, so are in conflict; precisely one of the visitors *owns* the process. If a process has no visitors, the next visitor becomes the owner. Thereafter, ownership may change in one of two ways: the current owner either deliberately relinquishes ownership (which does not necessarily entail rollback) or ceases to be visitor, in one of two ways: by rollback, triggered elsewhere in the transaction, or by commitment. If the latter, the current owner clone becomes the master and the clones of all the other visitors must be rolled back to their starts and re-cloned (this is the only way in which backtrack can be triggered). In every case, ownership passes to the oldest of the remaining visitors.

Commitment uses a stable property detection protocol (SPDP) derived from the termination detection algorithm of (Dijkstra and Scholten 1980), which is a form of reference count garbage collector. Specifically, the protocol detects that all the processes involved in a transaction have some *locally stable* property P. A locally stable property will never fail spontaneously at any process, but only by virtue of the process receiving another message from within the transaction.

To use the SPDP to detect transaction termination suggests taking P to be local termination. However, the local stability requirement means that P must also guarantee immunity from backtrack as the result of another transaction committing. This is the role of ownership. We guarantee that no clone belonging to the owning transaction of its host process is ever subject to backtrack from outside that transaction, and ownership is never confiscated. The SPDP will then detect termination if we take the property P to be local termination plus ownership.

To avoid deadlock, an owner which is unable to attain P must eventually back off and release its ownership. Moreover, the backoff must be transmitted to every clone of the transaction to allow the possibility of backtrack wherever other transactions may be waiting for ownership. This is accomplished by the originator of the backoff sending a special *backoff* message to all its acquaintance clones. Since these are the only messages that can invalidate P, they must also be SPDP messages.

Once termination has been detected by the transaction root, it disseminates a "second wave" of *commit* messages, following the paths of the original computation messages. On receipt of the commit message, a process commits the relevant transaction and transfers ownership as described above.

17.3.5 *An example*

Figure 17.4 shows active processes which are engaged in two transactions, with TIDs $T_1 < T_2$. Transaction T_1 and T_2 have been independently initiated by processes A and E respectively. Recall that each process is owned by exactly one transaction, although it may be hosting many, and each clone belongs to its parent transaction. Initial ownership of a process goes to the first visitor, and the transaction initiator always clones itself. In this example process C is hosting both T_1 and T_2. T_1 has reached process C before T_2, so owns C, which has cloned of itself for each transaction. We consider some possible outcomes in the above scenario.

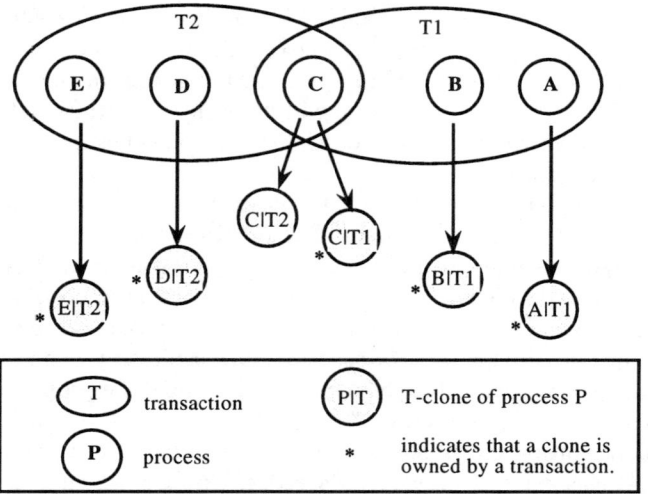

FIG. 17.4. A snapshot of two transactions, T_1 and T_2

- T_1 completes before T_2. Process A, the root of T_1, knows this when its SPDP reference count falls to 0. The clones $A|T_1$, $B|T_1$, and $C|T_1$ replace their masters. This invalidates clone $C|T_2$ and it must roll back to its start and be re-cloned from the updated C. Ownership of C then transfers to T_2, which runs to completion. Recall that a transaction cannot commit unless it owns all the processes it visits, but liveness ensures that all transactions commit eventually.
- T_1 has visited, prior to C, a process F already owned by another transaction that eventually commits. T_1 rolls back and ceases to be a visitor at C, whence ownership of C passes to T_2. T_2 then owns all its hosts and is able to commit. This outcome shows that even though T_2 is initially waiting on an older transaction at C, the eagerness of its clone may still be beneficial, although how beneficial is a decision that must depend on some suitable heuristics.
- T_1 backtracks to its start (that is, aborts itself) and commits by not changing C, thus releasing ownership. In this case ownership transfers to T_2 and there is no need for the work done by $C|T_2$ to be invalidated because the state of C is unchanged. This shows that backtrack can be optimized in certain special cases. Finally, we illustrate how the TID (age) ordering avoids deadlock and ensures liveness. Suppose T_2 later visits another process F, acquiring the ownership, and that subsequently T_1 also visits F before T_2 has committed. Then T_1 and T_2 wait on each other: T_2 for ownership of C and T_1 for ownership of F. Eventually, one or both of the transactions will start to back off and release ownership of all their host processes. However, T_1,

being the older transaction, will retain its ownership of C and acquire ownership of F. T_1 will therefore be able to commit.

17.3.6 Implementation issues

17.3.6.1 Communications In message-based computational models, it is usual to consider the communications layer as a black box that provides a channel with the desired *properties*—probably unique delivery, and perhaps monotonic (FIFO). Unfortunately, this may not be the most efficient approach where backtrack is involved. During the recovery that follows an undo, messages may be resent. These repeats should be effectively lost, otherwise they may be processed twice by the recipient. Similarly, the obvious way to tie an antimessage to its positive counterpart is to make the antimessage look like a "negative" repeat of the original. Messages must therefore contain sufficient information to distinguish repeats, and the backtrack protocol must act upon it.

Now consider how the channel service has been constructed in the first place. It will almost certainly be built upon an assumed unreliable and non-monotonic link, obtaining unique delivery by some sort of "repeat-until-ACKed" protocol. One effect of this is the possibility of copies of old messages suddenly popping out of the ether at a particular target process. So the channel protocol must be able to distinguish such repetitions from bona fide new messages, and it exercises this ability in providing the at-most-once part of its unique delivery service.

Thus, the channel protocol labels messages in order to separate them, whereupon the backtrack protocol relabels them in order to conflate them! This is just an example of the down side of modularity, but must be addressed by any system that claims good performance.

In the workstation-based implementation it is necessary to build a reliable lightweight datagram protocol as none of the protocols typically provided e.g. TCP, UDP, satisfy the communication requirements. In the case of an MIMD machine it is more likely to find a reliable datagram service already provided.

17.3.7 Stability

The stable process model provides two distinct functions:

- checkpointing
- recovery

Checkpointing needs the underlying stable store to provide failure-atomic transfer of the client and/or server virtual address regions to the stable storage medium. Although the process is the unit of stability, we shall see below that failure recovery requires the client and server components to be stabilizable separately. Retry requires that a checkpointed state can be reloaded as the current state.

17.3.8 *Failure recovery*

In principle, recovery from a failure — by which we mean a loss of information in volatile store — is just the recovery associated with a null backtrack. A process which temporarily fails is effectively undone back to its latest checkpoint, from which it recovers its current state. What complicates failure recovery is the fact that the log and input pool may also have been lost. In order to handle recovery by the backtrack mechanism, these backtrack data structures in effect the server state must be properly stabilized. We now derive two rules of recovery which make this precise.

- Once an incoming message has been ACKed, it is possible that no more copies of it will ever arrive again. Therefore an ACKed message in volatile store may be completely lost in the event of a failure. Furthermore, should a client run ahead into volatile input, even though this is unACKed and hence "in the ether", it might appear in a different order to a recovery computation, making this differ from the original. So the first rule of recovery must be ("consume" here means either ACK or act upon):

  R1: Never consume any message until it is stable. Given this rule, a sender can stabilize an output message in two ways: by writing it explicitly to stable storage, or by getting an ACK for it. Whenever we refer below to stabilizing an output message, we allow either of these possibilities.

- It is possible that a process might create a volatile section of output log — the *overrun*. Once an antimessage is consumed by a process, no more copies of the obsolete outputs will be generated. Should a failure strike during backtrack, those outputs may never be unsent. So we formulate the second rule of recovery as:

  R2: Never consume an antimessage until the overrun has been stabilized.

A traditional problem with failure recovery is the domino effect (Strom and Yemini 1985), whereby a cascade of backtrack throughout the system is triggered by one or more failures. The domino effect can only arise if global state information is destroyed by a failure, otherwise recovery is inevitable. The only potential source of information loss is an overrun, which could vanish in a failure. However, the rules specifically exclude any "information gaps", so the information persists somewhere in the system. An equivalent argument would be to observe that processes can always avoid overrunning without ever blocking.

17.3.9 *Other properties of the Warp model*

17.3.9.1 *Scalability*

Warp is a highly localized protocol. There are no explicit virtual time values and hence no need to calculate Global Virtual Time as was the case in Time Warp. Checkpointing is local so there is no need to carry out potentially expensive distributed snapshots. The protocol cost of a transaction is therefore linearly proportional to the number of sites it visits. There is, however, no mechanism at present which detects and compensates for hot spots such as an object that is in high demand from various transactions.

17.3.9.2 *Anchoring* As well as detecting termination, we would like to be able to detect *anchoring* of a transaction — immunity from further backtrack — irrespective of whether the computation has actually terminated. For interactive transactions, it can be important to get such a guarantee as soon as possible. The SPDP will detect anchoring if we take the property P to be simply ownership. It is still necessary to detect termination as well, in order that other ownership can be passed to other transactions. However, half the messages in each use of the SPDP are identical, so only 50 per cent more messages are needed to add the anchoring detection.

17.3.9.3 *Varimism* One of the novel features of the model of (Livesey 1991) is its 'varimism'. Tunable parameters of the system provide a spectrum between optimizm and pessimism. This feature is retained by the present model. Although we have implied above that clone computation is eager, making the whole approach optimistic, this need not be the case. Since the commitment protocol detects anchoring rather than computational termination, it is possible for a clone to be fully pessimistic by not executing until its parent transaction owns the host process. Allowing behaviours between these extremes (particularly with regard to how much a clone disseminates its parent before awaiting ownership) gives the spectrum of varimism.

17.4 Modifying the protocol for an MIMD architecture

17.4.1 *Non-volatile store issues*

The reader should now understand that the existing Warp implementation relies heavily on non-volatile storage for checkpointing, stability and virtual memory. The cost however of frequent accesses to disk from an MIMD node via a host computer is likely to be too high and must be minimized. We therefore propose to adopt a two-level store access strategy, making use of both volatile and non-volatile store. We also propose to use nodes with spare capacity for checkpointing and the creation of clones when necessitated by local memory shortages. In the event of no such nodes being available we have to go to disk and the performance degrades. This is a small move towards virtual memory but is carried out at the object level rather than the page level, does not depend on an MMU, and is less complex to implement. We therefore need to add some new management capabilities to the Warp part of a process which must now be able to establish how much free memory there is at its own node and which other nodes have free memory at any given time. In retrospect, our utilization of non-volatile storage in the existing implementation was partly influenced by the presence of a hardware MMU and operating system virtual memory support, that is, if store is already being paged to and from disk why not take advantage of it and control the situation.

17.4.1.1 *Failure recovery* We cannot see at present how to preserve the failure recovery features of the protocol described in Section 17.3.8. It depends on the message logs being stable before messages are actioned and the disk access delays would be intolerable. This has the unfortunate consequence of regarding the whole MIMD system as a single unit of failure when any node within it fails. It is possible to devise a scheme whereby nodes act as backups for each other, as in Isis (Birman 1991) and Argus (Liskov 1988) and we hope to address the fault-tolerance issue in the future.

17.4.1.2 *Example* If concurrent objects could not be cloned on the same node due to memory restrictions in the case of the example in Section 17.3.5 we would expect the clones to execute on other nodes. Similarly, checkpoints would be dumped to other nodes if there was a local memory shortage. In the event of no other nodes having spare capacity the objects would have to use disk accesses.

17.4.1.3 *Varimistic concurrency control* The varimistic concurrency control described in section 17.3.9.3 and in (Livesey 1991) provides an approach to adapting to a particular load and could be used to moderate the frequency of checkpointing for a particular object.

17.5 Status

17.5.1 *Design*

This includes formal verification of the model and the algorithms employed. We have formal proofs of the communication protocol, and the design of the detection algorithm is proceeding jointly with the development of a general proof of correctness of the interaction between the two time axes.

17.5.2 *Implementation*

The original implementation platform consisted of four autonomous Unix workstations connected by a dedicated 10Mb/s ethernet and 9.6K X25 links, providing both LAN and WAN operating conditions. We intend to experiment with the new two-level store strategy on a transputer-based Meiko Computing Surface (Meiko 1992). The Meiko CS Tools and CSN facilities provide a suitable infrastructure for prototyping and open up the possibility of a heterogeneous system that extends across Unix hosts and the transputer-based MIMD machine.

17.6 Related work

Timewarp, which can make heavier demands on an MIMD system than Warp, has been implemented in the form of an operating system, TWOS (Jefferson et al. 1987), which runs on various platforms including the BBN Butterfly GP1000 (Reiher 1992) and the CalTech/JPL Mark 3 Hypercube (Jefferson et al. 1987). These Timewarp implementations are distributed discrete event simulation platforms, however, and not suitable for general purpose parallel computing. There are many research systems which use transactions and atomicity as a structuring framework for distributed concurrency control, for example Argus (Liskov 1988). Argus uses locking, however, which brings deadlock problems. Furthermore it does not guarantee progress through the use of transactions. The Warp protocol is more than a basic transaction platform, it includes automatic redo and requires no separate deadlock detector or transaction manager to ensure that the system makes progress.

17.7 Summary

We have presented an approach to coherence in distributed object systems based on a transaction service that is supported by a uniform backtrack mechanism and integrated communication service. The Warp protocol is distributed, simple, deadlock-free, live, and fair. On machines with local disk it is also fault tolerant to volatile storage failures. The investigation into the feasibility of running the protocol on distributed memory MIMD machines has helped identify which features rely on specific types of system resources. We intend to experiment with a version of the protocol which utilizes a two-level store strategy on a transputer-based MIMD system.

REFERENCES

Acton, D. and Nufeld, G. (1992). Controlling concurrent access to objects in the Raven system. *Proc. 2nd International Workshop on Object Orientation in Operating Systems*, IEEE Computer Society Press, pp 148–152.

Birman, K. (1991). The Process group approach to reliable distributed computing. *Cornell University Department of Computer Science Research Report*.

Brown, A.L. and Rosenberg, J. (1991). Persistent object stores: an implementation technique. *Implementing Persistent Object Bases: Principles and Practice*, Morgan-Kaufmann, pp 199–212.

Chandy, K.M. and Lamport, L. (1985). Distributed snapshots: determining global states of distributed systems. *ACM TOCS*, **3**(1),63–75.

Chase, J., Issarny, V. and Levy, H. (1992). Distribution in a Single Address Space Operating System. *Proc 5th ACM SIGOPS Workshop*, Paper No. 12.

Chin, R.S. and Chinson, S.T. (1991). Distributed object based programming systems. *ACM Computing Surveys*, **23**(1),91–124.

Dasgupta, P. and Chen, R.C. (1991). Memory semantics in large grained persistent objects. *Implementing Persistent Object Bases: Principles and Practice*, Morgan-Kaufmann 226–238.

Dijkstra, E.W. and Scholten, C.S. (1980). Termination detection for diffusing computations. *Information Processing Letters*, **11**,1, 1–4.

Dollimore, J. et al.(1991). The Design of a system for distributing shared objects. *The Computer Journal*, **34**(6),514–521.

Hirotsu, T. and Tokoro, M. (1992). Object-oriented transaction support for distributed persistent objects. *Proc. 2nd International Workshop on Object Orientation in Operating Systems*, IEEE Computer Society Press, pp 13–25.

Jefferson, D.R. (1985). Virtual time. *ACM TOPLAS*, **73**,404–425.

Jefferson, D.R. et al. (1987). Distributed Simulation and the Time Warp Operating System. *Proc. of 11th ACM Symposium on OS Principles, SIGOPS Review*, **215**,77–93.

Lamport, L. (1978). Time, clocks, and the ordering of events in a distributed system. *Communications of the ACM*, **21**(7),558–565.

Liskov, B. (1988). Distributed Programming in Argus. *Communications of the ACM*, **32**(3),300–312.

Livesey, M.J. (1991). Distributed varimistic concurrency control in a persistent object store. *Implementing Persistent Object Bases: Prin-*

ciples and Practice, Morgan-Kaufmann 293–304.

Livesey, M.J. and Allison, C. (1992). A general purpose Time Warp toolkit. *Computer Science Research Report CS/92/1*, University of St. Andrews.

Mattern, F. (1990). Efficient distributed snapshots and global virtual time algorithms for non-FIFO systems. (Private communication).

MEIKO Ltd, (1992). *CS Tools for SunOs*.

Oszu, T.M. and Valduriez, P. (1991). *Principles of Distributed Database Systems*. Prentice-Hall, pp 277–281, 317–318.

Reiher, P.L. (1992). Experiences with Optimistic Synchronization for Distributed Operating Systems. *Proc. of Symposium on Experiences with Distributed and Multiprocessor Systems III*.

Schmuck, F. and Wylie, J. (1991). Experience with transactions in Quicksilver. *Proc. 13th ACM SIGOPS Symp. on Operating Systems Principles*, pp 239–253.

Strom, R.E. and Yemini, S. (1985). Optimistic recovery in distributed systems. *ACM TOCS*, **3**(3),204–226.

Wegner, P. (1987). Dimensions of object-based language design. *Proc ACM Conference on Object Oriented Programming Systems, Languages and Applications*, pp 168–182, ACM, New York.

Wilson, P.R. and Kakkad, S.V. (1992). Pointer swizzling at page fault time: efficiently and compatibly supporting huge address spaces on standard hardware. *Proc. 2nd International Workshop on Object Orientation in Operating Systems*, IEEE Computer Society Press, pp 362–377.

18

A FRAMEWORK FOR IMPLEMENTING HIGHLY PARALLEL APPLICATIONS ON DISTRIBUTED MEMORY ARCHITECTURES

Thierry Cornu

Swiss Federal Institute of Technology
EPFL DI, MANTRA Group
CH-1015 Lausanne, Switzerland
cornu@di.epfl.ch

Stéphane Vialle

Supélec Et. de Metz
2 rue Edouard Belin
F-57078 Metz cedex 3, France
vialle@ese-metz.fr

Abstract

We present here a massively parallel formalism, the Cellular Abstract Machine (CAM), and its implementation principles on a distributed memory multiprocessor architecture. The theoretical machine is more especially dedicated to connectionist applications and other massively parallel artificial intelligence applications, such as semantic networks. This paper briefly describes the CAM and gives an example of CAM program. Then we explain an implementation method on transputer-based architectures. We derive theoretical expressions predicting the implementation performances. We finally discuss briefly some practical issues of the actual multiprocessor implementation.

18.1 Introduction

The concept of an abstract or virtual machine has been widely used since the beginning of modern computer science history. Its first advantage is to hide from the developer the details of the real computer's architecture. The second advantage of the abstract machine concept is to increase software portability to a new computer architecture. Only the virtual machine itself has to be actually ported and not the languages and the applications running in it.

When it comes to parallel computing, the specific features of the hardware influence software development even more than in the case of sequential computing. Placement of the concurrent parts of the software on the different hardware components, load balancing, and communication optimization become major issues of the development. They sometimes induce programmers to re-parallelize parallel applications in a way different from their initial form. Eventually, in the course of solving a parallel programming problem, implementation constraints interfere with purely algorithmic considerations, compromising software reusability.

Thus in the case of parallel programming, the interest is obviously to find an abstract machine that would fulfil the following requirements:

- be portable on several multiprocessor architectures with reasonable efficiency;
- hide from the programmer most of the implementation constraints;
- still convey a precise idea for the programmer of how computation will be achieved.

In this paper we will present a virtual machine called Cellular Abstract Machine (CAM). This machine is more especially dedicated to explicitly distributed applications with a fine grain of parallelism and a high density of communication. Semantic networks based on marker passing mechanisms or neural networks are typical target applications for the CAM. We will first present the underlying computational model. We will also show that the CAM may be implemented with a reasonable efficiency on several MIMD multiprocessor architectures provided a sufficient parallel slackness may be achieved. Slackness means here that grain of the hardware architecture should be coarse enough compared to the one of the CAM application. The importance of parallel slackness for efficient implementation was first pointed out by Valiant (Valiant 1990). We will finally discuss briefly the implementation of the CAM on transputer networks of variable size.

18.2 The computational model of the CAM

18.2.1 *Overview of the CAM principles*

The underlying computational model of the CAM is an agent-based model. It belongs the family of actor- and object-based models. Its basic agents are called cells (from which, of course, the machine takes its name). A CAM program consists of a network of cells computing in parallel. The main features of the computational model are:

- The communication graph between cells is explicit. This means that a connection between two agents must be declared explicitly and lasts until explicitly destroyed. This is a first difference compared to classical actor models (Agha 1986) in which messages are sent using logical addresses, thus providing abstraction for the communication pattern. Explicit declaration of connection was a deliberate choice made for two distinct reasons. First, most target applications of the CAM (especially neural nets) do exhibit an explicit communication pattern. The form of this pattern is usually important to characterize the algorithms. Secondly, declaring communications explicitly gives the CAM programmers an idea of the communication burden implied by their programming style.

- The users declare cell types statically before run time. Thus cell types may be compiled in a simple way. Moreover, when it comes to implementing the CAM on distributed parallel architecture, dynamic migration of code will not have to be considered.

- The cells of a CAM program are dynamically created instances of the static pre-declared types. The pattern of connections between cells may also be created and modified dynamically. Dynamical configuration is a major feature of the CAM model. Most of the CAM potential applications require this feature. This is the case for semantic networks and for several neural net models (Platt 1991). Several other explicit parallel programming environments like the CSP-based Occam programming language do not support dynamic creation of processes.

- Finally, the main particularity of the CAM model is that it is globally synchronized. Every cell updates synchronously in a cyclic way, making the CAM model a synchronous MIMD model. This is probably the main difference from classical actor models of computation (Agha 1986) where concurrency is asynchronous. The effect of synchronism is to separate computation from communication in the model. Due to this feature the cellular model of computation is related to another more general computational model: the BSP model (Valiant 1989). BSP is also an MIMD synchronous formalism. Due to synchronism, the CAM is also related to the family of synchronous programming languages (Halbwachs et al. 1991, Boussinot and de Simone 1991, Le Guernic et al. 1991).

18.2.2 Concept of a cell

A cell may be seen as a basic abstract processor, with internal data storage, explicit connections to other cells, and an associated executable program responsible for periodically updating the output values of the connections to other cells. In the current programming language of the CAM, a program consists of a list of cell type declarations of which the actual cells of a program will be instances. These declarations include for each cell type the following elements:

- The cell type data structures, including the private variables names and those of the input and output channels to connect to other cells.

The same output channel may be connected to an unlimited number of input channels. Thus the communication pattern may be more general than just point-to-point connections.
- A set of transition rules specifying the actions to perform during each update, associated with each cell type. A transition rule consists of a precondition (head of the rule) and an action (body of the rule). A precondition is a Boolean expression testing the value of input channels and private variables. An action is a short sequential program. In the current programming language of the CAM, this program is expressed in a classical procedural style.

A cell update consists of a matching for all transition rules and a firing of at most one rule body. The corresponding action is described by a simple set of instructions in a procedural style.

18.2.3 Basic CAM cycle

Figure 18.1 gives the successive phases of a CAM cycle. Each cycle comprises three phases:
- The first phase is a computation phase, during which the outputs of the cells are updated according to the former values of their inputs and private variables. During this phase, cells may also issue requests for creating new cells, thus establishing and modifying connections between cells. Upon creation, each cell receives a unique identifier. This identifier serves as a logical name for the cell and will be used for further connection requests.
- The second phase is a management phase. During this phase, the requests for dynamic changes in the cellular topology are taken into account.
- During the last phase (the communication phase), the newly computed output of the communication channels between cells are actually sent to the receiving cells.

When a CAM program runs, only one cell, of a type called *main*, is initially created. It is up to the user to define this cell type. The initial cell will be the ancestor of every other cell of the program.

18.2.4 Programming languages for the CAM

The actual implementation of the CAM model consists of a superset of the C programming language, incorporating parallel programming primitives. These primitives mask the real multiprocessor architecture and provide facilities for creating cells, connecting channels, and so on. They are implemented *ad hoc* on each target hardware and are used as external calls of the C dialects available on different architectures.

Programming directly in this CAM native code is possible. However, it is not easy since the CAM instruction set exhibits a very low level of

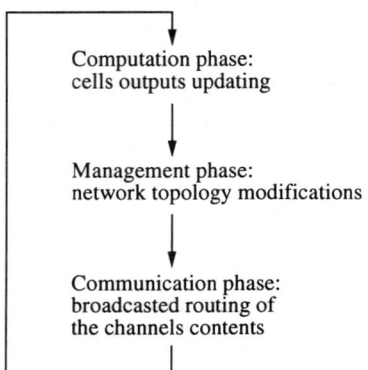

FIG. 18.1. Main stages of a CAM cycle

abstraction. Instead, dedicated higher-level languages have been designed, which follow the cellular computational model and compile easily into the CAM native code. So far two different versions of CAM dedicated high level languages are available, one of which is still under development (see §18.3). The next subsection presents an example of a program using the second version.

18.2.5 Example of programming style for the CAM

As an example of the CAM programming style, we will briefly explain how to create a feed-forward layered network of cells. This type of structure is very common in neural network applications. In Fig. 18.2, dotted lines represent the relations of filiation between cells and solid lines correspond to the actual communication channels between cells.

The basic idea of the program is that the first created cell (labelled M in the figure) will create layer cells (labelled either L for the ordinary intermediate layers, or I and F for the special case of initial and final layers) and connect them in the feedback direction (see Fig. 18.2.1). Each L cell (I, F, respectively) will first create during its first cycle (the second CAM cycle) the unit cells constituting the corresponding layer. Unit cells are labeled U for ordinary layers and I/O for user-interface cell types (see Fig. 18.2.2). Then each layer cell will be responsible for connecting its unit cells to the unit cells of the following layer. To enable the layer cells to perform these connections, the identifiers of the unit cells in the following layers are sent back through the connections between layer cells. Figure 18.2.3 shows the network state after this connection phase.

The listing corresponding to the declaration of the L cell type is given in Fig. 18.3. It is written in a dedicated language used on the CAM. The *initially* rule corresponds to the creation of the layer units during the first cycle of the L cell. Rule 1 then fires several times until all units of the

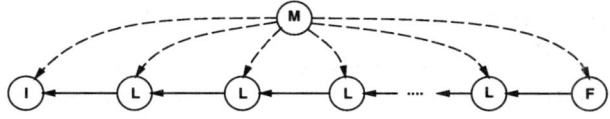

Network state after CAM cycle 1

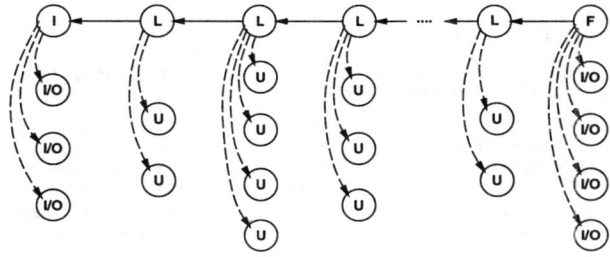

Network state after CAM cycle 2

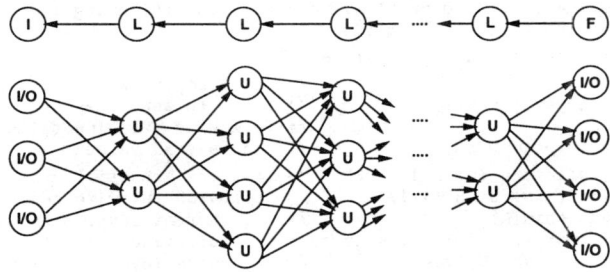

Network state after all layer connections

FIG. 18.2. Principle for creating a feed-forward layered structure

following layer are connected to the current layer and all identifiers of the units of the current layer have been sent to the preceding layer. Eventually, after fulfilling all connection tasks, rule 2 fires to eliminate the now useless L cell instance.

18.3 Implementation principles

Two different implementations of the CAM have been realized so far. The first one runs on a sequential machine and its principles are derived from the multiprocessor case. It is using the first dedicated high-level programming

```
typecell L_type (int l, int n) {
/* - l is the layer number.                                                  */
/* - n is number of units in the current layer.                              */
    in immat from_following_layer;/* Channel receiving unit cell identifiers */
                                  /* from following layer.                   */
    out immat to_prev_layer;      /* Channel sending unit cell identifiers of*/
                                  /* the current layer to the precedent layer*/
    immat neuron[DIM_LAYER_MAX];  /* Unit cells of the current layer         */
                                  /* identifiers table.                      */
    int i,j,k;
/* End of channels and variables declaration.                                */
/* Rule 'initially' is a special precondition automatically fired during     */
/* the first cycle of the cell life.                                         */
    initially ==> {
        for (i = 0; i < n; i += 1)
            neuron[i] = create U_type (l,i); /* Creates the n unit cells of the */
                                  /* layer and stores their logical          */
                                  /* identifiers in the array 'neuron'       */
        to_prev_layer = neuron[0]; /* Sends the identifier of the first      */
                                  /* unit cell to the previous layer.        */
        j = 1;                    /* Initialises the counter for             */
                                  /* identifiers sent.                       */
    }
/* Rule 1 : fires until all identifiers of unit cells have been sent to      */
/* the preceding layer and until all cells of the following layer have       */
/* been connected to the unit cells.                                         */
    (j <= n) || (from_following_layer != NULL) ==> {
        if (j < n)                /* If there are still cell identi-         */
            to_prev_layer = neuron[j]; /* fiers to send then send the        */
        else                      /* following cell identifier, else         */
            to_prev_layer = NULL; /* send a void value.                      */
        if (from_following_layer != NULL) /* If there are still units of the */
            for (k = 0; k < n; k += 1)   /* following layer to connect to    */
                connect chin[k]          /* the current layer then connect   */
                    of from_following_layer/* the following unit of the      */
                    oftype U_type          /* following layer to all units of*/
                    to chout               /* the current layer.             */
                    of neuron[k]
                    oftype U_type;
        j += 1;                   /* Increases counter for identi-           */
                                  /* fiers sent.                             */
    }
/* Rule 2 : fires when rule 1 does not and kills the cell.                   */
    (j > n) && (from_following_layer == NULL) ==>
        kill self;
}
```

FIG. 18.3. Declaration of the L cell type in a cellular language implemented on the CAM

language which may be considered as a simple prototype. The aim of this version was to validate the concept of cellular programming.

The second implementation is a parallel implementation of the CAM on a transputer network. Our choice of this type of architecture was motivated by its availability. The communication and management phases (see

IMPLEMENTATION PRINCIPLES 321

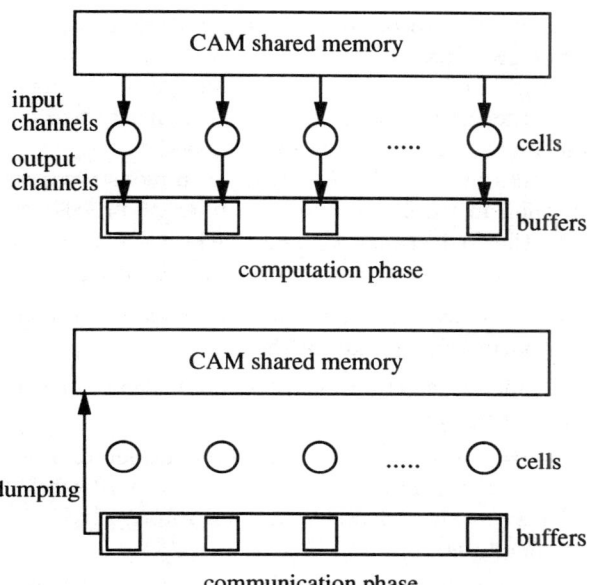

FIG. 18.4. Structure and general mechanism of the CAM

Fig. 18.1) have been operational on this version since 1992. An improved dedicated programming language derived from the prototype cellular language is currently under development for this version.

From now on we will deal with the implementation of the CAM on distributed memory architectures with fixed topology.

18.3.1 *Parallelization of the CAM cycle*

Each cell is implemented as a data structure, containing the cell's private variables and memory references on input and output channels. These channels are dynamically allocated in a logical shared memory area. If the cells at the two ends of a connection channel have been killed, the useless channels will be recycled by a garbage collector for a future connection operation. Moreover, according to the computational model (see Section 18.2), input channels at CAM cycle n will hold the output values computed at cycle $n-1$. Thus the global shared memory between cells will contain the channel values for the current CAM cycle. Simultaneously, output values for the next cycle will be stored in temporary buffers constituting the next version of the shared memory. At the end of each CAM cycle (during the communication phase) these buffers will be copied to the shared memory area. These principles are summarized in Fig. 18.4.

Within the computation phase of a CAM cycle, cells update independently from one another. Thus it is possible to map the cell pool on a pool of processors so as to perform the computation phase in parallel. As a result, the fine grain parallelism of the CAM will be simulated on a more coarse grain target architecture. The advantage of choosing coarse grain MIMD as the target architecture is that the explicitly parallel cellular network need not be re-parallelized in an other way.

On each processor a local CAM will be implemented, with integrally simulated parallelism. Cells running on different processors will be connected, due to the logical shared memory between the processors. The local CAM will have to be resynchronized twice during each cycle:

- a first time at the end of each computation phase (see Fig. 18.1) before initiating the update of the shared memory;
- a second time at the end of the communication phase (see Fig. 18.1) before initiating a new cycle.

In this way, the set of local CAMs may be considered as a unique CAM distributed among the different processors of the hardware.

The problems to solve for an actual multiprocessor implementation of the CAM are therefore:

- to map the cells on the processors optimally with respect to the balance of computational load;
- to map the logical shared memory of the CAM on the physical memory of the multiprocessor hardware;
- to find a way to resynchronize every processor, twice during each cycle;
- to find an optimal way to update the shared memory, at the end of each CAM cycle.

18.3.2 *Distributing the logical shared memory*

The CAM shared memory is locally duplicated on every processor so that computation phases can be performed locally. Updating the shared memory will then consist in routing the output buffers from each processor to every other.

As a first approach, the problem of mapping the cells in an optimal way has not been taken into account. Hence, in Section 18.4 we will first derive theoretical expressions of our implementation's performance for a network of cells *randomly* placed on the processor network. We make the hypothesis of a fair distribution of computing load among the processors.

The whole theoretical issue about implementing a CAM on a processor network with fixed topology can be divided into different subproblems, which may be treated independently up to a point:

- find an optimal topology for the processor network, not depending on any particular cellular program;

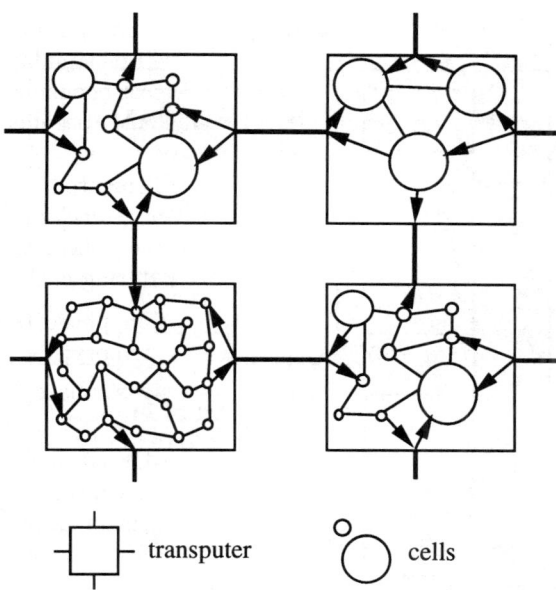

FIG. 18.5. Implementation principle on coarse grain MIMD distributed architecture

- find an optimal way of routing data for a given processor network topology, independently from any cellular network.

These two problems will be addressed in Section 18.4 with regard to performance evaluation.

18.3.3 Broadcasting data versus addressing data

For the channels' propagation phase, two basic implementation methods may be chosen: broadcasting or addressing. Communication between the different processes are said to be broadcast if the data produced by a processor are sent indistinctly to every other processor. On the other hand, if the new data are selectively sent only to the required processors, communication will be said to have been addressed. The two implementation options will need slightly different data structures for the local memory of each processor.

Broadcasting is the simpler communication method. The corresponding data structure is represented in Fig. 18.6. Each processor maintains a local copy of the shared memory of the application. During the computation phase, this copy is read and data are produced, corresponding to the current processor. Then during the channels propagation phase, the newly computed data are sent to every processor and local data are consequently updated.

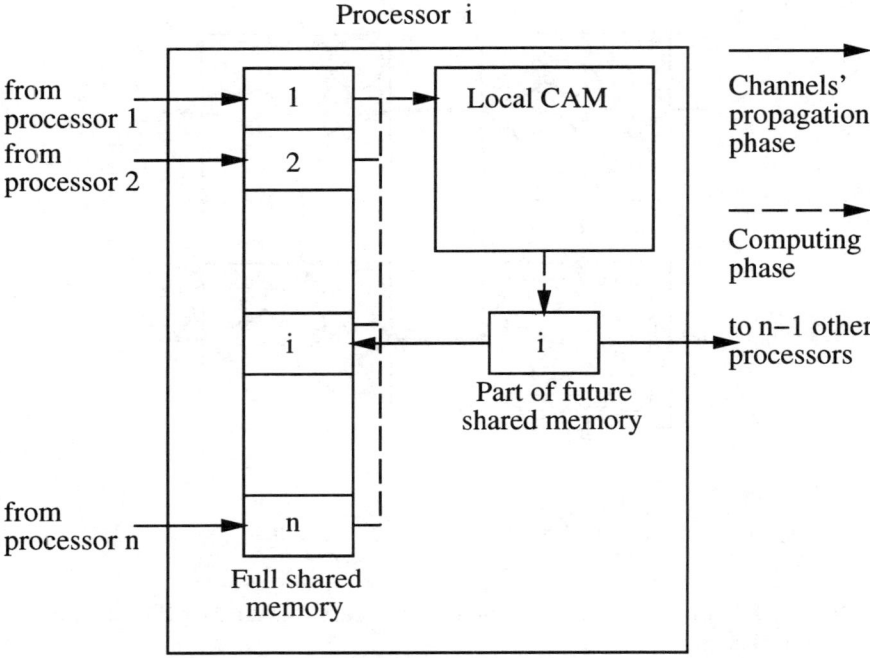

FIG. 18.6. Data structure in the local memory of a processor for a fully broadcast communication cycle

Addressing is a more elaborate communication mode, the aim of which is to reduce the amount of exchanged data compared with broadcasting. But the requirements are stronger: it is necessary to maintain, on each processor, the information necessary to distinguish, among the data computed by one processor, which part of the data is to be used by which other processor (Fig. 18.7).

When the shared data of an application are statistically destined to only one other specific processor, addressed communications will of course be preferred to broadcast communications, since only the useful data are to be transmitted. In addition, the local memory required on each processor for copying shared memory will be smaller. However, it is frequent in distributed parallel applications, and especially in our target connectionist applications, that the same data is sent to several different processors. A rough implementation of addressed communication then implies a partial duplication of some of the data. Either addressing or broadcasting may be more advantageous, depending on the expected characteristic of the cellular application. In Section 18.4 we will derive an expression of the average duplication rate below which addressing is more interesting than

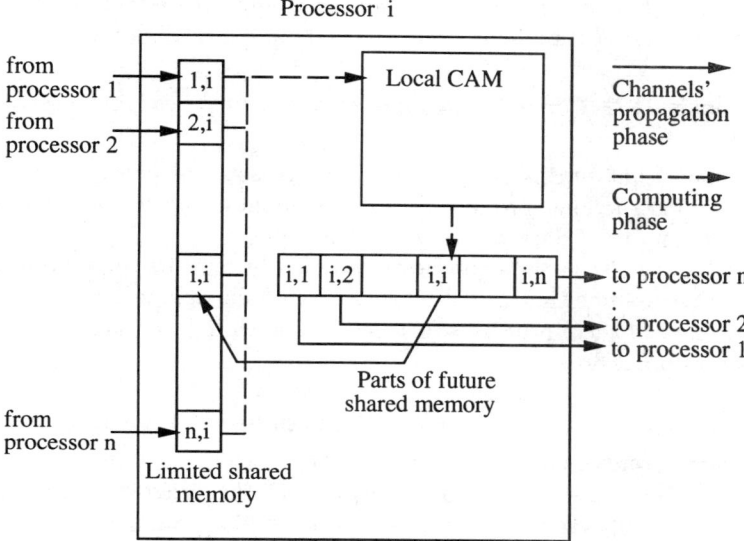

FIG. 18.7. Local data structure for addressed communications broadcasting.

18.4 Performance predictions

In this section we will give an estimate of the duration of a CAM cycle in our multiprocessor implementation. The estimate will depend on the number of processors of the architecture and on several parameters of the cellular program running on the CAM. From these calculations we will then derive expressions of speed-up and efficiency for our implementation method.

18.4.1 Duration of a CAM cycle

Duration of a complete cycle and hypothesis simplification The duration T_t of the total CAM cycle is the sum of the durations of the three phases defined in Fig. 18.1:

$$T_t = T_p + T_m + T_n \qquad (18.4.1)$$

where:
- T_p (processing) is the duration of the computation phase;
- T_m is the duration of the management phase;
- T_n (network) is the duration of the communication phase.

The calculation presented here is based on the following simplifying assumptions:

hypothesis 18.1 *The hardware architecture is homogeneous.*

This means that all processors are identical and are connected, through physical communication links, to the same number of neighbouring processors.

hypothesis 18.2 *Computation and communication load are evenly distributed.*

This means that during a cycle every processor will execute the same amount of instructions, produce the same amount of shared data, and send the same amount of data to every other processor.

In practice the latter hypothesis can clearly only be approximatively verified. We assume that it will be verified with enough accuracy provided the number of logical processes (the cells) is much higher than the number of physical processors.

Let us introduce several parameters for the subsequent calculation:

- Parameters of the cellular application:
  * M_S: size of the shared memory of the program. This is the cumulative size of all intercell connection channels;
  * α: the average number of local instructions to execute in order to produce one new byte of shared memory;
- Parameters of the hardware technology (for example, Transputers):
  * S_{proc}: computational speed of one processor of the architecture (in instructions per second);
  * F_b: the branching factor of the architecture; this is the number of neighbours to which a processor is directly connected through a physical communication link;
  * B_w: the unidirectional bandwidth of physical communication links;
- Parameter of the hardware size:
  * P: number of processors of the machine.

18.4.1.1 *Duration of the computation phase* Since we presumed data are evenly distributed, each processor will have to handle the same amount of the shared memory. The corresponding volume of data will thus be M_S/P bytes for each processor. The number of instructions each processor will have to execute will then be $\alpha.M_S/P$ (according to the definition of α). Eventually, if every processor works independently in parallel, the duration of the computation phase will be

$$T_p(P) = \frac{\alpha}{S_{proc}} \frac{M_S}{P} \qquad (18.4.2)$$

18.4.1.2 *Duration of the management phase* As a first approximation, we will not consider this phase in our estimate. Thus the subsequent estimation of speed-up and efficiency will be valid exclusively for cellular programs that have reached a purely static state (i.e. with no more cell creation or destruction). To extend our estimation to dynamic cellular programs, we would need to introduce additional parameters to characterize more precisely our applications (for instance, average number of cells created per cycle, ...). But as a first approximation and to keep our results as simple as possible, we will only examine the cases where

$$T_m = 0 \qquad (18.4.3)$$

18.4.1.3 Duration of broadcast communication phase Let us examine here the case of a broadcast communication phase (see Section 18.3.3 and Fig. 18.6). In the next paragraph we will see the case of the other implementation solution: addressed communication.

During a broadcast communication phase, each processor will have to send one data packet to every other. This packet will contain the M_S/P bytes of shared data produced during the preceding computation phase. The packet will first be sent to adjacent processors. Then it will be propagated by degrees to more distant processors. We will call *elementary transmission* the transmission of a packet from one processor to an adjacent processor. The time t_{trans} necessary for the elementary transmission of a data packet of size M_S/P is

$$t_{trans} = \frac{M_S}{P.B_w} \qquad (18.4.4)$$

according to the definition of B_w. In order for a packet from a processor to reach the $P-1$ other processors, $P-1$ elementary transmissions are necessary. As a whole, a communication phase will involve every P processors. So the number n_{trans} of elementary transmissions needed during a communication phase (broadcast) is

$$n_{trans} = P(P-1) \qquad (18.4.5)$$

On the other hand, according to the definition of F_b, the number of physical communication links in the whole architecture is:

$$n_{conn} = P.F_b \qquad (18.4.6)$$

For some kinds of processor topologies, it is possible to plan the succession of elementary transmissions in order to use all these links *optimally* (Cornu 1992) . In these cases, each physical link in the topology will support the same number n_{trans}/n_{comm} of elementary transmissions during the course of the phase. The total duration of the phase will be minimal in this case. Namely, we will have:

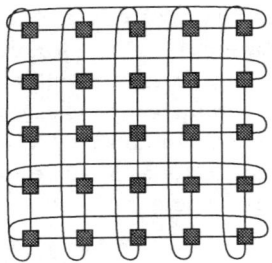
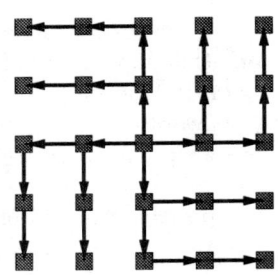

FIG. 18.8. Torus topology with an odd side

FIG. 18.9. Routing path inside the torus from any given processor to every other

$$T_{n-broad} = \frac{n_{trans}}{n_{comm}} t_{trans} = \frac{P-1}{P} \frac{M_S}{F_b.B_w} \qquad (18.4.7)$$

Several families of topologies, associated with transmission schemes, that reach that minimal duration are defined in (Cornu 1992). Among the optimal solutions for the transputers (i.e. for $F_b = 4$), we chose torus topologies with an odd side (Fig. 18.8). Several optimal schemes of elementary transmission are possible for these tori. Figures 18.9 and 18.10 explain the scheme actually chosen for our implementation. Figure 18.9 shows the routing path for data packets from one of the processors. Figure 18.10 defines the timing used along this path for propagating the data. When every processor of the architecture uses the same routing path and timing, elementary transmissions from any given processor do not collide and physical links are used optimally.

18.4.1.4 *Duration of addressed communication phase* We will treat here the other method for implementing communication (see Section 18.3.3 and Fig. 18.7). In this method, each processor, instead of sending one packet to every other processor, will send $P-1$ different packets each to a different processor. These new packets will be of smaller size. However, when the same data is to be sent to several other processors, it will have to be duplicated. Let us call γ the corresponding duplication rate, that is, the average number of different processors to which the same data is to be sent.

The calculation of the duration $T_{n-address}$ will not be conducted here exhaustively. As a matter of fact, it is largely inspired by the one for $T_{n-broad}$, presented above in detail. We will only give the result, that has been derived in (Cornu 1992). Once again we chose a torus topology with an odd side, that is, such that $P = (2r+1)^2$ where r is an integer. We will call r the radius of the topology. The routing path considered here is the same as for the former method (Fig. 18.9). There are more packets to route, however, so that the timing is be more sophisticated. The optimal time for an addressed communication phase is:

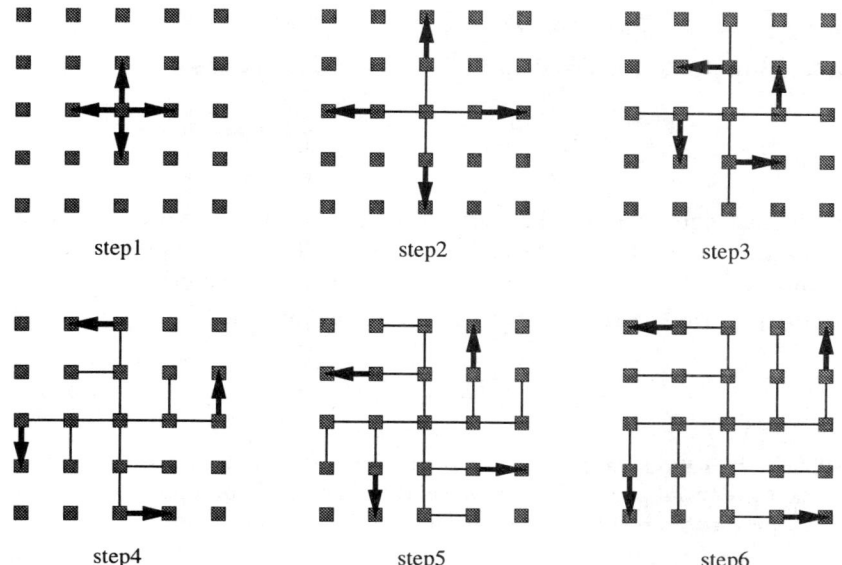

FIG. 18.10. Timing for the elementary transmissions along the routing path

$$T_{n-address} = \gamma \frac{r(r+1)}{2(2r+1)^3} \frac{M_S}{B_w} \qquad (18.4.8)$$

In contrast to eqn (18.4.7), eqn (18.4.8) is valid exclusively for the case of two-dimensional torus topologies. From the relation $P = (2r+1)^2$ and the two eqns () (with $F_b = 4$) and (18.4.8), it is clear that:

$$T_{n-address} \leq T_{n-broad} \iff \gamma \leq 2\sqrt{P} \qquad (18.4.9)$$

This expression defines the limit between the cases where addressed or broadcast routing performs better. The current CAM implementation uses broadcast routing exclusively. Addressed routing may be considered as a possible optimization for the future.

18.4.2 *Speed-up and efficiency*

In this section, we will derive expressions for speed-up and efficiency for our implementation method, as a function of the number of processors. These two parameters compare, in two different ways, the respective evolution of $T_p(P)$ et $T_n(P)$ as a function of P. Speed-up and efficiency are defined in the next paragraph. The calculation that follows is based on the values of T_n for *broadcast* communication.

18.4.2.1 *Basic expressions of speed-up and efficiency*

definition 18.1 *The speed-up $S(P)$ is defined as follows:*

$$S(P) = \frac{T_t(1)}{T_t(P)}, \qquad 0 \leq S(P) \leq P \qquad (18.4.10)$$

In the last equation, $T_t(1)$ is not meant to refer to the best possible sequential implementation of the CAM. Rather, $T_t(1)$ is the performance of a uniprocessor CAM using the multiprocessor implementation method.

definition 18.2 *Efficiency $e(P)$ is defined as follows:*

$$e(P) = \frac{S(P)}{P}, \qquad 0 \leq e(P) \leq 1 \qquad (18.4.11)$$

Efficiency gives a measure of the rate at which processors are used compared to their potential power. For instance a speed-up of 5 for a program running on 10 processors results in an efficiency of 50 per cent.

In order to simplify the following expressions, we will introduce here two new variables S_n and S_p defined as follows:

$$S_n = (F_b.B_w) \text{ in byte/second} \qquad (18.4.12)$$

$$S_p = (\frac{S_{proc}}{\alpha}) \text{ in byte/second} \qquad (18.4.13)$$

These two variables have a physical meaning:
- S_n gives the total communication bandwidth between one processor and the external world, considering all F_b links.
- S_p is the speed at which shared data may be locally produced on each processor. Since the coefficient α is involved in its definition, S_p depends to a large extent on the cellular application that the CAM is running.

If we keep the simplifying hypothesis of $T_m = 0$ and of a fairly distributed load, it is straightforward to derive expressions for $S(P)$ and $e(P)$ from eqns (18.4.1), (18.4.2), (18.4.7), (18.4.12) and (18.4.13). The results are:

$$S(P) = \frac{S_n}{S_p} \frac{P}{P - 1 + \frac{S_n}{S_p}} \qquad (18.4.14)$$

$$e(P) = \frac{S_n}{S_p} \frac{1}{P - 1 + \frac{S_n}{S_p}} \qquad (18.4.15)$$

The evolution of $S(P)$ and $e(P)$ is given in Fig. 18.11. The S_n/S_p ratio, which is the asymptotic maximum for the speed-up of our implementation,

PERFORMANCE PREDICTIONS 331

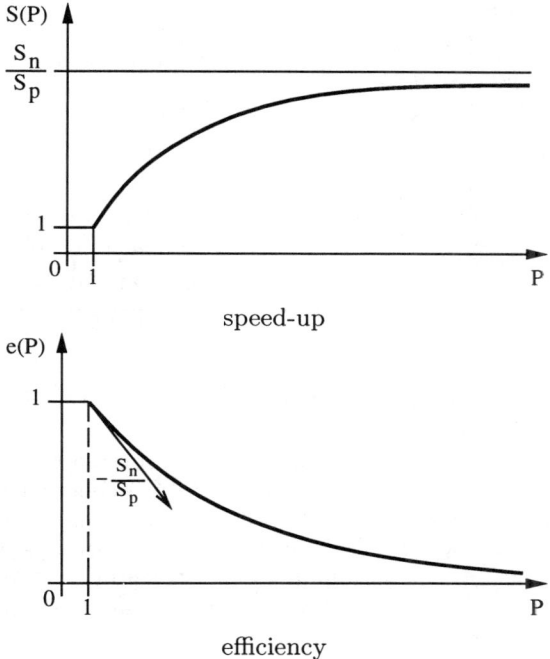

FIG. 18.11. Speed-up and efficiency as functions of the number of processors for broadcast user data and a static cellular network.

is also the slope of the efficiency curve as a function of P, at its starting point. Thus it determines how quickly the efficiency decreases when the number of processors grows.

It should be remembered that the expressions (18.4.14) and (18.4.15) only hold in the case of a fairly distributed load between the processors. If P grows indefinitely, the parallel grain of the cellular applications will no longer be finer than the grain of the architecture. Thus the curves of Fig. 18.11 are only valid for a number of processors P below a threshold (which depends on the application).

18.4.2.2 *Speed-up as a function of efficiency* From eqns (18.4.14) and (18.4.15), it is possible to derive the relation between speed-up and efficiency for a network of static cells:

$$S(e) = \frac{S_n}{S_p} - (\frac{S_n}{S_p} - 1)e, \qquad 0 \le e \le 1, \qquad 1 \le S \le \frac{S_n}{S_p} \qquad (18.4.16)$$

The resulting curve is a line, as shown in Fig. 18.12. This line illustrates the inevitable compromise that must be found between speed-up and efficiency of the implementation. On the curve $S = S(e)$ (Fig. 18.12) an operating

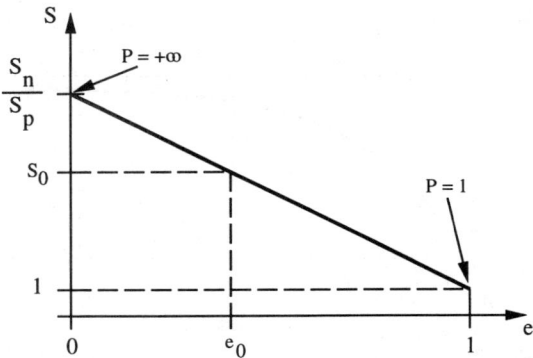

FIG. 18.12. Speed-up as a function of efficiency for architectures of variable size

point (let us call it (e_0, S_0)) may be chosen for any given type of cellular application. The corresponding convenient number of processors may then be derived from one of the two curves $S = S(P)$ or $e = e(P)$. Technical constraints will define a minimal desired speed-up and thus a minimal number of processors. Simultaneously, economical constraints will define a minimal tolerable efficiency and thus a maximal number of processors. If both constraints are not compatible, improvements will have to be envisioned. Any modification or optimization of the current implementation strategy, such as addressing user data instead of broadcasting, would be expected to modify the shape of the $S(e)$ curve. Hence, this representation of the speed-up as a function of the efficiency seems a useful tool in that it gives qualitative information on any implementation optimization. This criterion goes beyond the simple increase of the S_n/S_p ratio of the machine.

18.5 Practical parallel implementation

In the transputer implementation of the CAM, the communication phase has been implemented according to the principles presented in the last two sections. In order to produce an operational version, the management phase had also to be implemented. As already stated, the performance predictions presented so far do not cover this management phase.

In this section we will highlight several aspects of the practical implementation, most of which are linked to the implementation of the management phase.

18.5.1 *Management of the shared memory and placement of cells*

A distributed garbage collector is implemented over the shared memory. It recycles useless connections between cells that have been destroyed. This garbage collection contributes to minimizing the volume of data exchanged between the processors.

Cell placement takes place at the time of their creation. The target processor for a new cell is chosen by the processor issuing the creation request. For efficiency reasons, there is no cell migration. This would need to be updated through the network of the cells and their channels, which would not be efficient on our distributed message-passing architecture. The mapping heuristic is simply to balance the computational load of the processors. This very simple mapping method was presupposed in our theoretical performance predictions. The estimated load of a processor is based upon its number of cells and its number of output channels. At every cycle, each processor gets this information from the others, which allows it to decide which processors are most likely to receive new cells.

18.5.2 *Periodic synchronization of processors*

Transputers are resynchronized due to the message passings that take place during the management and communication phases. This communication uses a CSP-like synchronous protocol, which is the native communication protocol of transputers. In this protocol, communicating processes wait for each other at each communication appointment.

During each CAM cycle, processors exchange a predetermined number of messages in a predetermined order. Thus the communication phase is implemented in a completely deterministic way. The order of elementary transmissions is planned so as to avoid deadlocks. The management phase, which consists of several communication subphases for system data, is implemented in a similar way. As a result, transputers are resynchronized twice within each CAM cycle:

- First, after the computational phase: a processor will accept external communications only after its local computation phase is over.

- Second, at the end of a cycle: a processor will begin a new cycle only when the data from the last communication phase have been received from all other processors.

18.5.3 *Treatment of dynamic modification requests*

The management mechanism of requests (create, connect, kill...) is robust: it keeps the coherence of the cellular network, even in case of impossible or contradictory requests. For instance, an attempt to connect a dead cell can cause a warning, but does not create a wrong connection, neither does it waste shared memory. Simultaneous requests for the connection of one input channel to different output channels eventually create only one connection. Such a robustness is achieved by multiple system data routing between the involved processors. The management phase comprises three subphases, each one including routing and processing of subrequests.

This complex mechanism is necessary to detect and neutralize inconsistent requests.

In each processor, a cell manager uses an identifier table to list its cells, and each subrequest is intended for a unique processor cell manager. Thus we use an addressed routing for these data. It is implemented as a *store-and-forward* distributed routing (Ni and McKinley 1993). However, for small messages, routing control information becomes bigger than the messages themselves. Addressed routing issues more and smaller messages than broadcast routing. Thus it may happen to become less efficient. For this reason, system data can be broadcast rather than addressed, depending on the data volume to send. An automatic switching between broadcast and addressed modes has been implemented, which minimizes the system data routing delays (Gay 1992).

18.5.4 *Peripherals*

Input and outputs are managed by predefined cells, which can be created and connected like ordinary ones, except that those cells are automatically mapped onto the processors driving the *ad hoc* device.

18.6 Advancement of the CAM project and conclusion

The results achieved so far by the CAM project seem promising:

- A first CAM prototype has been running on sequential computers since 1990 (Cornu and Haton 1990). This prototype was associated with a first rough and simple programming language. This sequential version has been used to implement sample programs.
- Several programming examples are available so far, especially in the field of neural networks and parallel artificial intelligence. These applications have shed light on the possible advantages of the concept of cellular programming in the field of AI. The implementation of a semantic network on the CAM is presented in (Lallement *et al.* 1993). A multi-layered perceptron has also been realized on the CAM (Parmentier 1993), applied to classifying hand-written characters, and based on the algorithm of (McClelland *et al.* 1986).
- We presented in this paper a theoretical study of expected performances on a transputer network, demonstrating the viability of the CAM implementation method. In (Cornu 1992), different families of optimal processor topologies have been defined in addition to torus topologies.
- The communication kernel (broadcasting, addressing, automatic switching between broadcast and addressed mode, distributed garbage collection, creation, placement, destruction, and connections of cells) has been running on a transputer network since September 1992. Experiments have been conducted with this implementation using

benchmarks of transputers realized for this purpose (Gay and Vialle 1992). For an architecture with nine transputers, the performance prediction presented in this paper for the communication phase has been verified. Performance prediction still remains to be validated for architectures of different size.
- Upon this parallel CAM version, an enhanced programming language is currently under development, including stronger type checking, declaration of recursive (sequential) procedures, compilation in multiple separate files, and so on. The operational compiler on transputers is planned for the end of 1993.

Several research possibilities should be investigated in the future:
- Extensive programming experiments including real-size applications and evaluation of the parallel version of the CAM.
- Extensive benchmarking of the multiprocessor implementation of the CAM, with a growing number of processors.
- Porting the CAM implementation to other MIMD architectures.

18.7 Acknowledgements

We would like to thank Jonathan H. Covington, Yannick Lallement, Stéphane Monsallier and Jean-Marc Segrétain for their helpful comments on English style and presentation.

REFERENCES

Gul A. Agha. *Actors : A Model of Concurrent Computation in Distributed Systems*. MIT Press Series in Artificial Intelligence, Cambridge Mass, 1986.

F. Boussinot and R. de Simone. The Esterel language. *Proceedings of the IEEE*, **79**(9),1293–1304, 1991.

T. Cornu and J.P. Haton. An actor language for connectionism based on cellular automata. In *International Neural Network Conference, INNC90*, Paris, 1990.

T. Cornu. *Machine Cellulaire Virtuelle : Définition, Implantation et Exploitation*. PhD thesis, Université de Nancy I, CRIN INRIA Lorraine, Nancy, France, October 1992. In French.

S. Gay and S. Vialle. Analyse des divers parallélismes du transputer. *Lettre du transputer et des calculateurs distribués*, **15**,27–42, 1992. In French.

S. Gay. *Réalisation d'un Système de Communication Multiprocesseurs Optimisé*. DEA thesis, Université de Nancy I and Supélec Metz, France, 1992. In French.

N. Halbwachs, P. Caspi, P. Raymond, and D. Pilaud. The synchronous data flow programming language Lustre. *Proceedings of the IEEE*, **9**,1305 – 1320, 1991.

Y. Lallement, T. Cornu, and S. Vialle. An abstract machine for implementing connectionist and hybrid systems on multiprocessor architectures. In *Second International Workshop on Parallel Processing for Artificial Intelligence, PPAI-93*, Chambery, France, 1993.

P. Le Guernic, T. Gautier, M. Le Borgne, and C. Le Maire. Programming real time applications with signal. *Proceedings of the IEEE* **79**(9),1321–1336, sep 1991.

J.L. McClelland, D.E. Rumelhart, and the PDP Research Group. *Parallel Distributed Processing*, volumes 1 and 2. MIT press, 1986.

L.M. Ni and P.K. McKinley. A survey of wormhole routing techniques in direct networks. *Computer*, **26**(2),62–76, feb 1993.

F. Parmentier. *Implantation d'un réseau de neurones sur une machine cellulaire virtuelle*. DEA thesis, Université de Nancy I and Supélec-Metz, France, 1993. In French.

J. Platt. A resource-allocating network for function interpolation. *Neural Computation*, **3**(2),213–225, 1991.

Leslie G. Valiant. Bulk-synchronous parallel computers. Technical Report TR-08-89, Aiken Computation Laboratory, Harvard University,

Cambridge, MA 02138, USA, 1989.

Leslie G. Valiant. A bridging model for parallel computation. *Communications of the ACM*, **33**(8),103–111, 1990.